Mechanics Problems in Geodynamics Part II

Edited by
Ren Wang
Keiiti Aki

1996

Birkhäuser Verlag
Basel · Boston · Berlin

Reprint from Pageoph
(PAGEOPH), Volume 146 (1996), No. 3/4

The Editors:

Dr. Ren Wang
Department of Mechanics
Peking University
Beijing 100871
China

Dr. Keiiti Aki
Department of Earth Sciences
University of Southern California
Los Angeles, CA 90089-0740
USA

A CIP catalogue record for this book is available from the Library of Congress,
Washington D.C., USA

Deutsche Bibliothek Cataloging-in-Publication Data

Mechanics problems in geodynamics / ed. by Ren Wang; Keiiti Aki. –
Basel ; Boston ; Berlin : Birkhäuser.
NE: Wang, Ren [Hrsg.]
Pt. 2. – (1996)
   ISBN 3-7643-5412-7 (Basel . . .)
   ISBN 0-8176-5412-7 (Boston)

© 1996 Birkhäuser Verlag, P.O. Box 133, CH-4010 Basel, Switzerland
Printed on acid-free paper produced from chlorine-free pulp TCF ∞
Printed in Germany
ISBN 3-7643-5412-7
ISBN 0-8176-5412-7

9 8 7 6 5 4 3 2 1

Contents

PAGEOPH, Vol. 146, Nos. 3/4 (1996)

0033-4553/96/040405-02$1.50 + 0.20/0

Introduction

Ren Wang[1] and Keiiti Aki[2]

This topical issue is part II of the PROCEEDINGS of the IUTAM/IASPEI SYMPOSIUM on MECHANICS PROBLEMS in GEODYNAMICS held at Beijing on Sept. 5–9, 1994. Part I was published in Vol. 145 Nos. 3/4, 1995. As mentioned in the Introduction of that issue, it was truely an international gathering, bringing together scientists from 9 different countries and encompassing various different mechanics' aspects in this highly interdisciplinary subject of Geodynamics. In this part, we again have papers dealing with several subjects. The first five papers concern the global problems spanning different parts of the world. B. F. Chao and R. S. Gross investigated the coseismic effects on the spin rate, polar motion, and the low-degree gravitational field and computed the earthquake induced rotational and gravitational energy changes. S. I. Sherman pointed out systematic relationships between faulting and its evolution in a laboratory model and compared them with field data from the Baikal and other rift zones. Russian works written in Russian are introduced and well summarized. M. Liu presents an overview of the dynamic interactions between crustal shortening, extension and magmatism in the North American Cordillera and proposes scenarios for each stage of its evolution. Y. Shi, R. Allis and F. Davey give a finite-element modeling of the thermal region across the Southern Alps of New Zealand and point out the importance of frictional heating. T. N. Gowd, S. V. Srirama Rao and K. B. Chary considered the effects of continued convergence between India and Eurasia on the intraplate stresses and intraplate seismicity in the Indian subcontinent.

The next six papers concern composition and flow in the mantle. Y. B. Wang and D. I. Weidner, in an attempt to establish a connection between mineral physics and geodynamics, employed both the available mineral physics data on candidate lower mantle minerals and seismic data of the lower mantle in the estimation of $(d\mu/dT)p$, an important parameter in the dynamic analysis of the lower mantle. S. A. Weinstein tackles a very important question: why does a plate structure develop at all? He used a numerical model in a cylindrical annulus to investigate the coupling between surface plate motion and mantle convection. Z. R. Ye superim-

[1] 3131 Geoscience Building, Peking University, Beijing 100871, China.
[2]

poses the internally driven (free) and the plate driven (forced) convections which produced a surface velocity field that contains not only the poloidal component but also the toroidal component. Y. J. Chen considers both the effects of melt extraction and the decrease of melt production as the residual materials become increasingly difficult to melt on the rate of magma production beneath the mid-ocean ridges. These processes were usually neglected. In his next paper, he considers the dynamics of rifting and the thermal structure of mid-ocean ridges together with the underlying segmentation of mantle upwelling. Y. S. Fu, J. H. Huang and Z. X. Wei attempted to make use of isostatic anomalies to define density anomalies in the mantle under North China in a simplified model, without considering the lithospheric stretching there. The next five papers address the studies of earthquake mechanisms. Z. M. Yin and C. C. Rogers discuss the physical background of earthquake scaling relations, yielding some interesting results. R. K. Bhattacharyya gives a mathematical treatment of the reflection of plane waves from the free surface of an isotropic elastic half space with small random deviations from homogeneity. X. F. Chen and K. Aki present a new formulation for computing the stress field with given kinematic source parameters. Y. T. Chen and L. S. Xu *et al.* discuss the inversion of the source process of the 1990 Gonghe, China ($M_s = 6.9$) earthquake and the tectonic stress field in the NE Tibetan Plateau. Finally, L. B. Liu makes use of the Discontinuous Deformation Analysis (DDA) method in the estimation of long-term cross-fault survey and other geodetic data during the interseismic period in North China.

Finally, we wish to thank the reviewers for their patient and careful reviews. They are: Y. E. Cai, M. A. Chinnery, G. L. Chen, Y. J. Chen, Y. T. Chen, F. H. Cornet, P. Davies, C. Froidevaux, R. G. Gordon, A. T. Hsui, W. R. Jacoby, M. Kikuchi, M. Liu, T. Miyatake, J. W. Rudnicke, J. Y. Wang, H. S. Xie and R. S. Zeng. Special thanks are due to Dr. Renata Dmowska who not only assisted in organizing the successful Symposium, but, carried out most valuable editing of these Proceedings.

PAGEOPH, Vol. 146, Nos. 3/4 (1996)

0033–4553/96/040407–13$1.50 + 0.20/0

Seismic Excitation of the Polar Motion, 1977–1993

BENJAMIN FONG CHAO,[1] RICHARD S. GROSS,[2] and YAN-BEN HAN[3]

Abstract — The mass redistribution in the earth as a result of an earthquake faulting changes the earth's inertia tensor, and hence its rotation. Using the complete formulae developed by CHAO and GROSS (1987) based on the normal mode theory, we calculated the earthquake-induced polar motion excitation for the largest 11,015 earthquakes that occurred during 1977.0–1993.6. The seismic excitations in this period are found to be two orders of magnitude below the detection threshold even with today's high precision earth rotation measurements. However, it was calculated that an earthquake of only one tenth the size of the great 1960 Chile event, if happened today, could be comfortably detected in polar motion observations. Furthermore, collectively these seismic excitations have a strong statistical tendency to nudge the pole towards ~ 140°E, away from the actually observed polar drift direction. This non-random behavior, similarly found in other earthquake-induced changes in earth rotation and low-degree gravitational field by CHAO and GROSS (1987), manifests some geodynamic behavior yet to be explored.

Key words: Earthquake, polar motion, earth rotation.

1. Introduction

The earth's rotation varies slightly with time. The 3-D earth rotation variation can be conveniently separated into two components: (i) The 1-D variation in the spin rate, often expressed in terms of the length-of-day variation. (ii) 2-D variation in the rotational axis orientation, generically called the nutation when viewed from the inertial reference frame, or the polar motion as seen in the terrestrial reference frame.

There are two dynamic types of earth rotation variations (MUNK and MAC-DONALD, 1960). (i) The "astronomical variations" due to external luni-solar tidal torques that change the earth's angular momentum: Well-known examples include the tidal braking that causes the earth's spin to slow down over geological times, and the astronomical precession and nutation caused by torques exerted on the earth's equatorial bulge. (ii) The "geophysical variations" caused by large-scale

[1] Geodynamics Branch, NASA Goddard Space Flight Center, Greenbelt, Maryland 20771, U.S.A.
[2] Jet Propulsion Laboratory, California Institute of Technology, Pasadena, CA 91109, U.S.A.
[3] Beijing Astronomical Observatory, Chinese Academy of Sciences, Beijing 100080, China.

mass movement of internal geophysical processes under the conservation of angular momentum. These include tidal deformations in the solid earth and oceans, atmospheric fluctuations, hydrological variations, ocean currents, earthquake dislocations, post-glacial rebound, mantle tectonic movement, and core activities.

The present paper deals with the polar motion and its seismic excitation. Possible interactions between seismicity and earth rotation have been under consideration since the discovery of the polar motion a hundred years ago (e.g., LAMBECK, 1980). On one hand, a non-uniform earth rotation would give rise to a time-varying stress field inside the earth, which in turn may affect the triggering process of earthquakes. This effect with respect to the polar motion is in general an order of magnitude smaller than its tidal counterpart (LAMBERT, 1925), but having a longer time scale on the order of a year. CHAO and IZ (1992) conducted a statistical test in an attempt to correlate the occurrence of earthquakes with the contemporary polar motion based on nearly 10,000 earthquakes that occurred during 1977–1991; but only a weak correlation was found. A more definitive conclusion awaits further studies taking into account the tensorial nature of the seismic source mechanism and the stress field induced by polar motion.

On the other hand, the mass redistribution in the earth *as a result* of an earthquake faulting changes the earth's inertia tensor, and hence its rotation. Early simplistic earthquake faulting models considering only regional dislocations greatly underestimated the effect (MUNK and MACDONALD, 1960). After the 1964 Alaska earthquake it was recognized that the seismic dislocation, although decreasing with focal distance, remained non-zero even at teleseismic distances away from the fault (PRESS, 1965). When integrated globally (see equation (2) below) considering the size and source mechanism of the earthquake this displacement field can give finite effects (MANSINHA and SMYLIE, 1967). SMYLIE and MANSINHA (1971), DAHLEN (1971, 1973), RICE and CHINNERY (1972), and O'CONNELL and DZIEWONSKI (1976), among others, calculated the seismic excitation of polar motion based on realistic models. It was concluded that the great 1960 Chile and 1964 Alaska earthquakes should have produced significant changes in polar motion (see below). Unfortunately the accuracy of the polar motion record at that time was insufficient to yield any conclusive detection.

In the last three decades, the measurement of earth rotation has been progressively improved by three orders of magnitude in both accuracy and temporal resolution (e.g., EUBANKS, 1993, for a review). This is achieved through the advances in modern space geodetic techniques, primarily Very-Long-Baseline Interferometry (VLBI), Satellite Laser Ranging (SLR), and Global Positioning System (GPS). In fact, these techniques have measured the contemporary relative tectonic plate motions to within a few mm/year, and revealed coseismic, near-field displacements caused by recent earthquakes. The SLR technique has also greatly benefitted the determination of the earth's global gravitational field and detected slight temporal variations in the low-degree field. In the area of earth rotation measurement, the current accuracy is estimated to be within 1 milliarcsecond (1 mas = 4.85 × 10⁻⁹ radian, corresponding to about 3 cm of distance on the surface of the earth), while the formal errors are as low as 200 μas.

Taking advantage of these modern data as well as the seismic centroid moment tensor solutions made available in the Harvard CMT catalog (for a review, see DZIEWONSKI et al., 1993), SOURIAU and CAZENAVE (1985) and GROSS (1986) calculated the seismic excitation of polar motion using DAHLEN'S (1973) formulation for post-1977 major earthquakes. It was concluded that the polar motion excited by those earthquakes were all too small to be detected even in the modern earth rotation data. CHAO and GROSS (1987) developed complete formulae for calculating the earthquake-induced changes in earth's rotation and low-degree gravitational field based on the normal mode theory of GILBERT (1970). They made calculations for the 2,146 major earthquakes that occurred during 1977–1985. In addition to confirming the above conclusion, the result indicated that earthquakes have a strong statistical tendency to make the earth rounder and more compact. It also demonstrated a similar tendency for earthquakes to nudge the rotation pole towards the direction of about 150°E. CHAO and GROSS (1995) and CHAO et al. (1995) extended the calculation to include 11,015 major earthquakes for the period 1977–1993 to evaluate the corresponding rotational and gravitational *energy* changes.

The present paper will study the seismic excitation of polar motion in the framework of CHAO and GROSS (1987) but making use of the updated result of CHAO and GROSS (1995). The corresponding changes in length-of-day will only be presented in passing, as they are in general two orders of magnitude smaller (that is to say, polar motion is hundreds of times easier to excite than length-of-day by earthquakes). The physical reason is the following. The geophysical excitation acts against the "inertia" of the earth system. For length-of-day the inertia is the earth's axial moment of inertia C, whereas the inertia for polar motion is the difference between the axial and equatorial moments of inertia, $C-A$ (see equation 2), which is only $1/300$ of C.

2. Dynamics and Calculation

The excitation of the polar motion is governed by the conservation of angular momentum. The equation of motion is customarily expressed as (MUNK and MACDONALD, 1960):

$$m + \frac{i}{\sigma}\frac{dm}{dt} = \Psi \tag{1}$$

where m is the complex-valued pole location measured in radian; its real part is the x component (along the Greenwich Meridian) and the imaginary part the y

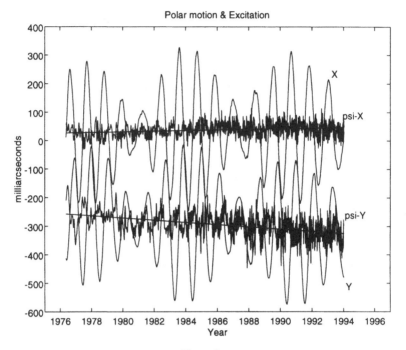

Figure 1
The x and y components of the observed polar motion and its excitation function (Ψ, obtained by deconvolution) for 1976.4–1994.0, in units of milliarcseconds. The straight lines are the least-squares fit to the excitation function.

component (along the 90°E longitude), σ is the frequency of the free Chandler wobble with a nominal period of 435 days and a Q value of 100, and Ψ is the complex-valued excitation function. Mechanically equation (1) is analogous to the excitation of a simple harmonic oscillator with a natural frequency σ.

The polar motion m traces out a prograde, quasi-circular path on the order of 10 m in the vicinity of the North Pole. Figure 1 shows the x and y positions of the pole at nominal daily intervals according to the "Space93" dataset (GROSS, 1994) derived from space geodetic measurements during 1976.4–1994.0. Besides a slow polar drift, the oscillation consists mainly of the annual wobble and the Chandler wobble. It is continually excited (otherwise it would decay away in a matter of decades); and the excitation function Ψ can be obtained numerically according to equation (1) in a process of deconvolution. The Ψ thus obtained is also given in Figure 1, together with the least-squares fitted straight lines representing the polar drift (see below). The geophysical problem is to identify and understand the sources of this "observed" Ψ. It is now known that a major source is the variation of the atmospheric angular momentum (e.g., CHAO, 1993; KUEHNE *et al.*, 1993). The problem is far from closed, and the earthquake dislocation remains a candidate excitation source.

The polar-motion excitation Ψ due to mass redistribution is given by

$$\Psi = 1.61(\Delta I_{zx} + i\, \Delta I_{yz})/(C - A) \qquad (2)$$

where I denotes the inertia tensor, the factor 1.61 takes into account the earth's non-rigidity and the decoupling of the fluid core from the mantle in the excitation process. Note that Ψ should also include an additional term due to mass motion, but that term is negligible in the case of abrupt seismic sources (CHAO, 1984).

With an abrupt step-function time history (compared to the considerably longer time scale of the polar motion), an earthquake faulting generates a co-seismic, step-function displacement field u (after the seismic waves have died away). Knowing the seismic moment tensor, u anywhere in the earth can be evaluated by the normal mode summation scheme of GILBERT (1970). The task, then, is to calculate the seismic Ψ according to equation (2), which consists of evaluating ΔI by a properly weighted integration of u over the globe. The reader is referred to CHAO and GROSS (1987) for details of the formulation and calculation method. This normal mode scheme is found to be extremely efficient.

Calculation has been conducted for 11,015 major earthquakes (with nominal magnitude greater than 5.0) that occurred during 1977.0–1993.6, using the seismic

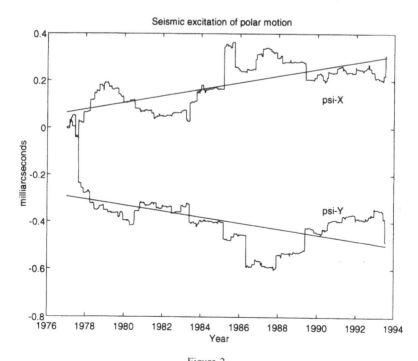

Figure 2

The x and y components of the seismic excitation of polar motion. The straight lines are the least-squares fit to the excitation function.

centroid moment tensor solutions published in the Harvard CMT catalog (e.g., DZIEWONSKI *et al.*, 1993). Smaller earthquakes, although numerous, have completely negligible effects. The adopted normal mode eigenfrequencies and eigenfunctions belong to the spherically symmetric earth model 1066B of GILBERT and DZIEWONSKI (1975). The net effect is then the accumulation of individual step-function contributions: $\Psi(t) = \Sigma_n \Psi_n H(t - t_n)$, $(n = 1, \ldots, 11{,}015)$. CHAO and GROSS (1995) have also calculated the corresponding changes in length-of-day and low-degree gravitational field.

3. Results and Analysis

Figure 2 shows the calculated polar-motion excitation function $\Psi(t)$ by the earthquakes. The starting value is arbitrarily chosen to be zero. Comparing Figure 2 with 1, it is obvious that the magnitude of the seismic excitation is insignificant during 1977.0–1993.6: It is of two orders of magnitude too small to explain the observed polar-motion excitation.

For the purpose of illustration, Table 1 lists the results for the following eight largest earthquakes in recent decades (with seismic moment M_0 exceeding 10^{21} Nm). These results have been reported elsewhere by CHAO and GROSS (1995).

Event I: May 22, 1960, Chile
Event II: March 28, 1964, Alaska, U.S.A.
Event III: August 19, 1977, Sumba, Indonesia
Event IV: March 3, 1985, Chile
Event V: September 19, 1985, Mexico
Event VI: May 23, 1989, Macquarie Ridge
Event VII: June 9, 1994, Bolivia
Event VIII: October 4, 1994, Kuril Is., Russia

The source mechanism of Events I and II, which occurred before the span of the Harvard catalog, are taken from KANAMORI and CIPAR (1974) and KANAMORI

Table 1

Magnitude and direction of polar-motion excitation by eight great earthquakes, and the corresponding length-of-day changes

Event (M_0, 10^{21} Nm)	I (270)	II (75)	III (3.6)	IV (1.0)	V (1.1)	VI (1.4)	VII (2.6)	VIII (3.9)
ΔLOD (μs)	−8.4	6.8	0.33	−0.10	−0.089	−0.059	0.192	−0.053
$|\Psi|$ (mas)	22.6	7.5	0.21	0.18	0.084	0.114	0.331	0.256
$\arg(\Psi)$ (°E)	115	198	160	110	277	323	122	129

(1970), respectively. Events VII (a deep-focused event) and VIII in 1994 are also outside our studied period. They have the largest seismic moment since Event III in 1977.

The following facts are observed: (i) The earthquake-induced ΔLOD is small: even the largest earthquakes (Events I and II) would only cause a ΔLOD that is barely discernable even with today's measurement precision. (ii) The direction of polar motion excitation, arg(Ψ), tend to cluster in the second quadrant (see below). (iii) Although the polar motion excitations by the latest 6 events have magnitude that were hardly detectable, Events I and II, if happened today, could be readily detected in the polar motion measurement. It should be noted, however, that it is in the polar motion *excitation* that an earthquake leaves its signature as a step function. In the polar motion path itself (which represents a time integration of the excitation, see equation (1)), the earthquake manifests itself as a "kink" in the path, and hence more difficult to detect. Further difficulties arise because of the abrupt nature of the earthquake source and the relatively "gappy" sampling of earth rotation in the past. Only recently has a virtually continuous monitoring of earth rotation become feasible by means of the GPS technique.

Figure 3 compares the power spectra of the observed and the seismic excitations for the same period of 1977.0–1993.6. The spectra are computed using THOMSON'S (1982) multitaper technique after removal of the mean values. The different spectral characteristic is evident. The frequency dependence f^n of the spectrum of the seismic excitation is found to follow $n = 2.0$, similar to a Brownian motion process. The power of the seismic Ψ is in general 40–60 dB lower than that observed. In particular, the power difference at the Chandler frequency band around 0.83 cycle per year is about 45 dB. We have also calculated and examined possible correlation between the two excitation functions in the Chandler band. Only a weak coherence at a moderate confidence level was found. These findings are consistent with GROSS (1986).

Despite the small magnitude, the seismic excitations Ψ collectively are interesting in their own right. Following CHAO and GROSS (1987), we shall now conduct statistical analyses to reveal their peculiar behavior.

Figure 4 shows the angular histogram ("rose diagram") of the arguments of the 11,015 Ψ's in thirty-six 10° increment bins. Apart from a concentration around 15°E, an abnormally large number of Ψ's cluster around 140°E. The distribution pattern is remarkably similar to the 2,146 Ψ's analyzed by CHAO and GROSS (1987), indicating that this pattern is robust with respect to time and number of earthquake samples. In fact, the statistical tendency of this angular anomaly is stronger now with the substantially greater number of samples: the normalized χ^2 found here is as high as 14.1, compared to 3.81 found in CHAO and GROSS (1987). Compared with, say, the 1% significant level of 1.64 or the 0.1% significant level of 1.90 for a random distribution (at 35 degrees of freedom), this asserts the extremely non-random nature of the distribution of Figure 4.

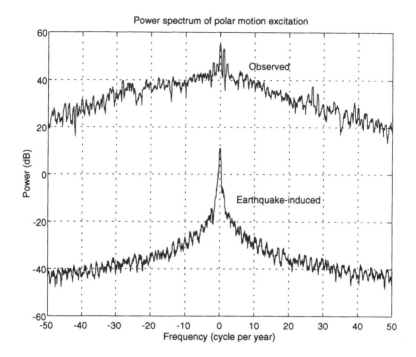

Figure 3

The (multitapered) power spectrum (in dB units of mas²/cpy) of the observed polar-motion excitation compared with that induced by earthquakes during 1977.0–1993.6. Mean values have been removed from the time series.

The preference of earthquakes in nudging the rotation pole towards ~ 140°E is also evident in Figure 2. The straight lines are the least-squares fit to the curves; their slopes give the velocity of the polar drift induced by earthquakes. This polar drift velocity vector (0.019 mas/yr, 131°E), as plotted in Figure 5, indeed points to that general direction. The magnitude of the vector is here magnified by 100 times, in order to be shown against other vectors:

The one labeled "O&D" (4.5 mas/yr, 148°E) is that similarly obtained using O'CONNELL and DZIEWONSKI's (1976) calculation for 30 great earthquakes during 1900–1964. So it appears that the earthquake-induced polar drift has continued its journey toward ~ 140°E at least since 1900 when fairly reliable seismic records had become available. It should be cautioned, however, that O'CONNELL and DZIEWONSKI may very well have greatly, and presumably to different extent, overestimated the sizes of their studied earthquakes (KANAMORI, 1976). Even though the directions of the polar excitation by *individual* earthquakes, which are determined by the source mechanisms according to plate tectonics, are realistic, the (vectorial) cumulative drift direction may be biased. However, closer examination of these individual directions indicates that this bias is not severe: Figure 6 plots the

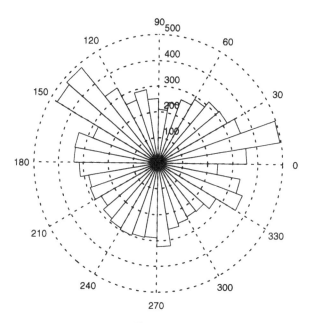

Figure 4

The angular histogram ("rose diagram") for the direction of the seismic excitation of polar motion by major earthquakes for 1977.0–1993.6, in 36 angular bins with respect to the terrestrial coordinate system.

rose diagram (in 20° bins) of O'CONNELL and DZIEWONSKI's 30 polar excitation directions and their cumulative polar drift. The directions have the same heavy clustering in the second quadrant, and hence the cumulative drift is largely unidirectional.

The other two vectors in Figure 5 are from polar motion observations; they agree well with each other. The one labeled "Space93" (3.84 mas/yr, −68°E) is simply obtained from the fitted line in Figure 1. The one labeled "Pole93" (3.22 mas/yr, −81°E) is similarly obtained from a polar motion dataset for 1900–1993, primarily based on the International Latitude Service data since 1900. It is seen that the observed polar drift directions are roughly opposite to those induced by earthquakes.

4. Conclusions

The direction of the seismic excitation depends on the focal location and source mechanism. Through computation, it is found that, at least since 1900, the seismic excitations collectively have a statistically strong tendency to nudge the pole towards ∼ 140°E, away from the actually observed polar drift direction. This non-

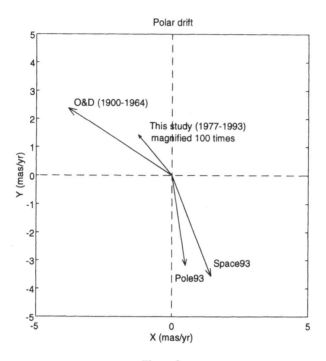

Figure 5
The calculated polar drift velocity induced by major earthquakes for 1977.0–1993.6, magnified 100 times, in comparison with O'Connell and Dziewonski's estimate (labeled "O&D") and the observed polar drift (labeled "Space93" and "Pole93").

random behavior is similar to other earthquake-induced changes computed for length-of-day and low-degree gravitational field, which indicate a strong tendency for the earthquakes to make the earth rounder and more compact (CHAO and GROSS, 1987). Specific questions arise: Why do earthquakes seem to "recognize" the existence of the rotation axis? Is there any as yet unseen, dynamic connection between the pole position and the occurrence and source mechanism of earthquakes? Or are they manifestations of some behind-the-scene geophyiscal processes? The dynamic reasons for such peculiar behavior of earthquakes are not clear in terms of the grand scheme of the plate tectonics. At the present time these remain mere speculations.

The magnitude of individual earthquake effect, on the other hand, depends largely on the seismic moment of the event. The rule of thumb is that the seismic moment of 10^{22} Nm would roughly produce 1 mas in polar-motion excitation and 1 μs in length-of-day change; the actual values depend on the focal location and seismic source mechanism. The polar motion excitations produced by the largest earthquakes during 1977–1993 were still an order of magnitude too small to be detected even with today's measuring accuracy of about 1 mas. However, it was calculated that the 1960 Chile event should have produced a discontinuity as large

(a)

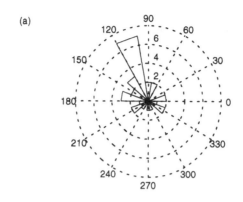

(b)

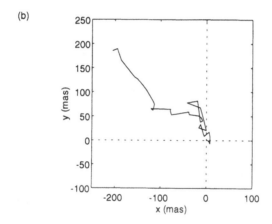

Figure 6

(a) The rose diagram for the direction of the seismic excitation of polar motion by 30 great earthquakes during 1900–1964 studied by O'Connell and Dziewonski, in 18 angular bins. (b) Their cumulative polar excitation in the terrestrial coordinate system.

as 23 mas in the polar-motion excitation function (but only 8 μs in ΔLOD). So an earthquake of only one tenth the size of that event, if happened today, could be comfortably detected in polar-motion observations, especially now that the GPS technique routinely provides sub-daily temporal resolutions. Further improvements in observation should allow detection of smaller (and hence potentially more numerous) earthquake signatures in the future.

Acknowledgments

The seismic centroid-moment tensor catalog is kindly provided via Internet by A. Dziewonski and his group at the Harvard University under support from the

National Science Foundation. The 1066B Earth Model is courtesy of R. Buland. The work of a co-author (RSG) described in this paper presents the results of one phase of research carried out at the Jet Propulsion Laboratory, California Institute of Technology, under a contract with the National Aeronautics and Space Administration (NASA). This study is supported by the NASA Geophysics Program and the Beijing Astronomical Observatory of the Chinese Academy of Sciences.

REFERENCES

CHAO, B. F. (1984), *On Excitation of Earth's Free Wobble and Reference Frames*, Geophys. J. Roy. Astron. Soc. *79*, 555–563.

CHAO, B. F., and GROSS, R. S. (1987), *Changes in the Earth's Rotation and Low-degree Gravitational Field Induced by Earthquakes*, Geophys. J. Roy. Astron. Soc. *91*, 569–596.

CHAO, B. F., and Iz, H. B. (1992), *Does Polar Motion Influence the Occurrence of Earthquakes*, EOS, Trans. Am. Geophys. Union *73*, 79.

CHAO, B. F. (1993), *Excitation of Earth's Polar Motion by Atmospheric Angular Momentum Variations, 1980–1990*, Geophys. Res. Lett. *20*, 253–256.

CHAO, B. F., and GROSS, R. S. (1995), *Changes of the Earth's Rotational Energy Induced by Earthquakes*, Geophys. J. Int., in press.

CHAO, B. F., GROSS, R. S., and DONG, D. N. (1995), *Global Gravitational Energy Changes Induced by Earthquakes*, Geophys. J. Int., in press.

DAHLEN, F. A. (1971), *The Excitation of the Chandler Wobble by Earthquakes*, Geophys. J. R. Astr. Soc. *25*, 157–206.

DAHLEN, F. A. (1973), *A Correction to the Excitation of the Chandler Wobble by Earthquakes*, Geophys. J. R. Astr. Soc. *32*, 203–217.

DZIEWONSKI, A. M., EKSTRÖM, G., and SALGANIK, M. P. (1993), *Centroid Moment Tensor Solutions for October–December, 1992*, Phys. Earth Planet. Int. *80*, 89–103.

EUBANKS, T. M., *Variations in the orientation of the earth*. In *Contributions of Space Geodesy to Geodynamics: Earth Dynamics* (eds. Smith, D. E., and Turcott, D. L.) (AGU, Washington, D.C. 1993), pp. 1–54.

GILBERT, F. (1970), *Excitation of the Normal Modes of the Earth by Earthquake Sources*, Geophys. J. R. Astr. Soc. *22*, 223–226.

GILBERT, F., and DZIEWONSKI, A. M. (1975), *An Application of Normal Mode Theory to the Retrieval of Structural Parameters and Source Mechanisms from Seismic Spectra*, Phil. Trans. R. Soc. London A-*278*, 187–269.

GROSS, R. S. (1986), *The Influence of Earthquakes on the Chandler Wobble during 1977–1983*, Geophys. J. R. Astr. Soc. *85*, 161–177.

GROSS, R. S. (1994), *A combination of Earth Orientation Data: Space 93*, IERS Technical Note *17*: Earth Orientation, Reference Frames and Atmospheric Excitation Functions, Obs. de Paris.

KANAMORI, H. (1970), *The Alaska Earthquake of 1964: Radiation of Long-period Surface Waves and Source Mechanism*, J. Geophys. Res. *75*, 5029–5040.

KANAMORI, H., and CIPAR, J. J. (1974), *Focal Process of the Great Chilean Earthquake May 22, 1960*, Phys. Earth. Planet. Int. *9*, 128–136.

KANAMORI, H. (1976), *Are Earthquakes a Major Cause of the Chandler Wobble ?* Nature *262*, 254–255.

KUEHNE, J., JOHNSON, S., and WILSON, C. R. (1993), *Atmospheric Excitation of Nonseasonal Polar Motion*, J. Geophys. Res. *98*, 19973–19978.

LAMBECK, K. *The Earth's Variable Rotation* (Cambridge Univ. Press, New York 1980).

LAMBERT, W. D. (1925), *The Variation of Latitude, Tides, and Earthquakes*, Proc. 3rd Pan-Pacific Science Congress, Tokyo, 1517–1522.

MANSINHA, L., and SMYLIE, D. E. (1967), *Effects of Earthquakes on the Chandler Wobble and the Secular Pole Shift*, J. Geophys. Res. *72*, 4731–4743.

MUNK, W. H., and MACDONALD, G. J. F., *The Rotation of the Earth* (Cambridge University Press, New York 1960).

O'CONNELL, R. J., and DZIEWONSKI, A. M. (1976), *Excitation of the Chandler Wobble by Large Earthquakes*, Nature *262*, 259–262.

PRESS, F. (1965), *Displacements, Strains, and Tilts at Teleseismic Distances*, J. Geophys. Res. *70*, 2395–2412.

RICE, J. R., and CHINNERY, M. A. (1972), *On the Calculation of Changes in the Earth's Inertia Tensor due to Faulting*, Geophys. J. R. Astr. Soc. *29*, 79–90.

SMYLIE, D. E., and MANISINHA, L. (1971), *The Elasticity Theory of Dislocations in Real Earth Models and Changes in the Rotation of the Earth*, Geophys. J. Res. *23*, 329–354.

SOURIAU, A., and CAZENAVE, A. (1985), *Reevaluation of the Chandler Wobble Seismic Excitation from Recent Data*, Earth Planet. Sci. Lett. *75*, 410–416.

THOMSON, D. J. (1982), *Spectrum Estimation and Harmonic Analysis*, Proc. IEEE *70*, 1055–1096.

(Received October 10, 1994, accepted April 30, 1995)

PAGEOPH, Vol. 146, Nos. 3/4 (1996)

0033-4553/96/040421-26$1.50 + 0.20/0

Faulting in Zones of Lithospheric Extension: Quantitative Analysis of Natural and Experimental Data

S. I. SHERMAN[1]

Abstract —Quantitative relationships between major fault parameters from geological observations and laboratory experiments are compared. Relationships are established between fault length, number, depth of fault penetration, amplitude of displacement and other characteristics. The width of destruction zones is estimated. Spacing between parallel faults of compatible length is evaluated. It is shown that there is a stable correlation between fault length and number, which is independent from the mode of loading the material under destruction. Destruction of the lithosphere is believed to occur according to the laws of deformation of Maxwell elasto-viscous body.

Key words: Fault, parameter, lithosphere, extension, stress, rift, experiment.

Introduction

The application of methods of quantitative analysis is becoming more widespread in geodynamics. This approach was initiated by NADAI (1963), GZOVSKY (1975), LEE (1984, 1991) and others. In the past two decades many reviews were presented concerning the problem (ARTYUSHKOV, 1979; TURCOTTE and SCHUBERT, 1982; MANDLE, 1988). Advances of mechanics being applied to geodynamic analysis can improve the latter and at the same time constrain a wide variety of conceptual hypotheses and models proposed in geology. Methods of mechanics and mathematical simulation describe geological structures and process numerically. However, it is not always possible to meet the requirement at the current state of knowledge. Besides, geodynamic processes are variable in time. It is often difficult to reconstruct their quantitative characteristics in retrospect. Physical modeling can provide numerical values of parameters of recent deep-seated geological structures. It allows the establishment of relationships between structures in depth and on the surface and to reveal the dynamics of their development. Moreover, physical modeling can provide statistical data on relationships between structures and processes which rarely take place in nature.

[1] Institute of the Earth's Crust, Siberian Branch of Russian Academy of Sciences, Lermonotov str. 128, Irkutsk, 664033, Russia.

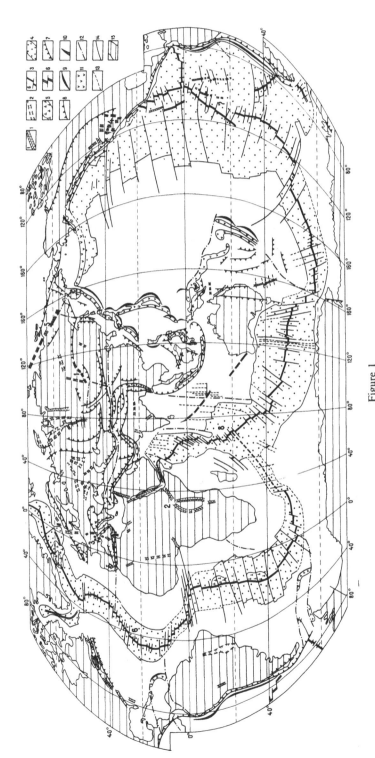

Figure 1

A world map of major extensional and compressional zones (SHERMAN and LYSAK, 1992). 1–6—extensional zones: 1—Cenozoic intracontinental rifts: (numbers on the map: 1—Baikal rift zone, 2—East African rift system, 3—Rhine-North African rift belt, 4—Cordillera rift system, 5—Moma rift zone); 2—pre-Cenozoic (paleorifts); 3–6—mid-oceanic rifts: 3—axes of Cenozoic mid-oceanic ridges (numbers on the map: 6—Middle Atlantic, 7—East Pacific, 8—Arabian-Indian); 4—oceanic crust formed in the Cenozoic; 5—Cenozoic volcanic plateaus; 6—pre-Cenozoic volcanic plateaus; 7–8—compressional zones: 7–8—intracontinental (7—Cenozoic, 8—pre-Cenozoic); 9—marine troughs; 10—onset volcanic arcs; 11—volcanic arcs; 12–14—strike-slip zones: 12—active in the Cenozoic, 13—passive in the Cenozoic; 14—transform faults.

Regularities of development and destruction of the lithosphere can be revealed more definitely if geological observations are complemented by physical modeling and numerical simulation. Based on the above approach, faulting in zones of lithospheric extension has been studied in an effort to deduce some quantitative regularities in the destruction of the lithosphere. Conditions of lithospheric extension are set up in the areas above ascending convective flows or result from tectonic forces acting in different directions in local lithospheric volumes. It is essential that the three principal stresses are not equal and that the maximum stress generated by the extensional tectonic forces is horizontal. Deformation and destruction of the lithosphere under the above discussed modes of loading produces a specific tectonic (geodynamic) regime, termed the rift regime, creating a complex of extensional structures among which grabens controlled by faults are the major ones. The objective of the study is to reveal qualitative and quantitative regularities of faulting in rift zones of the lithosphere, so as to compare the results of natural observation of destruction under extension and faulting with those obtained from laboratory models.

Quantitative Analysis of Faulting in Rift Zones of the Lithosphere: Data and Discussion

Extensional zones are widespread on the globe (Fig. 1). Their fault patterns are best studied in continental rift zones, the Baikal rift zone (BRZ) being a prominent one. It is an isolated system occupying a wide area of Central Asia with spacial relations neither with other continental rifts nor with oceanic rifts. Its relatively homogeneous Precambrian basement has been subjected to extension since the Oligocene. The formation of rift structures which started at that time has continued with varying intensity for more than 30 Ma. The most recent reactivation of the region took place in the Pliocene about 3.5 Ma ago. Tectonic movements are still intensive as evidenced by the activity of neotectonic processes and high seismicity. Seismicity is unevenly scattered in space, reflecting the recent state of the lithosphere, which can be regarded as instantaneous relative to its 30 Ma period of extension. In the central Baikal rift zone, where extension is maximum, seismic events occur mostly in the axial part of the zone and at its eastern boundary, though evidence of paleoseismic dislocations is more abundant on the western shore of the lake. The geometrical axis of the rift flanks is the symmetry plane of the area of seismicity of.these flanks, however, the density of earthquake foci is irregular.

Researchers generally believe that the BRZ is being formed by active rifting mechanisms and its energy source is represented by sublithospheric convection flow. It steadily affects the lithospheric base underneath the whole rift zone and the "cloud-like" irregular distribution of seismicity at the flanks or its historical migration, which is clearly registered in the central part of the zone, do not reflect

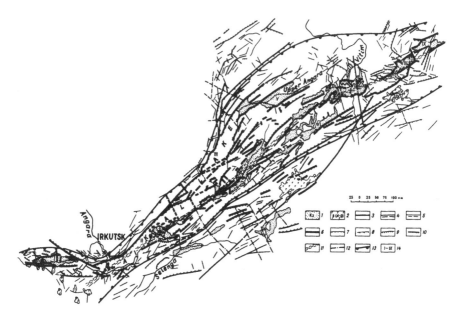

Figure 2

Fault system in the Baikal rift zone. 1—Cenozoic sediments; 2—Cenozoic basalts; 3—major faults; 4—regional faults; 5—local faults; 6—faults rejuvenated in the Cenozoic and Cenozoic faults; 7—normal faults; 8—oblique-slip faults; 9—thrusts; 10—tension (tear) faults; 11—zones of intensely fractured rocks; 12—faults defined by geophysical data but not verified by surface mapping; 13—boundaries of rift zones; 14—major depression in rift zone: I—Tunka; II—South Baikal; III—North Baikal; IV—Barguzin; V—Upper Angara; VI—Tsipa-Baunt; VII—Lower Muya; VIII—Chara; IX—Tokka.

the sublithospheric variations in the rate or in other characteristics of the flow. Seismicity, which is irregular in many parameters, reflects the ongoing irregular destruction of the base (lithosphere) which can be best studied from fault tectonics.

Figure 2 shows a map of major faults in the Baikal rift zone. Its analysis leads to the following conclusions. Normal faults are evidently predominant among other morphogenetic types of faults. For comparison, a schematic map of faults in the East African rift system and a more detailed fault map of the Kenya rift are shown (Fig. 3). The situations are identical in general. In the areas where the rift zone strike changes, as is seen at the BRZ flanks, normal faults are replaced by strike-slip faults with a normal component of displacement. These effects are well known. Two

Figure 3

Schemes of active faults in continental rift zones. a) East African rift system (after KAZMIN, 1987, modified). 1—faults, defined (solid line) and presumed (dashed line); 2—approximate boundary of extensional zone; b) Rhein rift (ILLIES, 1981); c) Rift Rio Grande (after EATON, 1987) (A) and scheme of fault density distribution (B).

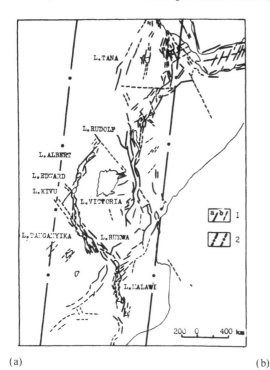

(a)

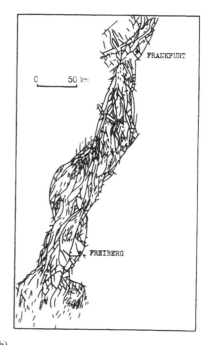

(b)

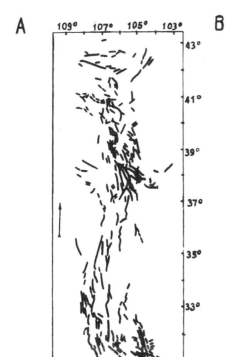

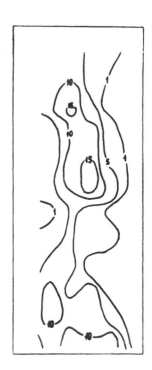

(c)

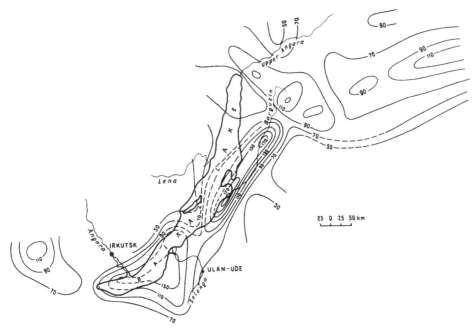

Figure 4
Fault density in the Baikal rift zone.

groups of faults at different scales can be distinguished: large-regional faults and small-local faults. In such a classification, the distinction is controlled by the fault length. Faults of a length less than half the thickness of the crust, i.e., 15–20 km, are regarded as local faults. Their depth of penetration (see below) does not exceed their length, and, consequently, they do not rupture the crustal layer. Faults longer than 15–20 km are considered regional ones. It has been established that the number of local faults is 10 times more numerous than that of regional faults.

Within the limits of the rift zone, the irregularity of the spatial distribution of faults of different scales is well observed. Large regional faults occur at the marginal parts of the extensional zone and control the formation of grabens and basins

Figure 5
Map of stress field in the Baikal rift zone (BRZ) from structural and seismic data. 1—orientation of the compressional (white) and tension (black) quadrants: 2—transregional faults (a—active; b—nonactive); 3—regional faults (a—active; b—nonactive); 4—normal faults (a) and strike-slip faults (b); 5—reverse faults (a) and reverse-slip faults (b); 6—thrust faults; 7—Cenozoic depressions. The scheme of regional tectonic stresses of the BRZ from geological structural and seismological data is given in the insert. 1—vector orientation of the regional stress field from geological structural data (a—horizontal; b—oblique; c—vertical). 2—vector orientation of the stress fields from seismological data for earthquakes of $M > 4$ (a—horizontal; b—oblique; c—vertical). 3—Cenozoic depressions.

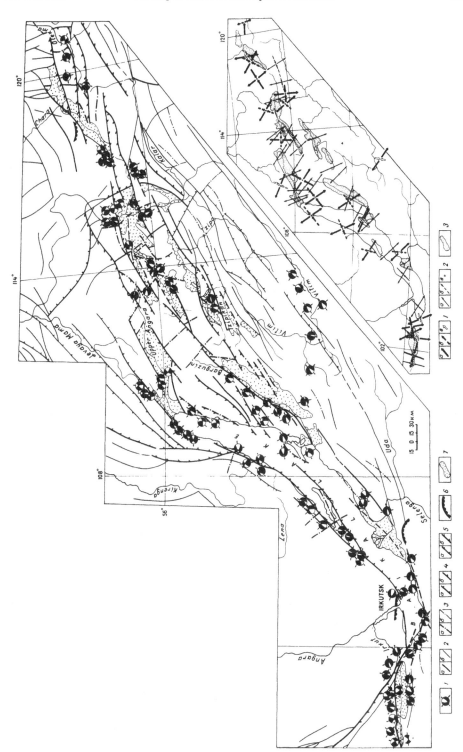

which are characteristic of areas of the most intense crustal extension and subsidence. This is most often associated with the formation of the extension "neck," caused by crustal thinning during the short quasi-plastic flow stage. It is reflected in the increased density of local faults and ruptures. Besides typical normal faults and strike-slip faults, pure tensional faults occur which compensate the intense extension of the crust.

Changes in fault density across the extensional zone is a regular phenomenon accompanied by changes in steepness of fault planes. The dip angle of normal faults generally decreases towards the geometric center of the extensional zone. Normal faults become less steep and compose a system of listric faults. There are exceptions when the dip angles of normal faults parallel to the extensional zone strike do not decrease towards the center of extension, as is observed at the Baikal rift zone flanks.

Fault maps show directions and types of faults. Fault age data are seldom mapped. However, mapping does not provide a qualitative indice of fragmentation of the crust and lithosphere, therefore information obtained from maps cannot always be considered complete.

Figure 4 displays the distribution of fault density for the BRZ. The map is based on geological materials differing in the degree of detailed representation. Fault density of the Baikal rift zone has been studied by the mid-scale geological maps (1:200000). The other zones (Fig. 3) are studied by maps of smaller scales. Nonetheless, the distributions of fault densities within continental rift zones are similar: areas of increased and decreased densities alternate along the rift zone strike. Comparison of this irregularity with the general geological state does not allow one to conclude that variations in fault density are related to the regional crustal structure. Furthermore, on an analogous pattern of fault density, variations were established by SEMINSKY (1986) while studying the internal structure—area of dynamic influence—of the Mongol-Okhotsk lineament. The irregular density of faults and ruptures along the extensional zone strike reflects general regular destruction which is attributed not only to the heterogeneous petrography and lithology of rocks.

The stress field of the Baikal rift zone has been studied by using structural methods. A scheme (Fig. 5) shows the results of these studies. In the central BRZ, stress fields of different types regularly alternate towards its terminations. Extensional stress in the central part gives way to extensional strike-slip and strike-slip stresses at the flanks. The consistent pattern of stress fields of different types is also observed across the strike of the zone, especially at its flank. At the northeastern flank, strike-slip stress fields, perpendicular to the axis, give way to extensional strike slip and strike-slip—extensional stress fields. Further out, even outside, at the structurally defined boundaries of the rift, one finds extensional stress fields. This pattern of stress fields is typical of the southwestern BRZ flank. Thus, local areas of other stress types occur within the zone of the lithosphere extension. It is

Table 1

Relationships between major parameters in the zones of lithospheric extension

Major parameters	Relationships between parameters		
	Natural data	Experimental data	General equation
Fault length L fault number N	$L = 151.4/N^{0.42}$	—	$a = f(L)$ $L = a/N^b$ $b \sim 0.4$
Fault length L and depth of fault penetration H	$H = 1.04L - 0.7$ (km) ($L < 30$ km) $H = 3.89L^{0.76}$ (km) ($L > 30$ km)	$H = (0.2 - 0.3)L + (0.04 - 0.03)$ (H, in cm)	$d \sim 1 \div 4$ $H = dL^n$ $n \sim 0.7 \div 1.0$
Strike-slip fault length L and horizontal displacement a	$a = 0.08L^{0.77}$ (km)	$a = 0.1L - 0.01$ (cm)	$k = 0.01 \div 0.08$ $a = kL^b$ $b = 0.8 \div 1.2$
Spacing between faults of the same direction M and compatible fault length L	$M = 0.29L + 1.74$ (Baikal rift zone) $M = 0.44L^{0.95}$ (continental rift zones)	$M = 0.31L - 0.01$ (cm) (elastic model) $M = 0.15L + 0.04$ (cm) (elasto-viscous model)	$k \sim 0.3 \div 0.5$ $M = kL^c$ $c \sim 0.5 \div 0.95$
Width of the zone of breakage M, thickness of deformation layer T, viscosity η and deformation rate V	—	$M = 1.03T + 0.0112 \lg \eta + 0.047 \lg V - 0.017$ (passive wing) $M = 1.43T + 0.0120 \lg \eta + 0.0025 \lg V - 0.042$ (active wing) $M = 2.70T + 0.003 \lg \eta + 0.028 \lg V + 1.153$ (zone of breakage)	

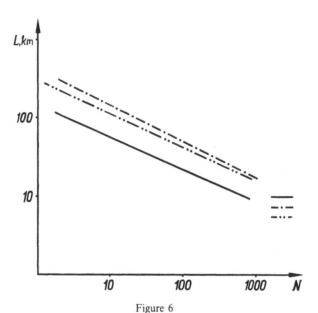

Figure 6
Fault length L versus fault number N. Baikal rift zone—solid line; East Siberian plate—dash-and-dot
line; Altai-Sayan region—dash-and-two dots line.

reasonable to suggest that they are related to changes in the structural conditions at
the flanks. Data on focal mechanisms of strong earthquakes ($M \geq 5.5$) are in good
agreement with the regional stress fields reconstructed by the geological and
structural analysis. This good agreement is due to the fact that in both cases stress
state is evaluated for the crustal volumes of the same order, the size of the volumes
eliminating possible effects of local structural inhomogeneities on estimates.

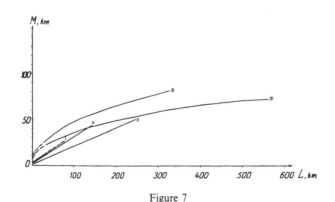

Figure 7
Mean spacing M between faults versus fault length L. I—Baikal rift zone; II—East African rift zone;
III—Altai-Sayan region; IV—Eurasia; V—continental rift zones.

Seismological data obtained from reconstructions of earthquake focal mechanisms in past decades provided information on the recent state of stresses in the BRZ. Geological stress field data, obtained from the analysis of tectonic fracturing and faulting characterize the state of stresses for an extended time since the Neogene. The good agreement of geological and seismological data is also indicative of the regional stability of the state of stress of the lithosphere during the latter part of the Cenozoic. However, local stress fields reconstructed from focal mechanisms of weaker earthquakes are more variable, principal stress orientations vary considerably; focal mechanisms showing local compression or strike-slip can occur in the background of the extensional field.

In other continental rift zones, e.g., in the Rhein rift zone, stress field reconstructions by seismological and geological methods are also in good agreement (AHORNER, 1975; GRINER and LORN, 1980). AHORNER (1975) described changes in orientations of the principal stress during the rift zone evolution since the Late Cretacious. Since the Middle Tertiary (35 Ma) to Late Quaternary, the regional stress fields of the Rhein rift system are stable and correlate with those reconstructed from focal mechanisms of strong earthquakes (AHORNER, 1975; ILLIES, 1975).

From these examples we conclude that the stress field in extensional zones of the lithosphere is stable in time; the period of its existence being compatible with the period of rift zone evolution. Variations of local stress fields are not attributed to local geology since temporal variations of the seismic regime are not compatible with the time of local geologic structures' formation. The latter cannot be regarded as factors controlling variations of local stress fields or the distribution of weak earthquake foci. Variations of local stress fields reflect irregularities of destruction of large lithospheric volumes. This suggestion is supported by physical modeling results (discussed below).

Modern methods of studying geotectonic regimes employ a quantitative description of structures and processes with increasing frequency. Numerical parameters of geologic structures in extensional zones, especially faults, can be relatively easily obtained. Quantitative parameters of faults are numerical values of length, number, depth of penetration, amplitude and direction of displacement, width of the fault influence zone, spacing between faults and density of fracture systems. Satisfactory relationships have been established between these parameters. The most interesting relationships of practical importance are those between fault length, their number, depth of fault penetration, amplitude of displacement, and width of the zone of fault influence among others. Many researchers have been and are engaged in studies of possible relationships between major fault parameters (RANALLY, 1977; SIBSON, 1985; HULL, 1988; MANDLE, 1988; WALSH and WATTERSON, 1988).

The established relationships between the major parameters of faults are listed in Table 1 and Figures 6–8. We discuss some of them in detail.

The qualitative regularity that large faults occur more rarely than small ones was known to practical geologists long ago. Research reveals that this regularity

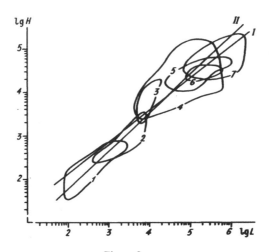

Figure 8

Relationship between fault depth H and fault length L (SAN'KOV, 1989). 1—depth of tin-bearing veins; 2—depth of gold-bearing bodies; 3—mean hypocenter depth; 4—size of aftershock area; 5—maximum hypocenter depth; 6—deep seismic sounding data; 7—gravimetric data.

reflects a property of crustal destruction that can be represented in a numerical form. Figure 6 exhibits plots of relationship $L(N)$ for the Baikal rift zone and other regions, given by the following equation

$$L = a/N^b \tag{1}$$

where L is fault length; N—number of faults; a—coefficient depending on the maximum length of faults analyzed; b—coefficient close to 0.4 which reflects the physical properties of rocks. It is most probably proportional to viscosity.

The above relationship allows us to conclude that during the formation of the fault pattern, that is during mega- and macro-destrution of rocks under natural conditions, general regularities of destruction are displayed. They are similar to the relationships between numbers of fractures of different lengths which were experimentally established for the Maxwell's body failure (KUZNETSOVA, 1969). Thus, deformation and faulting of the lithosphere under extension correspond to those of the Maxwell's body.

Relationship $L(N)$, which was established nearly 20 years ago, is in agreement with the fractal dimension studies which are widely used for quantitative estimation of complex geologic structures.

Figure 7 shows the relationship between fault length and density given by

$$M = 0.4 \, L^{0.95} \tag{2}$$

where M is the average spacing between faults; L—fault length. A physical explanation of this empirical relationship can be found by analyzing stress fields in the areas of interacting faults. It is expected that parameter M is dependent not only on L values, but other geologic parameters. However, geological observations do not allow the disclosure of other factors controlling changes in M.

An important fault parameter is depth of fault penetration. Different approaches to estimating depths of fault penetration are proposed in different branches of geology. The notion "depth of fault penetration" is a matter of debate. SAN'KOV (1989) proposed its definition and emphasized the "instability" of the parameter discussed. He claims that a numerical estimate of the depth of the lowermost fault boundary is considerably dependent on the method employed for its study. It is noteworthy that depth of fault penetration can be discussed only for periods of its initiation or reactivation since stresses are subsequently relaxed and the deep-seated fault segments can recover. Despite the complexity of the problem, depths of fault penetration have been studied in the Baikal rift zone and other regions (SHERMAN, 1977; LOBATSKAYA, 1987; SAN'KOV, 1989).

To estimate depths of fault penetration in the BRZ, a correlation between the average fault length and the average depth of earthquake hypocenters has been analyzed (SHERMAN et al., 1992). It is assumed that stresses are concentrated at fault tips, including the lower ones, and released by earthquakes. Computer processing of 1,258 fault length estimates and 2,146 earthquake hypocenter determinations revealed a correlation between the mean fault length and the mean hypocenter depth in the corresponding locations. The obtained regression equations are the following

$$H = 1.04L - 0.7 \text{ (km)}. \tag{3}$$

Here H is the average depth of hypocenters or of active penetration of faults; L is the average fault length. The equation is valid for L less than 25–30 km. With depth, the direct relationship is disturbed and the bilogarithmic ratio H/L changes (Fig. 8) (SAN'KOV, 1989). The regression equation is the following

$$H = 3.89L^{0.76}. \tag{4}$$

Ratio H/L decreases with depth.

A relationship between fault length and amplitude of displacement has been discussed in many papers (HERVE and CAILLEUX, 1962; KANAMORI and ANDERSON, 1975; MURAOKA and KAMOTA, 1983; RAMSEY and HUBER, 1983; WATTERSON, 1986; WALSH and WATTERSON, 1988; and others). For the BRZ, a correlation between length of shear ruptures and amplitude of displacement along them is given by

$$a = kL^b \tag{5}$$

where a is amplitude of displacement; L is length of ruptures; k and b are coefficients varying from 0.01 and 0.08 and from 0.8 to 1.2, respectively.

Geological Conclusions

—Normal faults are dominant over other types of faults;

—the spatial distribution of local and regional faults is irregular: local faults mainly occur at the central part of the rift zone, whereas regional faults evolve at its marginal parts;

—faults become selectively reactivated during rifting;

—density of faults is irregular;

—stress field patterns within the extensional zone are variable;

—changes in fault density and stress field variations are not always controlled by the structure of the basement;

—fault density depends on fault length;

—spacing between faults depends on their lengths;

—amplitude of displacement along a fault correlates with fault length;

—depth of fault penetration correlates with fault length;

—ratio of fault parameters is often valid for other geotectonic regimes.

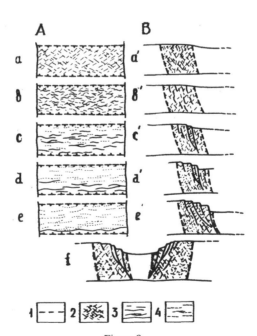

Figure 9

Scheme of deformation and dynamics of fracturing in extensional zone model. A —successive top views for different stages of deformation. B —vertical cross sections for different stages of deformation. 1—boundary of extensional zone; 2—preliminary joints; 3—active normal and tension faults; 4—passive normal and tension faults.

Modeling of Extensional Zones: Results and Discussion

The above quantitative correlations between fault parameters allow us to conclude that faulting in the lithosphere takes place according to rules or regularities. The established qualitative and quantitative characteristics of faulting in the zones of lithospheric extension have been compared with physical modeling results. The similarity of deformation in models and natural conditions was observed by the principle of self-similarity (SHERMAN et al., 1983):

$$\rho g L t / \eta = \text{const.} \qquad (6)$$

where ρ —density; g —acceleration of gravity force; L —linear dimensions; T —time; η —viscosity.

Physical models made from a brownish viscoplastic clay were deformed in a specially designed apparatus "Razlom" ("Fault"). Model viscosity was varied by changing the water content of the material. Experiments were undertaken by Bornyakov and Seminsky (SHERMAN et al., 1992).

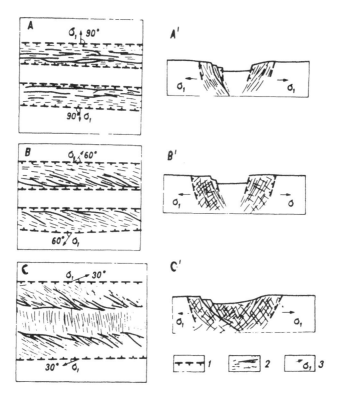

Figure 10

Dependence of extensional zone pattern on the vector of extension stresses (A, B, C —top views; A', B', C' vertical cross sections). 1 boundary of extensional zone; 2 active faults (solid line) and fractures which stopped to develop (dashed line); 3 – vector of extension stress.

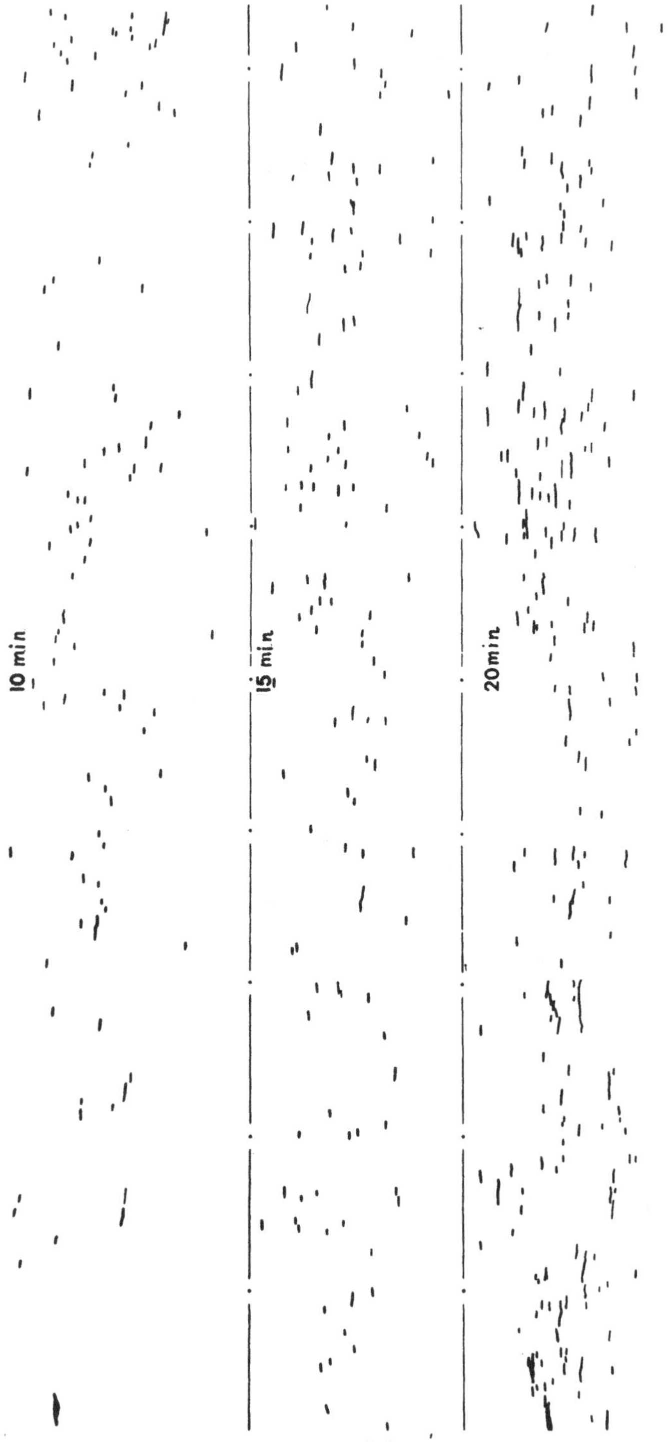
10 min
15 min
20 min

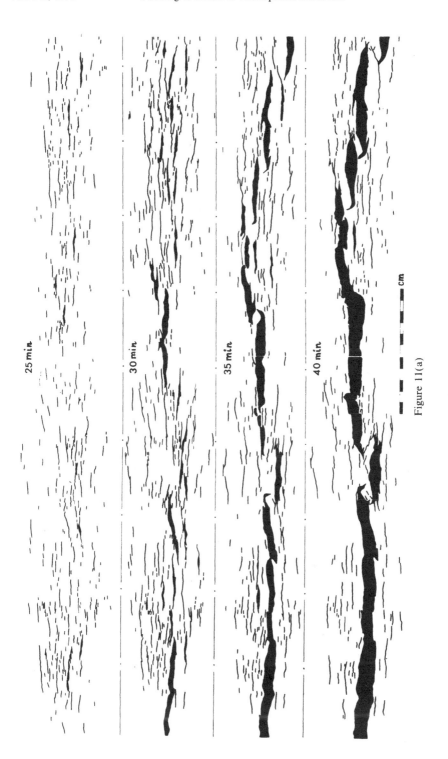

Figure 11(a)

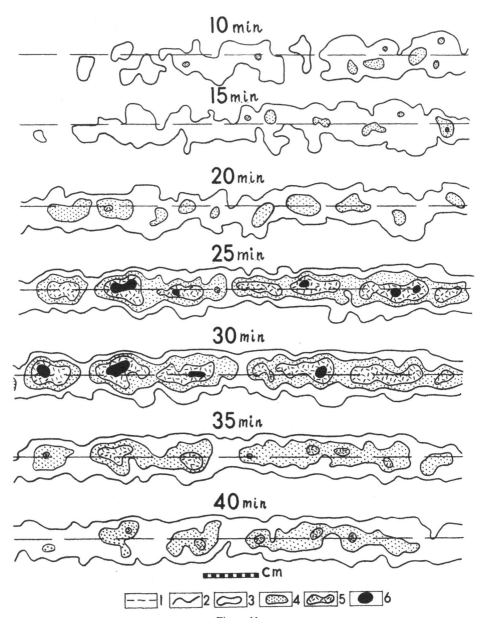

Figure 11
Fracturing at different stages of the extensional zone formation (normal fault) and fault density (results of modeling). A — top views of modeling faults; B — density of faults; 1 — border between movable parts; 2–6 — increase of density isolines.

The sequence of the fracture pattern formation in the extensional zone is shown in Figure 9. After deformation of the thinning neck," fracturing starts. At the initial stage, the process takes place in two linear zones bordering the neck. Fracturing starts by the formation of small joints. With the increase in deformation, the process spreads towards the central part of the zone. Large fractures evolve by joining the shorter ones. Small fractures which have not joined together cease development. Across the extensional zone strike, fracturing propagates from the periphery to the center. It has been observed that the dip angle of large fractures is variable. In the subsurface layers, it ranges from 60° to 85°. With depth, the fractures become less inclined, their dip angle being 50°–70°, to compose a system of listric normal faults. Younger fractures tend to occur closer to the graben axis and are more steeply dipping near the surface. The fracture pattern is considerably dependent on the viscosity and the rate of deformation. Models with $\eta = 10^{4-5}$ Pa · sec. and $V = 10^{-4}$ m/sec. produce results that correspond best to natural examples. The internal pattern of the extensional zone is predetermined by the direction of extension forces. Consequently smaller is the angle between the extension zone strike and the extension vector, the more favorable are conditions for the development of fractures obliquely to the extensional zone strike (Fig. 10).

The modeling provided estimates of the destruction zone width (m), depending on the model thickness h, deformation rate V and model material viscosity η as given by the following formula

$$m = 1.03\,h + 0.0122\,\lg\eta + 0.0047\,\lg V - 0.017. \qquad (7)$$

In the experiments with respect to the position of the moving wing (active part of the experimental apparatus) and that of the stable (passive) one, the symmetry of the extensional zone pattern relative to the axial lateral plane was disturbed. The destruction zone width in the active part was larger than that in the passive part. The total width of the destruction zone is given by

$$m = 2.70\,h + 0.003\,\lg\eta + 0.028\,\lg V + 0.153. \qquad (8)$$

For a series of experiments, the coefficient of correlation $r = 0.9$.

Experiments were undertaken to investigate the regularities of distribution of fault density within the limits of the extensional zone. To accomplish this goal, the density parameters of fault were estimated at regular time intervals during deformation. The results are shown in Figure 11. Notwithstanding the model material homogeneity and the stable mode of loading, fracturing is not uniformly distributed throughout the extensional zone. Fracture sets of maximum population are spaced at almost equal distances from each other. Profiles along the zone strike, constructed from the maximum fault densities, support the observation. With the increase in deformation, the degree of localization of maxima also increases thus leading to the decrease in the number of separate maxima and to simplification of the internal pattern of the extensional zone. During final stages of experiments, the

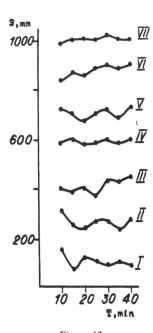

Figure 12
Along-strike migration of maxima of fault density in time. T—time; S—distance from maximum center
to the left side of the model.

extensional zone is represented by a series of maxima of the fracture density which
are bordered by individual large tension faults.

A set of experiments was performed using models of different thicknesses. It was
observed that the spacing between the maxima becomes larger if the deformed layer
thickness is increased. It drastically decreases with the increase in the rate of
deformation. Consequently, the irregular distribution of fault densities in natural
extensional zones reflects a specific feature of the lithospheric failure, and the
heterogeneous geology of the substratum is not important in the discussed process.

The experiments demonstrated that the occurrence of fracturing is not uniformly distributed over time and revealed that maxima of fracture density migrate
in time. Two sets of experiments were performed to investigate this phenomenon.
The first set provided data on changes in distribution of fracture density relative to
the extensional zone boundary at different stages of deformation. Figure 11 presents
the results of the study. Areas of higher density of fracture migrate in time in the
direction perpendicular to the model surface. Deviations of maxima from the model
axis can attain 30–40 percent of the total model width. The same phenomenon is
observed in the lateral direction. Figure 12 shows maxima of density of joints
corresponding to different stages of the extensional zone formation. Their centers
migrate relative to the terminations of the fault. Physical modeling suggests the

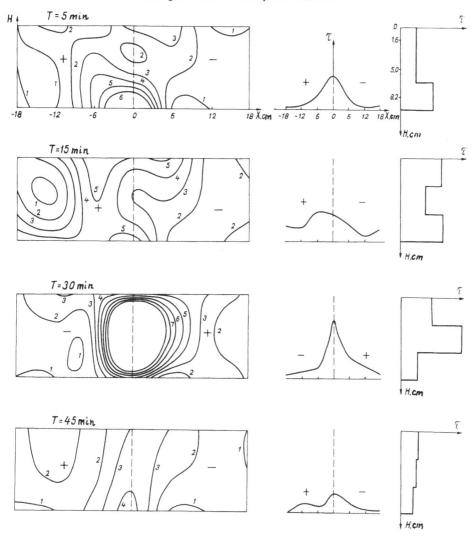

Figure 13

Approximate estimate of deformation rate in developing fault wings and epures of tangential stresses. The cross section is perpendicular to the axial fault plane (dashed). T—time; H—model thickness; X—distance from the fault plane in transverse direction; τ—tangential stresses. Isolines show deformation rates; "+" and "−" denote the fault wings which are under deformation at a given moment.

regional stability of areas of high fault density whereas locally maxima of fracture density migrate within the area of high-fault density both in time and space.

Migration of deformation fields in physical models was registered by sensing elements placed within the volume of the viscoplastic models. The experiments carried out by Buddo have been reported in SHERMAN et al. (1994). The experimental results support migration of deformation fields during the formation of the

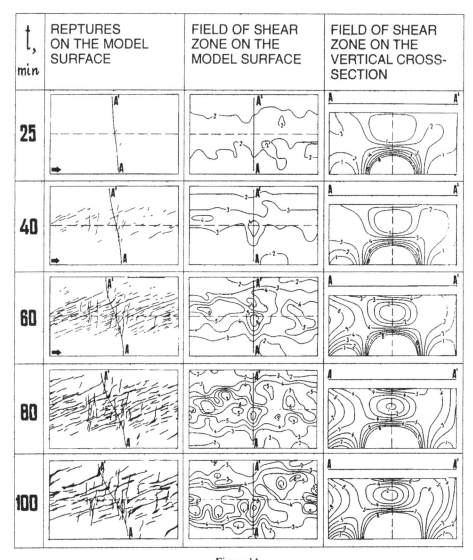

Figure 14
Relationship between fractures on the model surface and deformation fields on the surface and in the vertical model section for different stages of deformation.

extensional zone. Deformation fields migrate from one flank to the other, along and across the strike of the zone (Figs. 13 and 14).

The experiments supply evidence that the density of fractures is irregular within the zone of destruction and that local fields of deformation migrate in time. The same is typical of the areas of higher density of fractures. This conclusion is in agreement with the regime of seismic activity in the rift zones of the world.

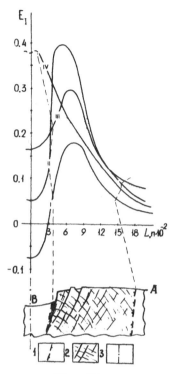

Figure 15

Distribution of elongation deformation E across the extensional zone strike, 1—boundaries of the zone; 2—fracture patterns; 3—axial line.

Distribution of areas of different types of deformation is shown in Figure 15. These are the data obtained by BORNYAKOV (1990). Data from four experiments with different boundary conditions are plotted. Curve I shows variations in the deformation of elongation (E) for the low viscosity models (10^4 Pa · sec.) at the low deformation rate ($V = 10^{-5}$ m/sec.). The higher values of E are registered at the marginal parts of the extensional zone. The intensity of deformations drastically decreases in the horizontal direction and the curve enters negative values. This indicates that in the central part of "extensional neck," conditions of extension are substituted by conditions of compression. Curve II is for the deformation rate up to 10^{-3} m/sec. In principle, the only distinction between the Ist and the IInd curves is that there is no transition to the area of compression. Curve III characterizes the conditions of deformation with the increased viscosity of the model material. Deformations of E across the strike of the extensional zone are irregularly scattered.

There are several extreme values, deformation decreases towards the center of the extensional zone. The experiments support the pattern of the extensional zone as deduced from the regional stress field analysis. Changes in local deformation fields and local stress have been experimentally registered. Thus, one more phenomenon typical of rift zones of the lithosphere was experimentally simulated.

Conclusions

1. Destruction of the lithosphere in extensional zones takes place according to certain regularities. It reflects a general property of the lithosphere which is subjected to failure as a physical body. The lithospheric destruction is best approximated by a self-similar Maxwell's body.

2. Under steady loading, destruction of the lithosphere occurs irregularly both in space and time. The interrupted course of the process is related neither with the nonsteady action of the internal endogenic energy source (as it is impossible) nor with the inhomogeneities of the lithosphere.

3. Seismicity reflects the ongoing destruction of the lithosphere and gives evidence of its intermittent mode which is, for the most part, independent of the geologic setting at the basement.

4. Relatively stable quantitative relationships between the major fault parameters are indicative of the regularities of faulting in the lithosphere under extension. They suggest that other relationships between structures caused by the lithospheric deformation and the processes synchronous to it can be established (Table 1).

5. The uneven destruction of extensional zones of the lithosphere requires a more thorough analysis in terms of fracture mechanics of mega-volume bodies.

Acknowledgements

The author is grateful for experiments and discussions with colleagues at the Laboratory of Tectonophysics: Drs. S. Bornyakov, K. Seminsky and V. Buddo. O. Baksheeva is thanked for his English version of the paper.

The author would like to express his appreciation to P. Davis for reviewing the original manuscript and kindly providing suggestions in the revised text.

To conclude, the author wishes especially to thank Mr. G. Soros and his International Science Foundation for providing a grant which allowed the author to participate in the International Symposium "On mechanics problems in Geodynamics" held September 5–9, 1994 in Beijing, China with a report represented in this publication. This research was supported by the Russian Fund of Fundamental Investigation, Grant 95–05–14211.

REFERENCES

AHORNER, L. (1975), *Present-day Stress Field and Seismotectonic Block Movements along Major Fault Zones in Central Europe*, Tectonophysics 29 (1–4), 233–249.

BORNYAKOV, S. A. (1990), *Dynamics of Structure Formation in a One-layer Model under Arbitrarily Active Mechanism of Extension*, Geologia i Geofizika *1*, 47–56 (in Russian).

EATON, G. P. (1987), *Topography and Origin of the Southern Rocky Mountains and Alvarado Ridge*, Continental Extensional Tectonics, Geol. Soc. Spec. Publ. *28*, 355–369.

GOLENETSKY, S. I. (1990), *Problems of the Seismicity of the Baikal Rift Zone*, J. Geodyn. *11*, 293–307.

GZOVSKY, M. V., *Fundamentals of Tectonophysics* (Nauka, Moscow 1975) 535 pp. (in Russian).

HERVE, I. C., and CAILLEUX, A. (1962), *Étude quantitative des Failles de Pechelbronn (Basin)*, Cahiers Geologiques, Paris, 68–70.

HULL, J. (1988), *Thickness-displacement Relationship for Deformation Zones*, J. Struct. Geol. *10* (4), 431–435.

ILLIES, J. H. (1975), *Recent and Paleo-intraplate Tectonics in Stable Europe and the Rhein Graben Rift System*, Tectonophysics *29* (1–4), 251–264.

ILLIES, J. H. (1981), *Mechanism of Graben Formation*, Tectonophysics *73*, 249–266.

JAIRI, F., and MIZUTANI, Sh. (1969), *Fault System of the Lake Tanganyika Rift at the Kigoma Area, Western Tanzania*, J. Earth Sci. Nagaya Univ. *17*, 71–95.

KANAMORI, H., and ANDERSON, D. L. (1975), *Theoretical Bases of Some Empirical Relations in Seismology*, Bull. Seismol. Soc. Am. *65*, 1071–1095.

KAZMIN, V. G., *Rift Structures of the East Africa Continental Division and the Ocean Onset* (Nauka, Moscow, 1987) 205 pp. (in Russian).

KAZMIN, V. G., BERHE, S., and WALSH, J. (1980), *Geophysical Map of the Ethiopian Rift and Explanatory Note*, Ethiop. Inst. Geol. Surv., 37 pp.

KUZNETSOVA, K. I., *Regularities of Failure of Elasto-viscous Bodies and Some Possibilities of their Application to Seismology* (Nauka, Moscow, 1969) 87 pp. (in Russian).

LEE, J. S., *Introduction of Geomechanics* (Science Press, Beijing, China 1984) 234 pp.

LEVI, K. G., *Neotectonic Crustal Movements in Seismoactive Zones of the Lithosphere* (Nauka, Novosibirsk 1991) 166 pp. (in Russian).

LOBATSKAYA, R. M., *Structural Zonality of Faults* (Nedra, Moscow 1987) 128 pp. (in Russian).

MANDLE, G., *Mechanics of Tectonic Faulting. Models and Basic Concepts* (Elsevier 1988) 407 pp.

NADAI, A., *Theory of Flow and Fracture of Solids*. Vol. 2 (McGraw-Hill Book Company Inc., New York 1963) 840 pp.

NUR, A. (1982), *Origin of Tensile Fracture Lineaments*, J. Struct. Geol. *4* (1), 31–40.

OTSUKI, K. (1978), *On the Relationship between the Width of Shear Zone and the Displacement along the Fault*, J. Geol. Soc. of Japan *84* (11), 661–669.

RANALLY, G. (1977), *Correlation between Length and Offset in Strike-slip Fault*, Tectonophysics *37* (4), T1–T7.

RUZHITCH, V. V., and SHERMAN, S. I., *Estimation of relationship between length and amplitude of ruptures*. In *Dynamics of the East Siberia Crust* (Nauka, Novosibirsk 1978), pp. 52–57 (in Russian).

SAN'KOV, V. A., *Depth of Fault Penetration* (Nauka, Novosibirsk 1989) 136 pp. (in Russian).

SEMINSKY, K. ZH. (1986), *Analysis of Distribution of Initial Fractures during the Formation of Large Faults*, Geologia i Geofizika *10*, 9–18 (in Russian).

SENGOR, A. M. C. (1990), *Plate Tectonics and Orogenic Research after 25 Years: A Tethyan Perspective*, Earth Sci. Rev. *27*, 1–201.

SHEMENDA, A. I. (1984), *Some Regularities of Lithosphere Deformation under Extension (from Modelling Results)*. Dokl. AN SSSR, 275, 2, 446–350 (in Russian).

SHERMAN, S. I., *Physical Regularities of Crustal Faulting* (Nauka, Novosibirsk 1977) 102 pp. (in Russian).

SHERMAN, S. I., BORNYAKOV, S. A., and BUDDO, V. YU., *Areas of Dynamic Influence of Faults* (Modelling Results) (Nauka, Novosibirsk 1983) 112 pp. (in Russian).

SHERMAN, S. I., and DNEPROVSKY, YU. I., *Crustal Stress Fields and Geological and Structural Methods of their Studies* (Nauka, Novosibirsk 1989) 157 pp. (in Russian).

SHERMAN, S. I., SEMINSKY, K. ZH., and BORNYAKOV, S. A. et al., *Faulting in the Lithosphere. Extensional Zones* (Nauka, Novosibirsk 1992) 228 pp. (in Russian).

SHERMAN, S. I., SEMINSKY, K. ZH., BORNYAKOV, S. A., ADAMOVICH, A. N., and BUDDO, V. Y., *Faulting in the Lithosphere. Compressional Zones* (Nauka, Novosibirsk 1994) 263 pp. (in Russian).

SIBSON, R. H. (1985), *A Note of Fault Reactivation*, J. Struct. Geol. *7* (6), 751–754.

STOYANOV, C., *Mechanism of Fracture Zone Formation* (Nedra, Moscow 1977) 144 pp. (in Russian).

TURCOTTE, D., and SCHUBERT, G., *Geodynamics, Applications of Continuous Physics to Geological Problems* (John Wiley and Sons, Inc., New York 1982).

WALSH, J. J., and WATTERSON, J. (1988), *Analysis of the Relationship between Displacements and Dimensions of Faults*, J. Struct. Geol. *10* (3), 239–247.

WATTERSON, J. (1986), *Fault Dimensions, Displacements and Growth*, Pure and Appl. Geophys. *124* (1–2,), 365–373.

ZHONG, J., SHAN, J., and WANG, Z. (1982), *The Experimental Study of the Formation of Grabens*, Sci. Geol. Sin. *2*, 161–178.

(Received September 8, 1994, revised March 25, 1995, accepted April 10, 1995)

PAGEOPH, Vol. 146, Nos. 3/4 (1996)

0033–4553/96/040447–21$1.50 + 0.20/0

Dynamic Interactions between Crustal Shortening, Extension, and Magmatism in the North American Cordillera

MIAN LIU[1]

Abstract —This paper examines the first-order dynamic interactions between crustal shortening, extension, and volcanism in tectonic evolution in the North American Cordillera. The protracted crustal compression in the Mesozoic and early Cenozoic (110–55 Ma) contributed to the subsequent Tertiary extension by thermally weakening the lithosphere and producing an overthickened (>50 km) and gravitationally unstable crust. In addition to post-kinematic burial heating, synkinematic thermal processes including conduction are shown significantly because of the long period of crustal contraction and the slow shortening rates (<4 mm/yr). The effects of shear heating were probably limited for the same reasons. Localized delamination of the lithospheric mantle may have contributed to the abundant plutonism and high crustal temperature in the southeastern Canadian Cordillera at the end of the orogeny. Most early-stage extension in the Cordillera, characterized by formation of metamorphic core complexes, resulted from gravitational collapse of the overthickened crust. Plutonism may have facilitated strain localization, causing widespread crustal extension at relatively low stress levels. Crustal collapse, however, was unlikely the direct cause of the Basin-Range extension, because the gravitational stresses induced by crustal thickening are limited to the crust; only a small fraction of the gravitational stresses may be transmitted to the lithospheric mantle. Nor could core complex formation induce the voluminous mid-Tertiary volcanism, which requires major upwelling of the asthenosphere. While the causes of the asthenospheric upwelling are not clear, such processes could provide the necessary conditions for the Basin-Range extension: the driving force from thermally induced gravitational potential and a thermally weakened lithosphere. The complicated spatial and temporal patterns of volcanism and extension in the Basin and Range province may be partially due to the time-dependent competing effects of thermal weakening and rheological hardening associated with intrusion and underplating of mantle-derived magmas.

Key words: Continental extension, crustal shortening, metamorphic core complex, Basin and Range.

1. Introduction

Tectonic evolution of continents involves many phases of crustal shortening, extension, and magmatism. Geological observations indicate close interrelations between these processes. For instance, many continental rifts developed in old thrust belts, and volcanism were common in extensional regimes. The purpose of this

---

[1] Department of Geological Sciences, University of Missouri, Columbia, MO 65211, U.S.A.

paper is to investigate dynamic interactions of these processes in tectonic evolution in the North American Cordillera, where a protracted phase of crustal compression during the Mesozoic and the early Cenozoic was followed by extensive continental extension and volcanism in the Tertiary. The resultant Basin and Range province in the western United States and Mexico is one of the most extended continental regimes in the world (Fig. 1).

While intensive studies in the past few decades have greatly refined our understanding of the Cordilleran tectonics, many fundamental questions, such as the causes of crustal extension, remain unclear. Much of the controversy is rooted in our incomplete understanding of the interrelations among various tectonic processes. CONEY and HARMS (1984) showed that formation of metamorphic core complexes, which characterizes the onset of Tertiary extension in the Cordillera, occurred mainly in a zone of overthickened crust in the hinterland of the

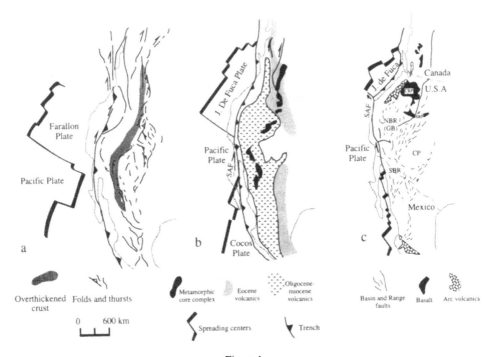

Figure 1
Tectonic setting in the North American Cordillera. (a) Early Tertiary: crustal shortening led to the development of the fold and thrust belt (FTB). Crust in the hinterland was thickened to more than 50 km. The gray curve shows the location of the present coastline. (b) Mid-Tertiary: metamorphic core complexes developed in the hinterland of overthickened crust; convergent plate boundary was progressively replaced by strike-slip motion along the San Andreas fault system (SAF). (c) Late Tertiary: widespread extension in the Basin and Range province. Thin lines depict block faults in the Basin and Range province. NBR, northern Basin and Range; GB, Great Basin; SRB, southern Basin and Range; SRP, Snake River Plain; CRP, Columbia River Plateau, CP; Colorado plateau. Modified from CONEY (1987).

Cordilleran fold and thrust belt (FTB) (Fig. 1). This observation indicates a link between crustal compression and extension, and gravitational collapse of overthickened crust has been proposed as the driving mechanism for extension in the Cordillera (CHEN and MOLNAR, 1983; CONEY and HARMS, 1984; SONDER et al., 1987; WERNICKE et al., 1987). However, the factors controlling crustal collapse, and hence the causes of the large spatial-temporal variations of core complex formation in the Cordillera, are not well understood (WERNICKE et al., 1987; LIU and FURLONG, 1993). Furthermore, not all core complexes occurred in the hinterland of the FTB. In the southern Basin and Range province, most core complexes developed in the midst of a deep-seated thrust belt (CONEY, 1980).

The relationship between core-complex formation and the younger (< 17 Ma) tectonomagmatism which is chiefly responsible for the Basin-Range structure is also questionable. Previous studies, either attributing Tertiary extension to plate interactions along the western margin of North America (ATWATER; 1970; LIPMAN, 1980; SEVERINGHAUS and ATWATER, 1990), or emphasizing local gravitational stresses associated with the thickened crust (CHEN and MOLNAR, 1983; CONEY and HARMS, 1984; SONDER et al., 1987; CHEN and MOLNAR, 1983), all treated core-complex formation and the Basin-Range extension as the results of common causes. Such hypotheses have considerable difficulty in explaining the vast differences in extension styles and magmatism associated with core complexes and the Basin and Range (CONEY, 1987).

Another major problem is the dynamic relations between magmatism and extension. The timing of core-complex formation is apparently correlated with the abundance of coeval magmatism (WERNICKE et al., 1987). However, correlation between the mid-Tertiary magmatism and crustal extension in the Great Basin (the northern Basin and Range) is controversial. In some regions the magmatism appears to be synextensional (GANS, 1987; GANS et al., 1989), while a number of recent studies argued that the mid-Tertiary volcanism and the Basin-Range extension were poorly correlated over time and space at a provincial scale (TAYLOR et al., 1989; BEST and CHRISTIANS, 1991; AXEN et al., 1993).

While understanding all details of the Cordilleran tectonics is very difficult, it is feasible to examine the first-order dynamic links between crustal shortening, extension, and magmatism. Results of such studies may shed light on many of the fundamental questions mentioned above. The following discussion begins with thermal-rheological effects of crustal shortening and their control on the subsequent crustal extension. I then examine the relationship between core-complex formation and the Basin-Range extension. I will show that crustal collapse may be the major cause of core complex formation but not for the Basin-Range extension. Finally, I discuss the thermal-rheological effects of intrusion of mafic magma in the crust and their roles in the Tertiary tectonomagmatism. I will show that thermal perturbations in the asthenosphere, indicated by the voluminous mid-Tertiary volcanism, may have provided the necessary conditions for the Basin-Range extension.

2. Crustal Shortening

Mesozoic and early Cenozoic (165–55 Ma) crustal compression in western North America involved a complicated history of subduction-related deformation and massive plutonism along the coastal margin (BURCHFIEL and DAVIS, 1975). Except for accreted terranes, contraction in the inland Cordillera occurred rather uniformly in a zone stretching from Canada to Mexico (STEWART, 1978; ELISON, 1991), telescoping more than 200-km crust and leading to the development of the fold and thrust belts. In the hinterland of the FTB the crustal thickness was nearly doubled to more than 50 kilometers (CONEY and HARMS, 1984; PARRISH et al., 1988).

The dynamic effects of crustal shortening on the subsequent extension may be twofold: (1) it produces an overthickened crust, which is dynamically unstable and tends to collapse under its own weight; (2) it thermally weakens the lithosphere. Here I focus on thermal-rheological effects associated with crustal shortening. Crustal collapse is discussed later.

Thermochronological evidence signifies that the crust was abnormally hot when core complexes developed (PARRISH et al., 1988). This is also indicated by the coeval plutonism closely associated with core-complex formation (ARMSTRONG and WARD, 1991). Theoretically, significant thermal weakening is necessary for major extension to occur, because the tensile strength of a normal continental lithosphere is about one order of magnitude higher than typical tectonic forces (SONDER et al., 1987; LYNCH and MORGAN, 1987). The question here is how much heat may be directly produced by crustal shortening.

Thermal effects of crustal shortening were the subject of numerous studies, mainly because of their importance in reconstructing the pressure-temperature-time paths of metamorphic rocks (OXBURGH and TURCOTTE, 1974; ENGLAND and RICHARDSON, 1977; ENGLAND and THOMPSON, 1984; SHI and WANG, 1988; RUPPEL et al., 1988). In general, major thermal processes associated with crustal shortening may include (1) thermal relaxation, (2) radiogenic heating, (3) frictional or shear heating, (4) heat advection by thrusts, (5) erosion, and (6) heat flux from the mantle. Many of these processes may occur simultaneously. The transient thermal evolution can be described as

$$\frac{\partial T}{\partial t} = \kappa \ \nabla^2 T + \frac{1}{\rho_c C_p} (A_r + A_s) - \vec{u} \cdot \nabla T - \dot{H} \tag{1}$$

where T is the temperature, $\vec{u}$ is the velocity vector, and the term $\vec{u} \cdot \nabla T$ represents thermal advection associated with crustal thrusting and erosion. The parameter κ is the thermal diffusivity, ρ_c is the crustal density, C_p is the specific heat, A_r and A_s are the volumetric radiogenic heating and shear heating, respectively, and $\dot{H} = (L/C_p)(\partial f/\partial T)(\partial T/\partial t)$ is the rate of temperature change due to partial melting, where f is melt fraction and L is the latent heat of fusion (LIU and CHASE, 1991).

Table 1

Model Parameters

Parameter	Definition	Value
Ar_0	Volumetric radioactive heating (surface value)	2×10^{-6} W/m^3
C_p	Specific heat	1000 J/kg. K
d	Characteristic length scale of radioactive heating	10 km
κ	Thermal diffusivity	10^{-6} m^2/s
L	Latent heat of fusion	420 kJ/kg
ρ_c	Density of the crust	2700 kg/m^3
ρ_m	Density of the mantle	3300 kg/m^3

Values of model parameters are given in Table 1. The time-dependent thermal evolution with various thermal processes and boundary conditions was simulated in a two-dimensional finite difference model (LIU and FURLONG, 1993).

2.1. Burial Heating and Lithosphere Weakening

During crustal shortening, heat is both advected by thrusting sheets and diffused by conduction. For crustal shortening at rates typical of plate motion (a few centimeters per year), heat advection by thrusting is much greater than thermal conduction, therefore thrusting may be regarded as instantaneous (OXBURGH and TURCOTTE, 1974). This assumption leads to a "sawtooth" geotherm (temperature variation with depth) across thrust belts immediately after crustal shortening (Fig. 2a). Thermal contrasts in the crust would be quickly erased by conduction, which is effective when thermal gradients are sharp, as indicated by Fourier's Law

$$q = -k \frac{dT}{dz} \tag{2}$$

where q is conductive heat flux, k is thermal conductivity, and dT/dz is the thermal gradient. Such processes are commonly referred to as thermal relaxation.

Notice that thermal relaxation involves no heat generation. Initially heat transfers from the thrust sheet to the underlying crust. In the absence of radiogenic heating, temperature of the lithosphere would approach a linear geotherm (Fig. 2a). In reality, crustal thickening also results in greater radiogenic heating because of the high concentration of radiogenic elements in the crust. Assuming an exponential distribution of radiogenic heating (LACHENBRUCH, 1970): $A = A_0 \exp(-z/d)$ (see Table 1), Figure 2a shows that the thrust sheet is heated slightly over its original temperature in 50 million years, and temperature increase in the underlying crust is considerably greater. The total effects of thermal relaxation and radiogenic heating in a thickened crust are sometimes referred to as burial heating.

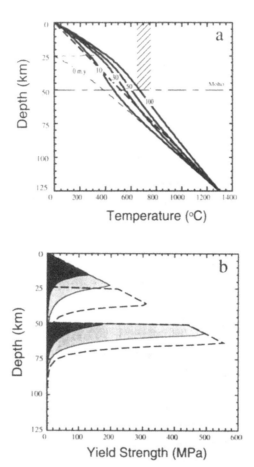

Figure 2

(a) Thermal effects of burial heating. The dashed "sawtooth" profile is the initial geotherm when a column of a "normal" lithosphere is buried under a 25-km thick thrust sheet. The geotherm is quickly smoothed by thermal relaxation. The dashed gray curve is the equilibrium geotherm in the absence of radiogenic heating. After approximately 10 my, temperature increase is mainly due to radiogenic heating. The hatched area indicates the range of solidus of wet granitic rocks. (b) Strength envelops of the lithosphere. Rheology structure of the lithosphere is approximated with a dry granitic crust and a dry olivine mantle (see Table 2). Dashed profile: the strength envelop immediately after crustal thickening; gray profile: after 50 my of thermal relaxation alone; dark profile: after 50 my of burial heating (thermal relaxation plus readiogenic heating). The vertically integrated lithospheric strength (equivalent to the areas enclosed by the strength envelops) are 1.5×10^{13} N/m (dashed curve); 1×10^{13} N/m (gray curve); and 3×10^{12} N/m (dark curve).

Associated with burial heating is mechanical weakening of the lithosphere, which results from burial heating and depression of the lithospheric mantle to a hotter regime under the thickened crust. The lithospheric strength may be represented by the vertically integrated yield strength, which is the minimum deviatoric stress required to cause either brittle or ductile deformation

$$S = \int_0^l [\text{Min}(\sigma_D(z), \sigma_B(z))] \, dz \tag{3a}$$

where l is the thickness of the lithosphere, σ_D and σ_B are the deviatoric stresses for ductile and brittle deformation, respectively

$$\sigma_D = \left(\frac{\dot{\varepsilon}}{A}\right)^{1/n} \exp\left(\frac{H}{nRT}\right) \tag{3b}$$

$$\sigma_B = \mu\sigma_n \tag{3c}$$

where $\dot{\varepsilon}$ is the strain rate, typically in the range between 10^{-14} to $10^{-16}\,\text{s}^{-1}$ for continental tectonics (unless otherwise specified, all calculations assume a strain rate of $\dot{\varepsilon} = 10^{-15}\,\text{s}^{-1}$). The rheological constants A and n, and the activation enthalpy, H, a.e lithology-dependent (Table 2). The parameters R is the gas constant, T is the absolute temperature, μ is the frictional coefficient (taken to be 0.65), and σ_n is the normal stress on the fault plane.

The weakening effects associated with burial heating are shown in Figure 2b. Immediately after an "instantaneous" crustal thickening, the integrated lithospheric strength is about $1.5 \times 10^{13}\,\text{N/m}$, one order of magnitude greater than typical tectonic forces (LYNCH and MORGAN, 1987). The lithosphere is then weakened by burial heating and by depression of the lithospheric mantle, which has a temperature-sensitive rheology, to a hotter regime. Note that these processes would take 30 to 50 million years to reduce the lithospheric strength to a level comparable to the tensile forces induced by crustal thickening, about $O(10^{12})\,\text{N/m}$. In the southeastern Canadian Cordillera, major crustal extension occurred only a few million years after the cessation of crustal compression (PARRISH et al., 1988; WERNICKE et al., 1987).

LIU and FURLONG (1993) argued that the instantaneous thrusting commonly assumed in thrust models may not be valid for the Cordillera where crustal compression was a long and slow process. The crustal shortening rates, even during the active periods, were only 2–4 mm/yr (ELISON, 1991). In such cases synkinematic thermal processes, including heat conduction and radiogenic heating, may be

Table 2

Flow Law Parameters

	A	H (kJ/mol)	n	References
Granite	10^{-88}	123	3	KIRBY and KRONENBERG (1987)
Diabase	$10^{-3.7}$	260	3.4	KIRBY and KRONENBERG (1987)
Olivine	$10^{3.28}$	420	3	RUTTER and BRODIE (1988)

important. This is clear from the Peclet number (*Pe*), which is essentially the ratio of heat advection to conduction: $Pe = vl/\kappa$, where v is the rate of thrusting, l is the thickness of thrust sheet, and κ is the thermal diffusivity. Take $l = 25$ km, $\kappa = 10^{-6}$ m^2/s, and $v = 4$ mm/yr, the Peclet number is only around 3, indicating that synkinematic thermal conduction is as important as heat advection by thrusting and therefore should be integrated through the period of crustal compression. LIU and FURLONG (1993) showed that synkinematic burial heating may have played a significant role in thermal evolution in the Cordillera. Nontheless, burial heating alone is unlikely to cause major crustal anatexis, which requires crustal temperature to be lifted over 650–750°C even for water-saturated source rocks (see Fig. 2a). Models predicting major crustal anatexis by burial heating usually assumed a constant mantle heat flux (e.g., ZEN, 1988). However, as burial heating and other thermal processes (see below) warm up the crust, thermal gradients near the crust-mantle boundary decrease (Fig. 2a), and so does mantle heat flux according to Eq. (2). LIU and FURLONG (1993) showed that mantle heat flux may be reduced by 40% with burial heating; the reduction can be much greater if strong shear heating occurs in the crust.

2.2. Shear Heating

Frictional or shear heating associated with crustal shortening has the potential to effectively weaken the lithosphere and cause significant crustal anatexis. This mechanism has been suggested as the major cause of magmatism in the Himalaya Main Central Thrust belt (MOLNAR and ENGLAND, 1990). The controversy of shear heating usually centers on the effective shear stresses. Heat flow along the San Andreas fault indicates that tectonic stresses in the fault are probably less than 10–20 MPa (LACHENBRUCH and SASS, 1978). On the other hand, higher stresses in the crust have been suggested by rock strength experiments (BYERLEE, 1978) and *in situ* field measurements (ZOBACK and HEALY, 1984).

The thrusting rates, however, are equally important but often overlooked. This is clear from the following equation

$$A_s = \frac{\sigma V}{\delta_z} \tag{4}$$

where A_s is the rate of volumetric shear heating, σ is tectonic stress in the shear zone, V is the thrusting rate, and δ_z is the thickness of the thrusting zone. When the thrusting rate is comparable with the typical plate velocity (30 mm/yr), shear heating is effective (Fig. 3a) and the lithospheric strength can be quickly reduced (Fig. 3b). However, for the low shortening rates in the Cordillera (ELISON, 1991), the effects of shear heating were limited even when a high tectonic stress of 100 MPa was assumed (Fig. 3a).

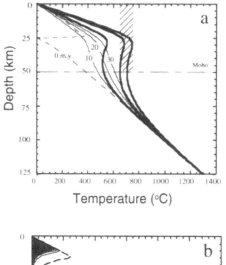

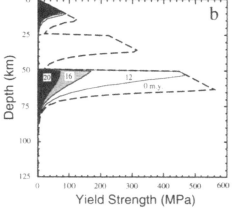

Figure 3

(a) Thermal effects of shear heating in addition to burial heating. Thin curves are for a thrusting rate of 5 mm/yr and a shera stress of 100 MPa; thick curves are for a thrusting rate of 30 mm/yr and a shear stress of 50 MPa. (b) Effects of shear heating on the lithospheric strength (corresponding to the case shown in thick curves in (a)). The integrated lithospheric strengths are 1.5×10^{13} N/m (0 my); 6×10^{12} N/m (12 my); 3×10^{12} N/m (16 my); and 2×10^{12} N/m (20 my).

2.3. Lithospheric Thickening and Thinning

Thus far our discussion has been limited to the crustal processes. However, crustal compression is only the surficial expression of the convergence of tectonic plates, and significant compressional deformation of the lithospheric mantle may occur simultaneously. Crustal thickening may be mirrored by thickening of the lithospheric mantle at depth (HOUSEMAN et al., 1981). On the other hand, significant lithospheric thinning may result from dynamic instabilities developed in a cold, thickened lithospheric mantle (BIRD, 1979; RANALLI et al., 1989).

Thickening of the lithosphere would significantly depress the geotherm and strengthen the lithosphere (Fig. 4). Thus regardless of whether or not lithospheric thickening occurred during Mesozoic crustal compression, the lithosphere was likely thinned rather than thickened at the end of the orogeny. With upwelling of the asthenosphere to a depth of 70 to 80 km, significant crustal anatexis, such as that associated with formation of the Valhalla core complex in the southeastern Canadian Cordillera (CARR, 1992; PARRISH et al., 1988), can be predicted without appealing to shear heating at extreme values (LIU and FURLONG, 1993). The causes of lithospheric thinning are open for speculation. The possible mechanisms include delamination of a thickened lithospheric mantle (BIRD, 1979) and downwelling

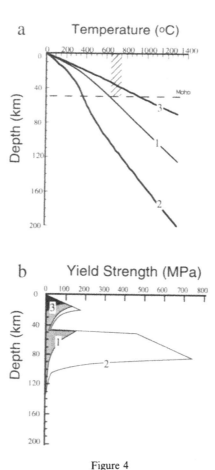

Figure 4

(a) Thermal effects of thinning and thickening of the lithosphere. For the "normal" geotherm (curve 1), the thickness of the lithospheric mantle is unchanged during crustal shortening. In the other two cases, the lithospheric mantle is either homogeneously thickened to a total lithospheric thickness of 200 km (curve 2) or thinned to a depth of 70 km (curve 3). All geotherms are after 50 my of the initial thermal event. The hatched area indicates the range of solidus of wet granitic rocks. (b) strength envelops corresponding to the geotherms in (a).

mantle material peeling off the overthickened thermal boundary layer (lithosphere) (HOUSEMAN *et al.*, 1981).

In summary, burial and shear heating associated with Mesozoic and early Cenozoic crustal compression may have contributed significantly to the high crustal temperature near the end of the orogeny. Because of the long and slow crustal contraction in the Cordillera, synkinematic thermal effects must be integrated over the period of crustal contraction. Nonetheless, at least in the southeastern Canadian Cordillera where magmatism was abundant near the end of the orogeny, significant thermal perturbations from the asthenosphere may be involved.

3. Tertiary Extension: Core-complex Formation and the Basin-Range Extension

Mesozoic and early Cenozoic crustal compression in the Cordillera was followed by wide-spread extension and volcanism in the Tertiary. Most studies treat the early stage extension, characterized by the formation of metamorphic core commmplexes, and the younger (< 17 Ma) Basin-Range extension as the consequences of common causes. Here I examine the core-complex formation and its relationship with development of the Basin and Range structure.

3.1. Core-complex Formation

Core complexes are high grade metamorphic rocks at middle crust that have been exposed along gently dipping extensional shear zones (CONEY, 1980). The occurrence of core complexes along the core zone of overthickened crust in the Cordilleran FTB indicates an origin of gravitational collapse (CONEY and HARMS, 1984).

A thickened crust at isostatic equilibrium is unstable and tends to collapse under its own weight (ARTYUSHKOV, 1973). The major deviatoric stress pulling apart the crust is the differential lithostatic pressure between the thickened crust and the surrounding crust (Fig. 5). For a constant crustal density, the gravitational potential depends mainly on h, the differential elevation between the thickened crust and the reference crust. Assuming an Airy-type compensation, h is related to crustal thickness and density by

$$h = \frac{(\rho_m - \rho_c)(a_t - a_r)}{\rho_m} \tag{5}$$

where a_t and a_r are the thickness of the thickened and the reference crust, respectively. The crustal thickness in the hinterland of the Cordilleran FTB reached 50 to 60 km when core complexes formed (CONEY and HARMS, 1984; PARRISH *et al.*, 1988). For a 35-km thick reference crust (i.e., the crust adjacent to the hinterland) and densities in Table 1, the relative elevation of the hinterland was

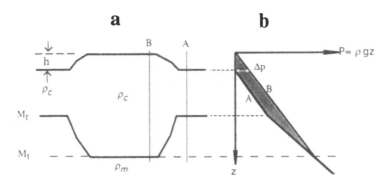

Figure 5
(a) A thickened crust at isostatic equilibrium. M_r, Moho under a reference crust; M_t, Moho under a thickened crust. (b) Pressure profiles across the reference crust (A) and the thickened crust (B). At a given depth, pressure under the thickened crust is greater than the reference crust. The differential pressure Δp tends to collapse the thickened crust. The total tensile force is the integration of Δp over the crust (equivalent to the shaded area). Notice that Δp vanishes below M_t.

about 2 km. The high elevation of the Cordilleran hinterland since Eocene was indicated by paleobotanic data (GREGORY and CHASE, 1992). The corresponding gravitational driving force, represented by the shaded area in Figure 5, is about 2×10^{12} N/m. Since the lithospheric strength in the hinterland may be reduced by thermal weakening to less than 2×10^{12} N/m (Figs. 3 and 4), significant extension may be predicted.

Major extension may occur even when the tensile force is smaller than the integrated lithospheric strength. Notice that the driving force for crustal collapse is limited to the crust (Fig. 5), whereas the rheology of continental lithosphere is intrinsically stratified. Because the lower crust may deform as ductile flow (BIRD, 1991; BLOCK and ROYDEN, 1990), crustal collapse may be mechanically decoupled from the lithospheric mantle. In such a case, crustal extension may not proceed for more than 20 million years or so, because the gravitational potential of a thickened crust diminishes during extension and conductive cooling associated with crustal extension increases the lithosphere strength (SHEN and LIU, 1994). The unroofing rates of the Valhalla core complex in the southeastern Canadian Cordillera was 1 to 2 mm/yr (LIU and FURLONG, 1993). At such rates, crust with an original thickness of 55 km may be reduced to a normal thickness of 35 km in 10 to 20 million years, which is consistent with the life span of 10 to 15 million years of crustal extension in this region (PARRISH et al., 1988).

Plutonism, which was commonly associated with core-complex formation, may facilitate more localized crustal extension at shallower levels by allowing ductile deformation to take place at a relatively lower stress level in the otherwise brittle regimes. LISTER and BALDWIN (1993) argued that this mechanism may explain the differential uplife of the footwalls during tectonic denudation of core complexes and the heterogeneity of $^{40}Ar/^{39}Ar$ apparent ages in some core complexes. Such a mechanism may be responsible for the widespread low-degree crustal extension in the Great Basin which spans a greater space and time than the few well-developed core complexes (AXEN et al., 1993).

3.2. The Basin and Range Extension

The Basin and Range province, one of the most extended continental regimes in the world, developed mainly since mid-Miocene (< 17 Ma) (ZOBACK et al., 1981; CONEY, 1987; WERNICKE et al., 1987). A number of studies linked the Basin and Range to the earlier core-complex formation as the consequence of the gravitational collapse of overthickened crust (CHEN and MOLNAR, 1983; CONEY and HARMS, 1984; SONDER et al., 1987). However, we have shown that the gravitational stresses associated with an overthickened crust are limited to the crust (Fig. 5). To cause whole-lithosphere extension, sufficient tensile stresses must be transmitted to the lithospheric mantle through the ductile lower crust. As a first approximation, the ductile layer may be approximated by a plane Poiseuille flow driven by lateral pressure gradient associated with topography. The shear stress that may be transmitted to the top of the lithospheric mantle is given by (TURCOTTE and SCHUBERT, 1982)

$$\tau = b\rho_c g \frac{dh}{dx} \tag{6}$$

where b is roughly the half thickness of the ductile channel, g is gravitational acceleration, and dh/dx is the topographic gradient, which is in the order of $1/100$ or smaller for the Cordilleran orogenic belt. Take b to be 5 km (BIRD, 1991), the maximum shear stress is 1.5 MPa. Clearly, such stresses are too small to cause significant extension in the lithospheric mantle.

The major phase of the Basin-Range extension occurred after the peak mid-Tertiary volcanism (see below). The thermal perturbations responsible for the voluminous volcanism may have provided the necessary conditions for the Basin and Range formation. As shown in Figure 6, with the topography being isostatically supported by an upwelled asthenosphere, gravitational tensile stresses are exerted on the entire lithosphere. For the same topographic head, the integrated gravitational tensile force is nearly twice that associated with an overthickened crust (cf. Fig. 5). The lithosphere would also be much weaker in this case. A major extension of the entire lithosphere is thus predictable. The accelerated extension and abundant mantle-derived magmatism in the Great Basin since mid-Miocene are consistent with such a scenario.

The present lithosphere in the Basin and Range province is abnormally hot and thin (~60 km), as indicated by the high flow (~90 mW/m²) (LACHENBRUCH and

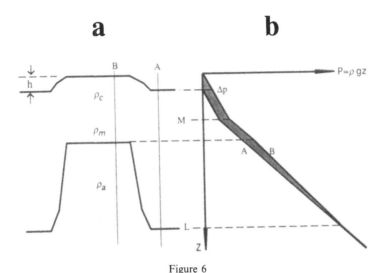

Figure 6
(a) Topography isostatically supported by thermal perturbations in the upwelled asthenosphere. *M*:
Moho; *L*: base of the reference lithosphere, also the depth of isostatic compensation in the model. (b)
Pressure profiles across the reference lithosphere (A) and the thinned lithosphere (B). Notice that the
differential pressure Δp is distributed across the whole lithosphere (cf., Fig. 5).

SASS, 1978), gravity studies (EATON *et al.*, 1978), and seismic data (ROMANOWICZ, 1979; PAKISER, 1985; SMITH *et al.*, 1989). Although the lithospheric structure before mid-Miocene is difficult to constrain, the voluminous mid-Tertiary volcanism in the Great Basin was indicative of strong thermal perturbations in the lithospheric mantle. It is demonstrated below that such conditions are also consistent with spatial-temporal patterns of tectonomagmatism in the Great Basin.

4. Magmatism: Thermal-rheological Effects

Whereas core-complex formation was clearly facilitated by plutonism (WERNICKE *et al.*, 1987; LISTER and BALDWIN, 1993; LIU and FURLONG, 1994), the dynamic link between the widespread mid-Tertiary volcanism and the younger extension (<17 Ma) remains unclear, partly because of the complicated spatial-temporal patterns of tectonomagmatism. One of the best studied yet controversial regions is the Great Basin. In some areas, Tertiary extension appeared to be synvolcanical (GANS 1987; GANS *et al.*, 1989). However, a number of recent studies suggest that, at a provincial scale, volcanism and extension generally do not correlate well over either space or time (TAYLOR *et al.*, 1989; BEST and CHRISTIANSEN, 1991; AXEN *et al.*, 1993). Volcanism culminated between 34 and 17 Ma, with an eruption of voluminous volcanic tuff in the center of the Great Basin (the so-called "ignimbrite flare-up") (Fig. 7). The synvolcanic extension was rather

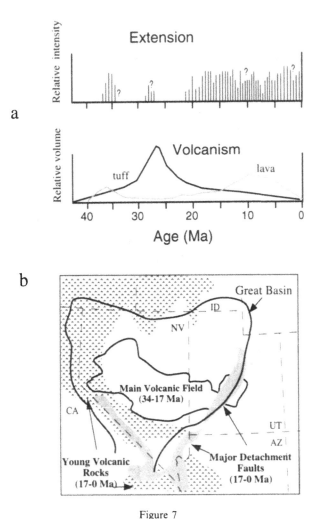

Figure 7

(a) Schematic temporal patterns of extension and volcanism in the Great Basin (adopted from BEST and CHRISTIANSEN (1991)). (b) Simplified map showing the spatial distribution of mid-Tertiary volcanism and the younger Basin-Range extension in the Great Basin (based on BEST and CHRISTIANSEN (1991), ARMSTRONG and WARD (1991), and WERNICKE et al. (1987)).

limited (BEST and CHRISTIANSEN, 1991). Most of the younger extension, which was chiefly responsible for the Basin and Range structure, occurred after the peak volcanism (< 17 Ma). Spatially, most of the younger Basin-Range extension did not occur in the main volcanic field in the central Great Basin, where the lithosphere was presumably hot and weak, but around its margins (Fig. 7b).

The poor correlations between volcanism and extension do not fit into the classical active or passive extension models (Sengor and Burke, 1978), and raise the question of whether these two processes were related to each other. Part of the answer may lie in the thermal-rheological effects associated with intrusion and underplating of mafic magmas, which were involved in the ignimbrite flare-up as the parental magmas for some of the silicic tuff (JOHNSON, 1991) and as the heat sources for crustal anatexis (HILDRETH, 1981; LIU and FURLONG, 1992).

Intrusion and underplating of mafic magmas have two competing effects on lithospheric strength and hence extensoin (LIU and FURLONG, 1994). On the one hand, heat advected by mafic magmas tends to thermally weaken the lithosphere. On the other hand, addition of mafic material to the crust increases the integrated strength of the lithosphere, because mafic material is considerably stronger than silicic rocks. The net effects of thermal weakening and rheological hardening are time-dependent. Depending on the thermal structure of the lithosphere and the mode of mafic intrusion, a large spectrum of spatial-temporal relations between volcanism and extension may be predicted (LIU and FURLONG, 1994).

Figure 8 exhibits a case that has a predicted spatial-temporal pattern of tectonomagmatism comparable with that which occurred in the Great Basin. The mafic intrusion, represented in the model by emplacement of a 4-km mafic sill with an initial temperature of 1200°C in the middle crust, was accompanied with upwelling of the asthenosphere to the base of the crust (Fig. 8). Transversing the volcanic field (the region intruded by the mafic sill) and the surrounding area, the minimum lithospheric strength is reached 5 to 10 million years after the mafic

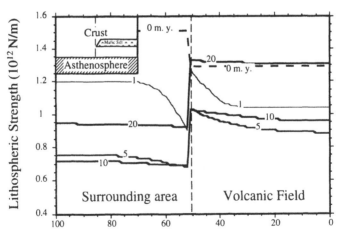

Distance From the Volcanic Center (km)

Figure 8

Predicted spatial-temporal evolution of lithospheric strength when mafic intrusion (a 4-km mafic sill emplaced at a depth between 17 and 20 km) is associated with upwelling of the asthenosphere to the base of the crust. The insert shows the model geometry. Note that the minimum lithospheric strength occurs 5 to 10 million years after the mafic intrusion; within such a time interval, the volcanic field becomes relatively stronger than the peripheral areas.

intrusion. Over such a period, the volcanic field has become relatively stronger than the surrouonding area. Assuming most extension occurs when the lithosphere is at its minimum strength and volcanism is closely linked to intrusion and underplating of mafic magma, Figure 8 implies a 5 to 10 my gap between peak volcanism and the major phase of extension, and development of most extension in areas near the margin of the main volcanic field. Such patterns are generally consistent with Tertiary tectonomagmatism in the Great Basin (cf., Fig. 7).

Two conditions are essential for the reversed lithospheric strength and the time gap between the peak volcanism and the major phase of extension in Figure 8: a thin lithosphere and relatively fast cooling of the intrusive magmas. The first condition implies that lithospheric strength is dominated by that of the crust. The gap of a few million years between mafic intrusion and the minimum lithospheric strength reflects the time needed for conductive thermal weakening of the crust. The second condition allows rheological hardening to overtake thermal weakening in a relatively short time. This is achieved in Figure 8 by emplacing the mafic sill at the relatively cold middle crust. Similar effects may result from intrusion of multiple magma bodies at various depths, which is probably a more realistic scenario.

5. Discussion and Summary

Although many details of the Cordilleran tectonics remain controversial, it is clear that dynamic interactions between crustal shortening, volcanism, and extension played a major role. Crustal shortening in the Mesozoic and early Cenozoic contributed to the high crustal temperature associated with core-complex formation. Synkinematic burial heating was important because of the low rates of crustal shortening. For the same reason, the effects of shear heating were probably limited. Consideration of multiple thrust zones predicts an even less effective shear heating, because thrusting rate on each fault is only a fraction of the average shortening rate: $V = U/n \cos \phi$, where U is the average shortening rate across the region, n is the number of thrusts, and ϕ is the thrusting angle, which is generally less than 30 degrees. Shear heating in such a system is thus small and diffusive. In the southeastern Canadian Cordillera, where crustal temperature was high and plutonism abundant, additional heat may be supplied by moderate lithospheric delamination. On the other hand, the relatively short life span (10 to 15 my) of crustal extension and the paucity of mantle-derived magmatism in this region (PARRISH et al., 1988) do not favor strong lithospheric thinning.

Core complexes in the hinterland of the Cordilleran FTB mainly resulted from gravitational collapse of overthickened crust. The extension was largely limited to the crust and mechanically decoupled from the lithospheric mantle. Magmatism may have facilitated the crust-mantle decoupling and allowed localized ductile deformation in the otherwise brittle upper crust. These processes may be responsible for the widespread low-degree crustal extension in the Great Basin (AXEN *et al.*, 1993).

Gravitational collapse of overthickened crust, however, was unlikely the major cause of the Basin and Range development, because the gravitational tensile stresses associated with an overthickened crust are limited to the crust; only a small fraction of the tensile stresses may be transmitted through the ductile lower crust to the lithospheric mantle. Nor can core-complex formation directly cause the ignimbrite flare-up, as significant decompressional partial melting requires upwelling of asthenosphere to a depth less than 50 km (McKENZIE and BICKLE, 1988). Such conditions require major lithospheric extension for more than 30 million years, even in a progressive stretching model that assumes maximum thinning near the bottom of the lithosphere (LIU and FURLONG, 1993). LEEMAN and HARRY (1993) circumvented this problem by suggesting that mid-Tertiary volcanism in the Great Basin was derived from partial melting of basaltic material locked in the lithospheric mantle. However, such a model still needs mechanisms to extend the whole lithosphere.

The model consistent with the general patterns of volcanism and the Basin-Range extension is one that involved significant thermal perturbations in the subcrustal mantle during the ignimbrite flare-up. With significant upwelling of the asthenosphere, the complicated spatial-temporal patterns of tectonomagmatism in the Great Basin may be explained by the competing effects of thermal weakening and rheological hardening associated with mafic intrusion. Such deep thermal perturbations may provide the essential conditions for the Basin and Range formation: (1) a sufficiently weak lithosphere and (2) a driving mechanism for whole-lithosphere extension. With the topography being isostatically supported by an abnormally hot upper mantle, whole-lithosphere extension may be induced by the gravitational potential in a manner similar to crustal collapse. The change from convergence between the North American and the Farallon plates to the transform boundary along the San Andreas fault around 25–29 Ma (ATWATER, 1970) may have relaxed compressional stresses at the western margins of the North American plate, thus facilitating the Basin-Range extension.

The nature of thermal perturbations beneath the Basin and Range province is not clear. Subduction-related processes (LIPMAN, 1980) may fit the general pattern of southward migration of the mid-Tertiary volcanism, but it is difficult to explain the strong thermal perturbations witnessed by the ignimbrite flare-up in the Great Basin. Mechanisms such as "deblobing" (HOUSEMAN *et al.*, 1981; RANALLI *et al.*, 1989) or delamination (BIRD, 1979) of a thickened lithosphere can provide the thermal pulses for the volcanism and extension, however these hypotheses are difficult to test.

Acknowledgments

This work is supported by ACS-PRF grant 27925-G2 and the Research Council of the University of Missouri. I thank Y.J. Chen and R. Gordon for helpful comments, and Professors R. Wang and K. Aki for inviting me to the IUTAM Symposium on Geodynamics.

REFERENCES

ATWATER, T. (1970), *Implications of Plate Tectonics for the Cenozoic Tectonic Evolution of Western North America*, Geol. Soc. Am. Bull. *81*, 3513–3536.

ARMSTRONG, R. L., and WARD, P. (1991), *Evolving Geographic Patterns of Cenozoic Magmatism in the North American Cordillera: The Temporal and Spatial Association of Magmatism and Metamorphic Core Complexes*, J. Geophys. Res. *96*, 13,201–13,224.

ARTYUSHKOV, E. V. (1973), *Stresses in the Lithosphere Caused by Crustal Thickness Inhomogeneities*, J. Geophys. Res. *78*, 7675–7690.

AXEN, G. J., TAYLOR, W., and BARTLEY, J. M. (1993), *Space-time Patterns and Tectonic Controls of Tertiary Extension and Magmatism in the Great Basin of the Western United States*, Geol. Soc. Am. Bull. *105*, 56–72.

BEST, M. G., and CHRISTIANSEN, E. H. (1991), *Limited Extension during Peak Tertiary Volcanism, Great Basin of Nevada and Utah*, J. Geophys. Res. *96*, 13,509–13,528.

BIRD, P. (1991), *Lateral Extrusion of Lower Crust from under High Topography, in the Isostatic Limit*, J. Geophys. Res. *96*, 102,755–102,866.

BIRD, P. (1979), *Continental Delamination and the Colorado Plateau*, J. Geophys. Res. *84*, 7561–7571.

BURCHFIEL, C. B., and DAVIS, G. A. (1975), *Nature and Controls of Cordilleran Orogenesis, Western United States: Extensions of an Earlier Synthesis*, Am. J. Sci. *275-A*, 363–396.

BLOCK, L., and ROYDEN, L. H. (1990), *Core Complex Geometries and Regional Scale Flow in the Lower Crust*, Tectonics *9*, 557–567.

BRACE, W. F. (1972), *Laboratory Studies of Stick-slip, and their Application to Earthquakes*, Tectonophysics *14*, 189–200.

BYERLEE, J. D. (1978), *Friction of Rocks*, Pure and Appl. Geophys. *116*, 615–626.

CARR, S. D. (1992), *Tectonic Setting and U-Pb Geochronology of the Early Tertiary Ladybird Leucogranite Suite, Thor-Odin-Pinnacles Area, Southern Omineca Belt, British Colombia*, Tectonics *11*, 258–278.

CHEN, W-P., and MOLNAR, P. (1983), *Focal Depths of Intracontinental and Intraplate Earthquakes and their Implications for the Thermal and Mechanical Properties of the Lithosphere*, J. Geophys. Res. *88*, 4183–4214.

CONEY, P. J. (1980), *Cordilleran Metamorphic Core Complexes: An Overview*, Mem. Geol. Soc. Am. *153*, 7–31.

CONEY, P. J., *The regional tectonic setting and possible causes of Cenozoic extension in the North American Cordillera*. In *Continental Extensional Tectonics* (Coward, M. P., Dewey, J. F., and Hancock, P. L., eds.) (Geological Society Special Publ. *28*, 1987) pp. 177–186.

CONEY, P. J., and HARMS, T. A. (1984), *Cordilleran Metamorphic Core Complexes: Cenozoic Extensional Relics of Mesozoic Compression*, Geology *12*, 550–554.

DICKINSON, W. R., and SNYDER, W. S. (1979), *Geometry of Subducted Slabs Related to the San Andreas Transform*, J. Geol. *87*, 609–627.

EATON, G. P., WAHL, R. R., PROSTKA, H. J., MAHEY, D. R., and KLEINKOPF, M. D., *Regional gravity and tectonic patterns: Their relation to late Cenozoic epeirogeny and lateral spreading in the western Cordillera*. In *Cenozoic Tectonics and Regional Geophysics of the Western Cordillera* (Smith, R. B., and Eaton, G. P., eds.) (Mem. Geol. Soc. Am. *152*, 1978) pp. 51–91.

ENGLAND, P. C., and RICHARDSON, S. W. (1977), *The Influence of Erosion upon Mineral Facies of Rocks from Different Metamorphic Environments*, J. Geol. Soc. Lond. *134*, 201–213.

ENGLAND, P. C., and THOMPSON, A. B. (1984), *Pressure-temperature-time Paths of Regional Metamorphism I. Heat Transfer during the Evolution of Regions of Thickened Continental Crust*, J. Petrol. *25*, 894–928.

ELISON, M. W. (1991), *Intracontinental Contraction in Western North America: Continuity and Episodicity*, Geol. Soc. Am. Bull. *103*, 1226–1238.

GANS, P. B. (1987), *An Open-system, two-layer Crustal Stretching Model for the Eastern Great Basin*, Tectonics 6, 1–12.

GANS, P. B., MAHOOD, G. A., and SCHERMER, E. (1989), *Synextensional Magmatism in the Basin and Range Province: A Case Study from the Eastern Great Basin*, Geol. Soc. Am. Spec. Paper 233, 58 pp.

GREGORY, K. M., and CHASE, C. G. (1992), *Tectonic Significance of Paleobotanically Estimated Climate and Altitude of the Late Eocene Erosion Surface, Colorado*, Geology 20, 581–585.

HILDRETH, W. (1981), *Gradients in Silicic Magma Chambers: Implications for Lithospheric Magmatism*, J. Geophys. Res. 86, 10,153–10,192.

HOUSEMAN, G. A., McKENZIE, D. P., and MOLNAR, P. (1981), *Convective Instability of a Thickened Boundary Layer and its Relevance for the Thermal Evolution of Continental Convergent Belts*, J. Geophys. Res. 86, 6115–6132.

JOHNSON, C. M. (1991), *Large-scale Crustal Formation and Lithosphere Modification beneath Middle to Late Cenozoic Calderas and Volcanic Fields, Western North America*, J. Geophys. Res. 96, 13,485–13,508.

KIRBY, S. H., and KRONENBERG, A. K. (1987), *Rheology of the Lithosphere: Selected Topics*, Rev. Geophys. 25, 1219–1244.

LACHENBRUCH, A. H. (1970), *Crustal Temperature and Heat Production: Implications for the Linear Heat-flow Relation*, J. Geophys. Res. 75, 3291–3300.

LACHENBRUCH, A. H., and SASS, J. H., *Models of an extending lithosphere and heat flow in the Basin and Range province*. In *Cenozoic Tectonics and Regional Geophysics of the Western Cordillera* (Smith, R. B., and Eaton, G. P., eds.) (Mem. Geol. Soc. Am. 152, 1978) pp. 209–250.

LEEMAN, W. P., and HARRY, D. L. (1993), *A Binary Source Model for Extension-related Magmatism in the Great Basin, Western North America*, Science 262, 1550–1554.

LIPMAN, P. W., *Cenozoic volcanism in the western United States: Implication for continental tectonics*. In *Studies in Geophysics: Continental Tectonics* (National Academy of Sciences, Washington, D.C. 1980) pp. 161–174.

LISTER, G. S., and BALDWIN, S. L. (1993), *Plutonism and the Origin of Metamorphic Core Complexes*, Geology 21, 607–610.

LIU, M., and CHASE, C. G. (1991), *Evolution of Hawaiian Basalts: A Hotspot Melting Model*, Earth. Planet. Sci. Lett. 104, 151–165.

LIU, M., and FURLONG, K. P. (1992), *Cenozoic Volcanism in the California Coast Ranges: Numerical Solutions*, J. Geophys. Res. 97, 4941–4957.

LIU, M., and FURLONG, K. P. (1993), *Crustal Shortening and Eocene Extension in the Southeastern Canadian Cordillera: Since Thermal and Mechanical Considerations*, Tectonics 12, 776–786.

LIU, M., and FURLONG, K. P. (1994), *Intrusion and Underplating of Mafic Magmas: Thermal-rheological Effects and Implications for Tertiary Tectonomagmatism in the North American Cordillera*, Tectonophysics 237, 175–187.

LYNCH, H. D., and MORGAN, P., *The tensile strength of the lithosphere and the localization of extension*. In *Continental Extensional Tectonics* (Coward, M. P., Dewey, J. F., and Hancock, P. L., eds.) (Geological Society Special Publ. 28, 1987) pp. 53–66.

McKENZIE, D., and BICKLE, M. J. (1988), *The Volume and Composition of Melt Generated by Extension of the Lithosphere*, J. Petrol. 29, 625–679.

MOLNAR, P., and ENGLAND, P. (1990), *Temperatures, Heat Flux, and Frictional Stress near Major Thrust Faults*. J. Geophys. Res. 95, 4833–4856.

OXBURGH, E. R., and TURCOTTE, D. L. (1974), *Thermal Gradients and Regional Metamorphism in Overthrust Terrains with Special Reference to the Eastern Alps*, Schweiz. Min. Petr. Mitt. 54, 641–622.

PEACOCK, S. M. *Thermal modeling of metamorphic pressure-temperature-time paths: A forward approach*. In *Metamorphic Pressure-Temperature-Time Paths* (Spear, F. S., and Peacock, S. M., eds.) (AGU, Washington, D.C. 1989) pp. 57–99.

PAKISER, L. C., *Seismic exploration of the crust and upper mantle of the Basin and Range province*. In *Geologists and Ideas: A History of North American Geology* (Drake, E. T., and Jordan, W. M., eds.) (Geol. Soc. Am., Continnian Special 1, 1985) pp. 453–469.

PARRISH, R. R., CARR, S. D., and PARKINSON, D. L. (1988), *Eocene Extensional Tectonics and Geochronology of the Southern Omineca Belt, British Columbia and Washington*, Tectonics 7, 181–212.

RANALLI, G., BROWN, R. L., and BOSDACHIN, R. (1989), *A Geodynamic Model for Extension in the Shuswap Core Complex, Southeastern Canadian Cordillera*, Can. J. Earth Sci. *26*, 1647–1653.

ROMANOWICZ, B. A. (1979), *Seismic Structure of the Upper Mantle beneath the United States by Three-dimensional Inversion of Body Wave Arrival Times*, Geophys. J. Roy. Astro. Soc. *57*, 479–506.

RUTTER, E. H., and BRODIE, K. H. (1988), *The Role of Tectonic Grain Size Reduction in the Rheological Stratification of the Lithosphere*, Geol. Rundschau *77*, 295–308.

RUPPEL, C., ROYDEN, L., and HODGES, K. V. (1988), *Thermal Modeling of Extensional Tectonics: Application to Pressure-temperature-time Histories of Metamorphic Rocks*, Tectonics *7*, 947–957.

SENGOR, A. M. C., and BURKE, K. (1978), *Relative Timing of Rifting and Volcanism on Earth and its Tectonic Implications*, Geophys. Res. Lett. *5*, 419–421.

SEVERINGHAUS, J., and ATWATER, T., *Cenozoic geometry and thermal state of the subducting slabs beneath western North America*. In *Basin and Range Extensional Tectonics near the Latitude of Las Vegas, Nevada* (Wernicke, B. P., ed.) (Geol. Soc. Am. Mem. *176*, 1990) pp. 1–22.

SHEN, Y. Q., and LIU, M. (1994), *Dynamic Links between Core Complex Formation and the Basin and Range Development*, EOS Trans., AGU *75*, 678.

SHI, Y., and WANG, C.-Y. (1988), *Two-dimensional Modeling of the P-T-t Paths of Regional Metamorphism in Simple Overthrust Terrains*, Geology *15*, 1048–1051.

SONDER, L. J., ENGLAND, P. C., WERNICKE, B., and CHRISTIANSEN, R. L., *A physical model for Cenozoic extension of western North America*. In *Continental Extensional Tectonics* (Coward, M. P., Dewey, J. F., and Hancock, P. L., eds.) (Geological Society Special Publ. *28*, 1987) pp. 187–201.

STEWART, J. H., *Basin and Range structure in western North America: A Review*. In *Cenzoic Tectonics and Regional Geophysics of the Western Cordillera* (Smith, R. B., and Eaton, G. L., eds.) (Mem. Geol. Soc. Am. *152*, 1978) pp. 1–31.

THOMPSON, G. A., CATCHINGS, R., GOODWIN, E., HOLBROOK, S., JARCHOW, C., MANN, C., MCCARTHY, J., and OKAYA, D., *Geophysics of the western Basin and Range Province*. In *Geophysical Framework of the Continental United States* (Pakiser, L. C., and Mooney, W. D., eds.) (Geol. Soc. Am. Mem. *172*, 1989) pp. 177–203.

TAYLOR, W. J., and BARTLEY, J. M. (1992), *Prevolcanic Extensional Seaman Breakaway Fault and its Geological Implications for Eastern Nevada and Western Utah*, Geol. Soc. Am. Bull. *104*, 255–266.

TAYLOR, W. J., BARTLEY, J. M., LUX, D. L., and AXEN, G. J. (1989), *Timing of Tertiary Extension in the Railroad Valley-Pioche Transect, Nevada: Constraints from $^{40}Ar/^{39}Ar$ ages of Volcanic Rocks*, J. Geophys. Res. *94*, 7757–7774.

TURCOTTE, D. L., and SCHUBERT, G., *Geodynamics* (John Wiley & Sons, New York 1982) 237 pp.

WERNICKE, B., CHRISTIANSEN, R. L., ENGLAND, P. C., and SONDER, L. J., *Tectonomagmatic evolution of Cenozoic extension in the North American Cordillera*. In *Continental Extensional Tectonics* (Coward, M. P., Dewey, J. F., and Hancock, P. L., eds.) (Geological Society Special Publ. *28*, 1987) pp. 203–221.

WILSON, J. M., MCCARTHY, J., JOHNSON, R. A., and HOWARD, K. A. (1991), *An Axial View of a Metamorphic Core Complex: Crustal Structure of the Whipple and Chemehuevi Mountains, Southeastern California*, J. Geophys. Res. *96*, 12,293–12,311.

ZEN, E-an (1988), *Thermal Modeling of Stepwise Anatexis in a Thrust-thickened Sialic Crust*, Trans. Roy. Astro. Soc. Edinburgh *79*, 223–235.

ZOBACK, M. D., and HEALY, J. H. (1984), *Friction, Faulting, and "in situ" Stress*, Ann. Geophys. *2*, 689–698.

ZOBACK, M. D., ANDERSON, R. E., and THOMPSON, G. A. (1981), *Cenozoic Evolution of the State of Stress and Style of Tectonism of the Basin and Ranges Province of the Western United States*, Phil. Trans. Roy. Soc., London Serial *A 300*, 407–434.

(Received September 8, 1994, revised April 6, 1995, accepted May 20, 1995)

PAGEOPH, Vol. 146, Nos. 3/4 (1996)

0033–4553/96/040469–33$1.50 + 0.20/0

Thermal Modeling of the Southern Alps, New Zealand

YAOLIN SHI,[1] RICK ALLIS,[2] and FRED DAVEY[2]

Abstract — Finite-element modeling of the thermal regime across the Southern Alps of New Zealand has been carried out along two profiles situated near the Franz Josef and Haast valleys. The modeling involves viscous deformation beneath the Southern Alps, including both uplift and erosion, and crustal/lithospheric thickening, as a result of crustal shortening extending to 20 mm/y of a 25-km thick crust. Published uplift rates and crustal thickness variations along the two profiles are used to constrain the modeled advection of crustal material, and results are compared with the recent heat flow determinations, 190 ± 50 mW/m^2 in the Franz Josef valley and 90 ± 25 mW/m^2 in the Haast valley. Comparisons of the model with published K-Ar and fission track ages, show that the observed heat flow in the Franz Josef valley is consistent with observed zircon fission track ages of around 1 Ma, if the present-day uplift rate is close to 10 mm/y. Major thermal differences between the Franz Josef and Haast profiles appear to be due to different uplift and erosion rates. There is weak evidence that frictional heating close to the Alpine fault zone is not significant. The modeling provides explanations for the distribution of seismicity beneath the Southern Alps, and predicts a low surface heat flow over the eastern foothills due to the dominant thermal effect of crustal thickening beneath this region. Predicted temperatures at mid-crustal depth beneath the zone of maximum uplift rate are 50–100°C cooler than those indicated in previously published models, which implies that thermal weakening of the crust may not be the main factor causing the aseismicity of the central Southern Alps. The results of the modeling demonstrate that the different types of reset age data in the region within 25 km of the Alpine fault are critical for constraining models of the deformation and the thermal regime beneath the Southern Alps.

Key words: Thermal modeling, Southern Alps, fission track age.

Introduction

The Southern Alps of New Zealand mark the zone of continental collision between the Pacific and Australian plates (Figure 1). For much of the last 40 Ma, the plate boundary through the region has been predominantly strike-slip, with around 600 km of displacement occurring on the Alpine fault zone (WALCOTT, 1978; 1979). During the last 10 Ma, compression normal to the plate boundary has increased, resulting in some 60–70 km of shortening across the boundary zone in the region of the Southern Alps (WALCOTT, 1979, pers. comm. 1994). Most of the

[1] Graduate School, USTC, Academia Sinica, Beijing 100039, China.
[2] Institute of Geological and Nuclear Sciences, Wellington, New Zealand.

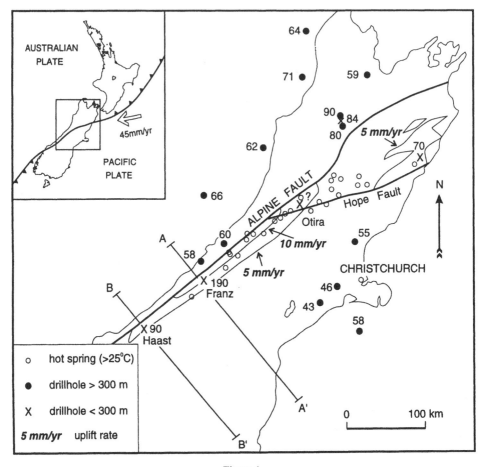

Figure 1

Central and northern South Island, New Zealand, showing the Alpine fault, available heat flow data (values in mW/m², from FUNNELL and ALLIS, in preparation), hot spring locations, and the 5 and 10 mm/y uplift contours. A-A' and B-B' mark the Franz Josef and Haast profiles which are the subject of the tectono-thermal modeling in this paper.

shortening has probably occurred during the last 5 Ma, with deformation concentrated within the Pacific plate as thickening, uplift and erosion. In the area of maximum uplift rate, the topography may now have reached an equilibrium profile, with uplift and erosion rates approximately equal (WELLMAN, 1979). Here, Wellman has proposed that the bulk of the crust is deflected upwards on a curving Alpine fault, with the remainder of the Pacific plate crust and lithospheric mantle accumulating at depth in a broad root zone beneath the Alps. North of the Southern Alps, the surface location of the plate boundary shifts eastwards to the Hikurangi trough, where Pacific plate oceanic crust subducts beneath the North Island and the northern part of the South Island. The southern extension of the plate boundary zone into the Fiordland subduction zone beneath the southwest South Island involves a transition from predominantly shortening of the Pacific plate continental crust in the Southern Alps, to predominantly shortening of the Australian plate oceanic crust beneath Fiordland (ALLIS, 1986).

Present-day uplift and erosion in the Southern Alps is concentrated in a 150-km long strip, about 20 km in width, adjacent to the central portion of the Alpine fault (Figure 1). Uplift rates for the South Island have been summarized by WELLMAN (1979), based on the elevations and ages of uplifted marine benches and river terraces together with the tilt of stranded beaches. Most of the uplift occurs along the axis of the Southern Alps rather than against the Alpine fault which is approximately 10 km to the west. This is believed due to fault drag, or shearing, on the eastern, uplifted side of the Alpine fault (J. ADAMS, 1979, 1980). The highest uplift rates exceed 10 mm/y about 5 km southeast of the central portion of the Alpine fault. The high erosion rates in the Southern Alps have contributed to the exhumation of garnet-bearing, amphibolite schists adjacent to the Alpine fault, with progressively lower grade metamorphic rocks occurring to the east. Age dating, using K-Ar and fission track analysis methods, confirms a pronounced apparent younging of rocks towards the Alpine fault, with apparent ages of around 1 Ma or less occurring in the area of highest uplift and erosion (ADAMS, 1981; ADAMS and GABITES, 1985; KAMP, 1992; KAMP and TIPPETT, 1993; TIPPETT and KAMP, 1993). Interpretation of the fission track annealing trends has led Kamp and Tippett to conclude that a maximum of nearly 16 km of erosion may have occurred adjacent to the Alpine fault during the present phase of crustal shortening, that the main phase of uplift was underway by 5 to 8 Ma (depending on location) and may have increased greatly at 1.3 Ma, and that the maximum uplift and erosion rates are close to 10 mm/y.

The exceptionally high erosion rate in the Southern Alps has profound implications on the thermal regime in the crust. ALLIS et al. (1979) showed, from one-dimensional considerations, that the near-surface temperature gradient could be in the range of 70–180°C/km for erosion rates of 10–20 mm/y, with the thermal regime dependent on both the uplift rate and the original depth (temperature) at the start of the uplift path. Two-dimensional (2-D) modeling of the thermal regime, assuming vertical uplift trajectories ranging up to 10 mm/y and a 25 km base to the uplift zone, confirmed near-surface temperature gradients increasing to 200°C/km after about 2 million years, with the brittle-ductile transition rising to less than 5 km depth in the high uplift zone (KOONS, 1987). Supporting evidence for high temperatures at shallow depths are two high heat flow determinations (FUNNELL and ALLIS, in preparation); a metamorphic component in hot spring waters (ALLIS et al., 1979); fluid inclusion properties in cross-cutting quartz-calcite veins (Craw, 1988; Craw et al., 1994); low seismicity implying a weak crust (EVISON, 1971; HATHERTON, 1980; ANDERSON and WEBB, 1994); and the anomalously young age dates mentioned above.

The purpose of this paper is to present the initial results of finite-element modeling of the Southern Alps thermal regime in which the new heat flow data are used as a constraint. The results can also be compared with the recently published fission track data. In contrast to the previous modeling of ALLIS *et al.* (1979) and KOONS (1987), the deformation beneath the Southern Alps is represented as a Newtonian viscous model with a prescribed erosion history at the surface, and with lower crust and mantle able to be deflected downwards to form a root zone. The model therefore combines both the effects of uplift which enhance the surface heat flow, as well as the effects of crust and mantle thickening which tend to decrease the surface heat flow.

Heat Flow Measurements

Although the thermal regime in the major offshore basins adjacent to the South Island is now relatively well determined, few heat flow measurements have been made within the Southern Alps themselves. Offshore, on the continental platforms of the Taranaki basin to the northwest, and the Great South basin to the southeast, a heat flow of $60 \pm 5 \, mW/m^2$ has been observed (FUNNELL *et al.*, 1995; FUNNELL and ALLIS, in preparation). In these regions, the only significant deformation in the last 100 Ma has been a late Cretaceous period of rifting followed by passive margin subsidence. However in the parts of these basins closer to the present landmass of New Zealand (i.e., eastern and southern Taranaki basin, northwestern Great South basin), the heat flow has been influenced by the development of the Pacific-Australian plate boundary during the last 30 Ma The Southern Alps represent an extreme example of the deformation on this boundary, with heat flow expected to be greatly enhanced due to the large amount of erosion during the last few million years.

Figure 1, illustrates the distribution of heat flow determinations in the vicinity of the Southern Alps. Most of the data are derived from bottom hole temperature measurements in oil exploration drillholes (corrected for the drilling disturbance), together with thermal conductivities derived from matrix lithologies and porosities determined from wireline log data (method described by FUNNELL *et al.*, 1995). The uncertainties in each determination depend largely on the quality of the bottom hole temperature data, but are typically better than $\pm 5 \, mW/m^2$. In the northern part of the South Island, the observed heat flow ranges from 58 to 85 mW/m^2. The high heat flow of 80–85 mW/m^2 in this area occurs in a basin which is known to have had over 3 km of erosion in the last 5 Ma, and the heat flows here are consistent with this amount of erosion (FUNNELL and ALLIS, in preparation). The heat flow values around Christchurch at the east coast range from 43 and 46 mW/m^2 to the southwest, to 58 mW/m^2 in the offshore well. The two low values of heat flow are based on wells with large thicknesses of Quaternary gravels, and the apparently depressed heat flows are considered to be a consequence of large-scale movement of groundwater.

Reliable heat flow measurements in Alpine regions are difficult to obtain because of the potentially large magnitude of perturbing effects such as steep topography and water circulation. Despite these potential problems, 3 holes between 200 and 300 m deep were drilled in the Southern Alps to investigate whether the expected high heat flow anomaly could be confirmed. The results of this drilling programme are discussed in detail in FUNNELL and ALLIS (in preparation), and are briefly summarized here. The location of the holes is shown in Figure 1, and the observed temperature profiles are shown in Figure 2.

The northernmost hole, Otira, was sited alongside the Hope fault, about 10 km southeast of the Alpine fault. Major fracturing was encountered during the drilling. The temperature was found to be almost constant over much of the hole, and indicative of a downflow of water from about 50 m depth. After unsuccessful attempts to grout the fractures, the hole was abandoned for conductive heat flow measurements.

The Franz Josef hole is 4 km from the Alpine fault, and is situated in the region of highest uplift and erosion of the Southern Alps (>10 mm/y erosion rate; WELLMAN, 1979). Drilling encountered unfractured schist, but the effect of the schist was to cause the hole to deviate significantly from vertical. The geotherm profile was corrected for the observed tilt. Above about 100 m depth, the curvature in the temperature profile is caused by the warming of the ground surface since the retreat of the Franz Josef glacier about 25 years before the measurements. Below about 150 m depth, the gradient is 95°C/km, and when combined with the mean conductivity of 3.4 W/mK, yields an observed heat flow of 320 mW/m². The schist has a strongly anisotropic thermal conductivity on a centimeter scale (Figure 2b), but on an outcrop scale the mean conductivity is considered to yield a representative value. The observed heat flow in the hole is enhanced by a large topographic effect due to the shape of the glacial valley. When corrected for this, the heat flow reduces to 190 ± 50 mW/m². Most of the uncertainty is related to the topographic correction, which was estimated by a 3-D terrain model (after KAPPELMEYER and HAENEL, 1974), by a 2-D finite-element approximation at right angles to the glacial valley which included the effects of erosion, and by a 2-D analytical approximation to the valley topography (after LEES, 1910). The greatest uncertainty arises from the assumption that the thermal regime is conductive. The possible effects of the high uplift and erosion rates causing an overestimation of the topographic correction were investigated in the finite-element modeling, but these transient effects were found to be negligible. The lack of significant loss zones encountered during drilling, the linearity of the observed temperature profile, the high-temperature gradient in the region but lack of abundant hot springs, are all consistent with a predominantly conductive thermal regime. The 25% uncertainty in the heat flow value is considered realistic for the site characteristics.

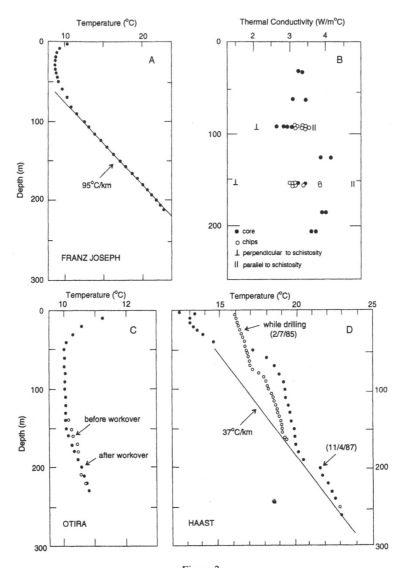

Figure 2

Temperature profiles from the 3 drillholes in the Southern Alps, and the thermal conductivity measurements on core retrieved from the Franz Josef. The Otira hole contained a downflow and was unsuitable for heat flow purposes. The upper part of the Franz Josef hole shows the warming effects due to the retreat of a glacier from the site 30 years before the drilling and the measurements. The Haast hole sustained internal upflows during the course of drilling due to a positive head; the temperature gradient is based on the inflow temperatures, which indicate a conductive regime despite the inflows. The three wells are marked in Figure 1.

The Haast drillhole is situated towards the southern end of the Southern Alps, about 6 km east of the Alpine fault. The uplift rate here is about 5 mm/y (WELLMAN, 1979). The well encountered progressively higher overpressures while being drilled, reaching over 6 m of artesian head at the total depth of 270 m. Temperature measurements during and after drilling showed the effects of upflow around the casing, despite efforts to grout off interzonal flow. Based on the temperature of the inflow points, the *in situ* temperature profile appears to be linear, with a gradient of 37°C/km. Assuming the heat flow to be predominantly conductive, the observed heat flow is 110 mW/m², which reduces to 90 ± 25 mW/m² with a topographic correction. If convective upflow of water is also occurring, the observed conductive component of the heat flow will be a minimum estimate of the total heat flow. Although the possibility of large-scale convection in the schist cannot be excluded, there are no hot springs in this part of the Southern Alps, despite the relatively high conductive heat flow. In the modeling that follows the thermal regime here is assumed to be conductive.

To summarize the above, the background heat flow on the undisturbed continental crust adjacent to the Southern Alps is considered to be 60 ± 5 mW/m². Within the convergent plate boundary zone where crustal thickening, uplift and erosion has occurred, the heat flow appears to rise to a maximum of 190 ± 50 mW/m² in the area of greatest erosion rate adjacent to the Alpine fault (~10 mm/y at the Franz Josef site). Further south at the Haast site, where the erosion rate is around 5 mm/y, the heat flow is 90 ± 25 mW/m². These values are used as the main thermal constraints in the modeling.

Numerous hot springs occur along the northern part of the Southern Alps, particularly in the sector between Franz Josef and extending to 50 km north of the Hope fault (Figure 1). Although ALLIS *et al.* (1979) considered them to be due to elevated heat flow associated with the high erosion rates, the frequency of springs adjacent to the central portion of the Hope fault raises doubt over this explanation. The uplift and erosion rate here is considerably lower than further to the southwest or northeast. FUNNELL and ALLIS (in preparation) suggest that anomalously high permeability in a brittle upper crust is the primary cause of the hot spring occurrences. Major fractures may be open to at least 3 km depth in the region to the north of the Hope fault, allowing rapid upflow of warm water to the surface. Further south, in the region of highest uplift and erosion rate, the high temperature gradient (~60°C/km below the topographic effects) undoubtedly contributes to the occurrence of the hot springs. Good data on the flow and temperatures of the springs is generally lacking. However high flow rates (>1 kg/s) appear to be rare, and in the high uplift zone of the Southern Alps, the springs are considered to represent a minor component of the total heat flow.

Modeling

Heat conduction and heat advection are two major mechanisms of heat transport in tectonically active regions, as governed by the basic equation:

$$cp\frac{\partial T}{\partial t} - cpU\,\nabla T = K\,\nabla^2 T + Q$$

where c is the specific heat, ρ is density, T is temperature, t is time, U is the velocity of rock mass, K is the thermal conductivity, and Q is any heat source.

In this paper, a simple two-dimensional numerical model is constructed, based on a conceptual tectonic model proposed by WELLMAN (1979). The advective flow pattern is initially calculated assuming a Newtonian viscous rheology, followed by calculation of the thermal evolution for the last 10 Ma. Results are compared with other observations, such as heat flow measurements, fission track and K-Ar ages, seismic data, etc., along two southeast-trending cross sections; the Franz Josef profile to the north and the Haast profile to the south (Figure 1).

The finite-element mesh employed in both the dynamic and thermal calculations used 1170 nodes and 1100 elements in total (Figure 3a). Each element usually was sized 3 km × 2.5 km in regions of major interest. The Australian and Pacific plates are divided by the Alpine fault which is assumed to dip 45°SE, as suggested by the foliation study of SIBSON *et al.* (1979) and WELLMAN (1979). The normal thickness of the crust is assumed to be 25 km (WOOD, 1991; REYNERS and COWAN, 1993), but southeast of the Alpine fault, the crust has been thickened to around 40 km, as suggested by gravity data (WOODWARD, 1979). In the modeling discussed here, the crust is assumed to be a flat layer, 25 km thick at 10 Ma. The crust of the Australian plate is assumed to remain rigid, while the crust of the Pacific plate undergoes progressive compression and thickening and finally takes its present configuration as shown in Figure 3. The earthquake focal depth distribution in regions southeast of the Alpine fault is characterized by a seismic upper crust, an aseismic lower crust and a seismic upper mantle (REYNERS, 1987), which implies a stratified rheology for the lithosphere. In our model, a low viscosity, ductile, lower crust is sandwiched between a high viscosity, upper crust of 10 km initial thickness, and a high viscosity mantle. The viscosity contrast of the three layers is assumed to be 100:1:100. Tests on the effects of these viscosity assumptions suggested only a minor influence on the results discussed below.

The boundary conditions of the dynamic modeling are constrained by plate tectonic observations of the late Cenozoic compression in the Southern Alps. The Pacific crust is assumed to have been pushed from the east at a rate of 2.67 mm/y

---

Figure 3

a. The finite-element mesh used for the numerical modeling. b. Schematic cross section of the compression and uplift modelled on the Franz Josef profile. The uplift profile was taken from WELLMAN (1979), and it was assumed that 70% of the present-day 20 mm/y shortening is occurring as uplift and erosion, with the remainder being accommodated as crustal thickening. The shape of the root is controlled by the deformation which is shown in 3c. The Haast cross section is similar, but has a peak uplift value of 6 mm/y, and a larger crustal root because of an assumed 50:50 slpit between uplift and thickening. c. Resultant flow pattern on the Franz Josef profile.

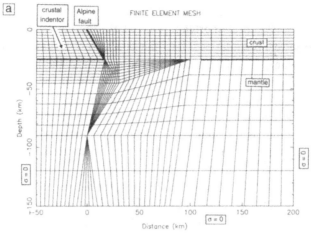

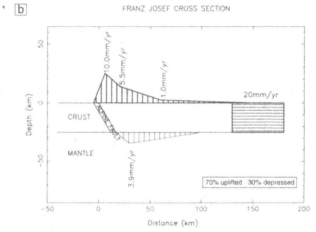

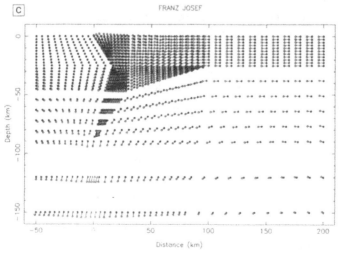

from 10 Ma to 5 Ma, with the movement then increasing linearly to be 20 mm/y at present. This produces a total shortening of 70 km. The deformation rates at the other (upper and lower) boundaries follow the same trend in changing with time. A current rate of 20 mm/y is also assumed for the motion at the lower boundary of the model for downgoing lithosphere beneath the region of the Alpine fault. The modeled deformation at the base of the crust and in the upper mantle is known to be an oversimplification, however the emphasis in this modeling is the uplift zone in the upper crust and its thermal implications. The time-scale is too short for thermal anomalies induced in the upper mantle by the compressional tectonics to be affecting the upper crustal thermal regime.

Observed uplift rates (WELLMAN, 1979) are applied as the upper boundary conditions, and assumptions about these conditions have a major influence on the resulting thermal regime. The Franz Josef profile contains a higher rate of uplift and erosion, therefore, less thickening of crust. Our model assumes 70% of the compressed crust has been uplifted and eroded, and 30% has been underplated and has accumulated as a crustal root. The maximum, present-day uplift rate is 10 mm/y, occurring about 6 km from the Alpine fault (Figure 3b). The Haast profile has a lower uplift and erosion rate (maximum of 6 mm/y), and the crust reaches greater thickness (the model assumes 50% of the compressed crust is uplifted and eroded, and 50% is depressed to thicken the crust). The different assumptions for the two profiles were chosen after consideration of both the present-day uplift rate variation along the profiles, and the crustal thickness inferred from the negative Bouguer gravity anomaly in the southern part of the Southern Alps.

Recent plate motion model (DEMETS *et al.*, 1990) suggests that the present-day convergent rate at Alpine fault could be 12 mm/y rather than the 20 mm/y assumed in the modeling. A 12 mm/y convergence rate requires most of the shortening on the Franz Josef profile to be occurring as uplift and erosion, and makes the formation of a significant crustal root difficult. It also means that, for the estimate of total shortening of 60–70 km over the last c. 5 Ma, there has been very little increase of the convergence rate during that time. If the above figures of 12 mm/y and 60–70 km are true, the geological evidence for an increase in the uplift rate within the last 5 Ma (WELLMAN, 1979) implies that early convergence has occurred predominantly as crustal thickening, with more recent convergence manifest as uplift and erosion. In our model, the uplift and crustal thickening occur simultaneously as the convergence rate increases uniformly over the last 5 Ma. The effects of these assumptions will be discussed later.

A steady-state geotherm is assumed for the initial temperature. The initial temperature profile is constrained mainly by the background heat flow observations of c. 60 mW/m^2 in basins away from the Neogene plate boundary zone. This implies a temperature about 800°C at the depth of 40 km, which is consistent with the observed seismic upper mantle (REYNERS and COWAN, 1993). Radioactive heat generation is assumed to be a uniform $0.8 \, \mu W/m^3$ for the entire crust, with no radiogenic heat source in the mantle. The surface temperature is assumed zero. Fixed temperatures are given for both the right and the left side boundary. Constant heat flow is assigned to the lowest boundary of the model, at a depth of 150 km, and has little effect on the crust thermal evolution in a time duration of 10 m.y.

An example of the flow pattern produced by the model is shown in Figure 3c, and this provides the basis for advective thermal computation. In order to obtain a stable and accurate solution, the upwind method is applied for heat advection calculations (SHI and WANG, 1987).

The thermal modeling assumes a uniform thermal conductivity of 3 W/mK for the crust. Basement rocks exposed in the South Island range from greywacke in the east, to schistose rock with progressively higher metamorphic grade as the Alpine fault is approached. Characteristic thermal conductivities range from c. 2.5 W/mK for greywacke to c. 3.4 W/mK in the 10–20 km wide zone of outcropping garnet-amphibolite schists which have been exhumed from mid- to lower-crustal depths (PANDEY, 1981; FUNNELL and ALLIS, in preparation). Although an average value is adequate for this modeling, care is needed when comparing the predicted heat flow or temperature gradient with that observed in the two drillholes. This is because the two holes are located in schist with a conductivity of 3.4 W/mK, and because of the strong advection of heat in this locality, the thermal regime is dominated by the temperature difference between the base of the advection zone (10–20 km depth), the surface temperature, and the average thermal diffusivity $(K/\rho c)$. Therefore the observed temperature gradient (corrected for topographic effects) and the modeled temperature gradients should be compared, rather than the heat flows. Alternatively, the observed heat flows should be scaled by a factor of 3/3.4 to be comparable to the modeled heat flow. The latter has been used in Figures 4–7, and this adjustment reduces the observed values to $170 \pm 50 \, mW/m^2$ at 4 km from the Alpine fault in the Franz Josef profile, and to $80 \pm 25 \, mW/m^2$ at 6 km from the fault on the Haast profile. This 10% downward adjustment in the heat flow is significantly less than the uncertainties, and it has little effect on the conclusions resulting from the modeling. In Figures 11 and 12, the observed temperature gradients, corrected for topography, are directly compared with gradients inferred from the trends in reset ages.

Results

Many models have been tested and analyzed on both the Franz Josef and Haast profiles, and four of them are presented here for discussion. The two examples for each profile are with and without a frictional heat source in the Alpine fault zone.

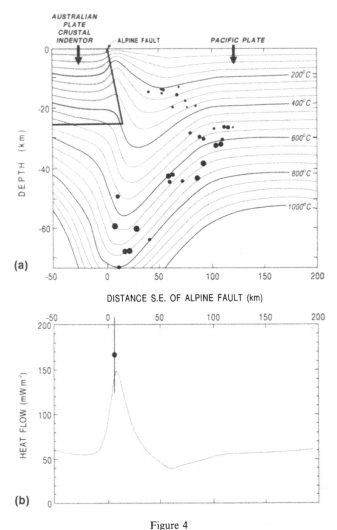

Figure 4

Present-day thermal structure and surface heat flow of Model 1, the Franz Josef profile with no frictional heating. The heat flow point is from the drillhole on the profile. The seismicity in the vicinity of the profile (from REYNERS, 1987) is superimposed. Note the band of seismicity lying between the 500 and 800°C isotherms. Distances are relative to the surface position of the Alpine fault. The assumed rigid indentor of Australian plate crust is marked for reference purposes.

Model 1—Franz Josef Cross Section without Frictional Heating

The isotherms and surface heat flow are shown in Figure 4. At 4 km southeast of the surface trace of the Alpine fault the calculated heat flow is about 130 mW/ m², which is lower than the observed heat flow value of 170 ± 50 mW/m², although it is within the uncertainty estimates. Explanations for the possible difference include the modeled uplift and erosion rate at the site being about 20% too small (~ 10 mm/y compared to 8 mm/y in the model); the average thermal diffusivity being about 20% smaller than assumed; frictional heating occurring on the Alpine fault zone (model 2); the uplift path being modified in such a way that uplift occurs from greater depth (and temperature), and with less lateral heat loss; and the observed heat flow being too high due to local groundwater flow, since a hot spring is located 2.5 km from the heat flow measurement site. Some of these factors are discussed later in the paper after a comparison of the model predictions with other thermally sensitive data.

The overall pattern of the isotherms down to 60 km depth displays depressed temperatures, similar to that occurring at subduction plate boundaries. However, the rapid uplift and erosion near the Alpine fault produces elevated temperatures in the upper crust. Assuming a stratified rheology with crustal rocks above a temperature of 350°C ductile and aseismic, and mantle lithologies brittle and prone to earthquakes up to temperatures of 800°C (CHEN and MOLNAR, 1983; RANALLI and MURPHY, 1987; HYNDMAN and WANG, 1993), the pattern of the isotherms can be used to infer lateral variations in strength along the profile. Based on the above concepts, our thermal calculations suggest a brittle crust of about 15 km thick beneath the east coast, thickening to about 25 km at a distance of 40 km SE from the Alpine fault, and reducing to 10 km adjacent to the fault. The results also suggest significant changes in the depth range of brittle mantle. The maximum depth of the seismogenic zone increases from about 40 km beneath the east coast, to nearly 70 km beneath the Alpine fault. This is in agreement with the observations of REYNERS (1987; superimposed on Figure 4). Using data from a local microearthquake network located close to the line of the Franz Josef profile, Reyners found a maximum focal depth of 30–40 km in central South Island, which increased to 73 km in the region 45 km southeast of the Alpine fault, with a possible further reduction of focal depth close to the fault, in agreement with the geotherm trend shown in our calculation. This pattern of west-dipping isotherms in the upper mantle is common to all four models. Crustal seismicity reduces towards the Alpine fault (HATHERTON, 1980), which may be related to the rise in upper crustal temperatures in the high uplift and erosion zone (ALLIS et al., 1979).

Model 2—Franz Josef Cross Section with Frictional Heating (Figure 5)

For slow sliding of a fault, where pore pressure and shear strength on the fault do not change rapidly, the amount of frictional heating can be calculated by $Q_f = \tau_s * V$, where τ_s is the shear stress on the fault, and V is the long-term sliding velocity between the two walls of the fault. The maximum shear stress on the fault is limited by the friction (or BYERLEE) law (BYERLEE, 1978):

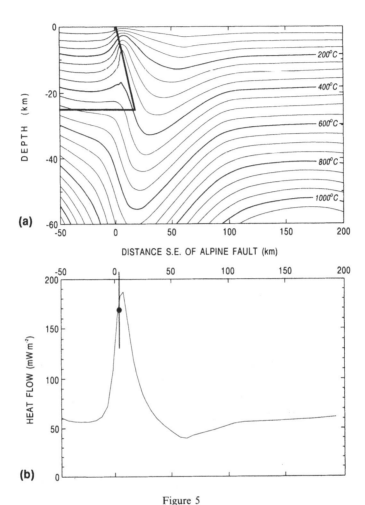

Figure 5
Present-day thermal structure and surface heat flow of Model 2, the Franz Josef profile with frictional heating on the Alpine fault extending to 10 km depth. Labelling is the same as in Figure 4.

$$\tau_s = f*(\sigma_n - P_p)$$

where f is the frictional coefficient, usually about 0.8 for rocks; P_p is pore pressure, and σ_n is the normal stress acting on the fault. Here, intermediate value of τ_s is assumed, with $f = 0.8$, σ_n being the lithospheric load, and the pore pressure being hydrostatic. The shear heating is assumed to occur on the Alpine fault down to 25 km depth.

The calculated isotherms are raised by about 50°C in a narrow zone within several kilometers of the fault. The calculated surface heat flow at the Franz Josef heat flow measurement site is 170 mW/m², in agreement with the observation. However, we believe the fitting of the single heat flow value may be fortuitous, and

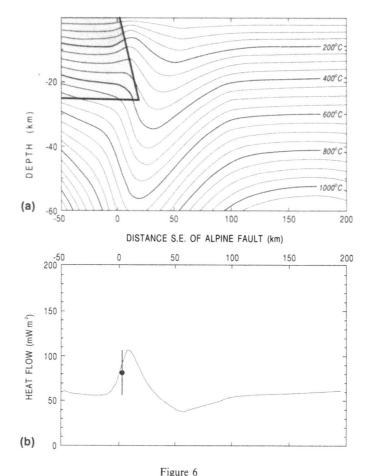

Figure 6
Present-day thermal structure and surface heat flow of Model 3, the Haast profile with no frictional heating. Labelling is the same as in Figure 4.

this is not sufficient to support a claim that Model 2 fits the observation better than Model 1. It could be argued that pore pressures in the Alpine fault zone will be close to lithostatic rather than hydrostatic, and therefore the fault zone is weak. Alternatively it has been argued that the fault zone is a high stress regime and extensive shear heating is occurring (SCHOLZ et al., 1979). Comparisons with the apparent age trends provide some additional insight (discussed below).

Model 3—Haast Cross Section without Frictional Heating (Figure 6)

This model differs from Model 1 of the Franz Josef cross section due to a relatively lower rate of uplift and erosion (6 mm/y at present), and a greater crustal root suggesting increased thickening. This is supported by geological observation

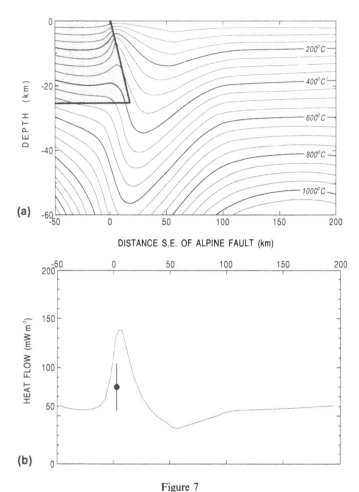

Figure 7
Present-day thermal structure and surface heat flow of Model 4, the profile with frictional heating.
Labelling is the same as in Figure 4.

(WELLMAN, 1979) and gravity data (WOODWARD, 1979). The calculated heat flow
($100 \, \mathrm{mW/m^2}$), 6 km SE of the Alpine fault is on the high side, but in reasonable
agreement with the observation of $80 \pm 25 \, \mathrm{mW/m^2}$.

Model 4—Haast Cross Section with Frictional Heating (Figure 7)

Frictional heating causes the calculated heat flow 6 km SE of the Alpine fault to
be about $130 \, \mathrm{mW/m^2}$, significantly higher than the observed value.

Comparison with Fission Track Data

Because of the sparsity of heat flow measurements in the studied region, and the difficulties of obtaining reliable values in mountainous terrain, heat flow modeling alone is unlikely to provide a strong constraint on the tectonic regime. However, numerous fission track and K-Ar age data are now available from outcrop around the Southern Alps and these potentially provide more stringent constraints. Past interpretations of the age data have usually relied on estimates of the vertical temperature gradient in the crust to relate closure or annealing temperatures of the age determination to depths (e.g., C. ADAMS, 1979; TIPPETT and KAMP, 1993). Numerical modeling of the uplift paths provides a substantially more powerful tool for interpreting the age data. In the discussion below, the modeling results from the Franz Josef profile are compared with published age data and preliminary conclusions are made. These results suggest that further refinement of the modeling and a comparison with more detailed age data along the length of the Southern Alps are desirable.

The trajectories of rocks now exposed at the surface at distances extending to 20 km from the Alpine fault can be traced back as shown in Figure 8a. With the age, location and temperature variation known from our calculations, the thermal and uplift histories of rock samples at any specific location can therefore be derived. The thermal history can then be applied to calculate apparent ages, based on the dating method being considered. In this paper we simplify the age analysis by assuming that the annealing temperature for fission tracks in apatite is 120°C, and 240°C for zircon fission tracks, and the retention temperature of Argon (K-Ar dates) is 350°C for coarse grained muscovite (CLIFF, 1985; HURFORD, 1986; TIPPETT and KAMP, 1993; C. ADAMS, pers comm., 1994). More elaborate methods of calculating fission track ages using the thermal history are also available (e.g., WILLETT, 1992)

Figure 8b converts the uplift paths in Figure 8a into temperature-time space for the no-friction Franz Josef model, with the assumed temperature thresholds for the dating methods superimposed on the figure. The figure is useful for visualising the trends in apparent ages for the different uplift paths occurring along the profile. For example, the model predicts that rock presently outcropping 15 km from the Alpine fault began its uplift path at an elevation which is shallower, and therefore cooler, than the Ar closure temperature, therefore its K-Ar age will reflect its pre-uplift history. However its zircon fission track age is predicted to be around 8 Ma, and its apatite fission track age is predicted to be around 0.6 Ma. The effects of conductive cooling due to lateral heat flow into the stable block west of the Alpine fault can be seen in the uplift path for rock presently outcropping at 2.5 km from the fault. This effect is negligible 5 km from the fault.

TIPPETT and KAMP (1993) suggest a cooling path for rock within 5 km of the Alpine fault in the vicinity of the Franz Josef profile, based on reset zircon ages,

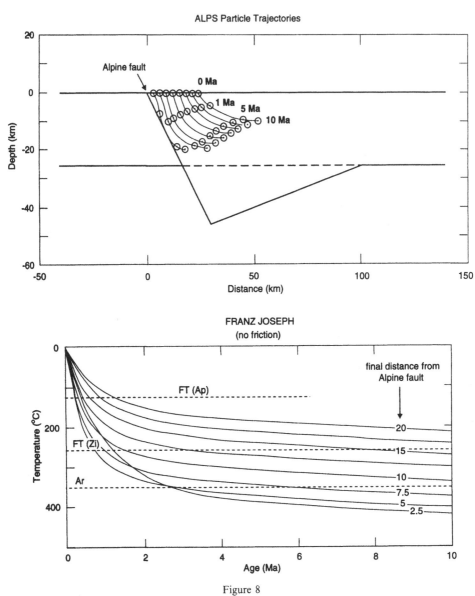

Figure 8

a. Trajectories of rock movement on the Franz Josef profile for present-day outcropping rock at 2.5 km increments from the Alpine fault. Circles mark the points on each path at 1, 5, and 10 Ma. b. The temperature-time history for each of the paths shown in 8a. Horizontal dashed lines are the assumed temperature thresholds which result in ages being set during the uplift and cooling (FT (Ap) and FT(Zi) are apatite and zircon fission track thresholds; Ar is K-Ar (muscovite) threshold).

K-Ar ages in biotite and muscovite, and estimate of schist cooling since the start of the late Cenozoic uplift (at c. 7 Ma). Their temperature-time data are superimposed on our model curves for cooling adjacent to the Alpine fault, with and without the

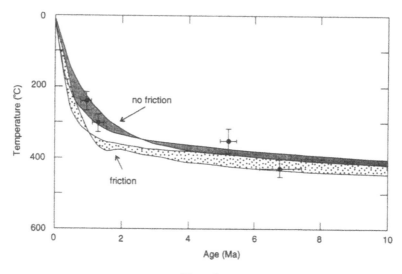

Figure 9

Modelled temperature-time histories for present day outcrop at 2.5 and 5 km from the Alpine fault on the Franz Josef profile. Light stippling is Model 1 (no friction), heavy stippling is for Model 2 (frictional heating). Points with error bars are data from dating methods and the schist metamorphism inferred by TIPPETT and KAMP (1993).

presence of friction, in Figure 9. If the size of the error bars is accepted (1σ, TIPPETT and KAMP, 1993), then based on the reset fission tracks points (<5 Ma), the no-friction model appears to be a slightly better fit than the model including frictional heating on the fault. The acceleration in cooling on the uplift path shown in Figure 9 could be misinterpreted as indicating an increase in the regional uplift and erosion rate in the last 2 Ma, especially if 1-D (vertical) uplift paths are being considered. In fact, the increase in uplift rate is a consequence of the curvature in the uplift path as the Alpine fault is approached (Figure 8a), and the general shape of the cooling curve in Figure 9 would occur even if the crust is in a steady state of uplift and erosion.

TIPPETT and KAMP (1993) assume that a negligible change in the thermal regime occurred prior to about 1.3 ± 0.3 Ma, because of low apparent uplift and erosion rates in the initial phase of Late Cenozoic compression (~1 mm/y; based on trends in Figure 9). Our modeling suggests that this assumption may not be valid. The cooling curves on Figures 8a or 9 predict that c. 100°C of cooling could have occurred on all paths between 10 Ma and 1 Ma, for rock now at the surface within 20 km of the Alpine fault. Simple one-dimensional thermal calculations, using an uplift and erosion rate of 1 mm/y for 5–10 Ma, confirm the model predictions (e.g., POWELL et al., 1988).

The predicted trend in reset zircon fission track ages within 20 km of the Alpine fault along the Franz Josef profile are compared with the compilation of ages

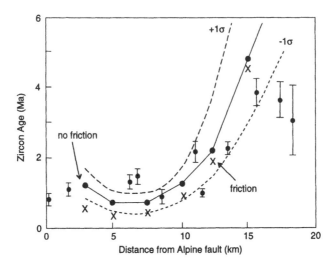

Figure 10

Comparison between the modelled zircon age pattern within 20 km of the Alpine fault on the Franz Josef profile and the observed age trend (data from Figure 16, TIPPETT and KAMP, 1993). Solid circles linked by line are the model ages without friction, crosses are with frictional heating. The dashed lines are the spread in model ages (without friction) assuming a $\pm 1\sigma$ of 25°C on the threshold temperature of 240°C (after HURFORD, 1986). Error bars on TIPPETT and KAMP data are also $\pm 1\sigma$.

presented by TIPPETT and KAMP (1993) in Figure 10. The predicted ages are based on the 240 ± 25°C ($\pm 1\sigma$) fission track closure temperature (HURFORD, 1986). The model data fits the trend reasonably, particularly when error bars are taken into consideration. The uncertainties in the model predictions increase greatly with distance from the Alpine fault because of the ± 25°C uncertainty in the closure temperature, and the relatively slow rate of temperature change at lower uplift rates. The uncertainties in the observed zircon ages also increase with distance from the fault by a similar amount.

The effect of friction on the fault reduces the model ages by between 0.5 and 1 Ma. For present-day rock outcropping within about 10 km of the fault, the model without frictional heating appears to be a marginally better fit to the data; at greater distances the increasing uncertainties in zircon ages make it impossible to distinguish between models with and without frictional heating. Supporting evidence for a lack of frictional heating may also be contained in the K-Ar ages within 4 km of the Alpine fault. The compilation of TIPPETT and KAMP (1993) shows K-Ar ages of 4.5–6 Ma, which compare with model ages of c. 3 Ma with no frictional heating, and 1–2 Ma with frictional heating. Within 2.5 km of the fault, there is the suggestion in Figure 10 that the modeled uplift rates, with no frictional heating, result in zircon ages significantly higher than observed. This could mean that the assumed decrease in uplift rate near the Alpine fault due to drag (Figure 3b) is not as great as modeled. Alternatively, some frictional heating could be

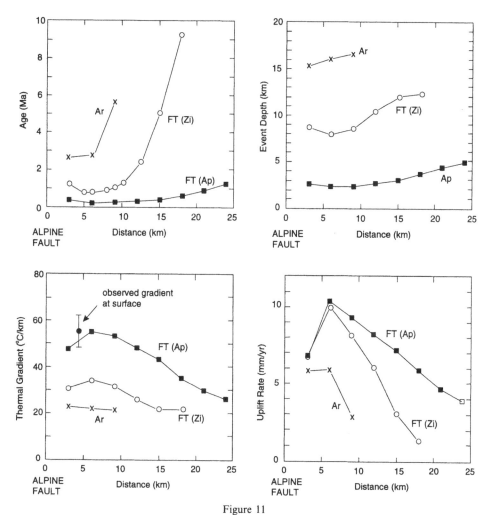

Figure 11

Composite of the thermal, age, and tectonic trends derived from modeling the Franz Josef profile with no frictional heating. The event depth is the depth when the various ages are set on the uplift path; the thermal gradient is derived from the temperature threshold and the event depth; uplift rate is derived from the event depth and the reset age. Age labels are the same as in Figure 8.

affecting uplift paths within 1–2 km of the Alpine fault. More reset age data are needed to resolve this.

For present-day rock outcropping around 15 km from the fault, Figure 10 suggests that the model ages are increasing more quickly than the observations, although the differences are probably within uncertainties. Inspection of Figures 8a and 8b reveals that at 5 Ma, these rocks were 40 km from the surface position of the Alpine fault and at about 10 km depth. The zircon ages for these rock paths at this time were very sensitive to the initial thermal regime and the uplift rate during the

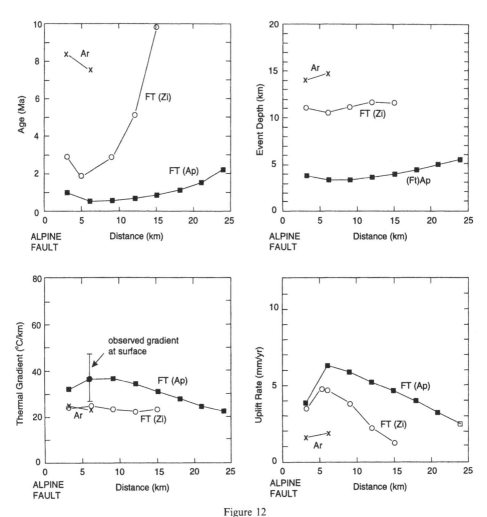

Figure 12

Composite of the thermal, age, and tectonic trends derived from modeling the Haast profile with no frictional heating. Refer to Figure 11 caption and text for more details.

early phase of the late Cenozoic compression. If these differences between the models and the zircon data are real, then they could be due to the uplift path for outcrop, presently 15 km from the Alpine fault, originating from a greater depth than that modeled, the initial thermal regime being slightly warmer than the 60 mW/m² (~20°C/km) assumed, or the early phase of uplift and erosion (between 5–10 Ma) being slightly greater than that assumed. More detailed age data is needed from rock outcropping in the zone 10–30 km from the Alpine fault in order to resolve some of these questions.

Figures 11 and 12 demonstrate the "windows" of the uplift histories represented by the different types of reset ages, and the potential for discriminating between the very recent uplift history and the early phase of the uplift and erosion episode. The same threshold temperatures of 120, 240 and 350°C are assumed for apatite and zircon fission tracks ages, and K-Ar muscovite ages, respectively, as previously used. The event depth refers to the depth at which the rock cools below the threshold temperature, the age provides the time when this occurred, the thermal gradient is the threshold temperature divided by the event depth, and the uplift rate is the event depth divided by the age. Both the thermal gradients and the uplift rates represent average values over differing amounts of space and time since the uplift episode began. The examples in Figure 11 are for the Franz Josef profile with no frictional heating, those in Figure 12 are for the Haast profile with no frictional heating. The differences between the two figures give an indication of the sensitivity of the age data to uplift and erosion rates (Franz Josef profile has a peak uplift rate of 10 mm/y; Haast a peak rate of 6 mm/y). The most notable differences in ages are within 10 km of the fault, with apatite fission track ages being <0.5 Ma at Franz Josef compared to between 0.5–1.0 Ma at Haast; zircon fission track ages being c. 1 Ma compared to 2–3 Ma; and K-Ar (muscovite) ages being 3–6 Ma compared to 7–8 Ma. The temperature gradients inferred from the apatite fission track ages and the event depth peak at around 55°C/km between 5–10 km from the Alpine fault on the Franz Josef profile, compared to between 35–40°C/km on the Haast profile.

The apatite ages provide the best indicator of the present-day tectono-thermal regime, representing average rates within the last 0.5 Ma and within the upper 5 km of crust, for outcrop within some 20 km of the Alpine fault. The predicted temperature gradient derived from the apatite age appears to be in reasonable agreement with the observed gradients at the two heat flow sites, as was found when the models predicted the surface heat flows in Figures 4 and 6. The uplift rate derived from the apatite ages is also a good match to the present-day uplift rate imposed as a surface boundary condition to the model.

The zircon fission tracks ages produce generally lower uplift rates and temperature gradients because of the longer time period represented. Within about 5 km of the Alpine fault on both profiles (Figures 11, 12), the zircon-derived uplift rate is close to the present-day surface uplift rate because of the near-vertical uplift path from the event depth over the last 1 Ma. The temperature gradients derived from the zircon fission tracks are lower than the present-day regime close to the fault mainly because of thermal lag effects. In the region 10–20 km from the Alpine fault, where the top of the totally reset zircon ages is now being exhumed, the zircon ages indicate the start of the uplift period (5–10 Ma) and the thermal regime at that time (i.e., ~20°C/km). These pre-uplift conditions apply to depths of 10 km and distances of c. 50 km from the present position of the Alpine fault.

The reset K-Ar (muscovite) profile is restricted to outcrop within 10 km of the Alpine fault on the two profiles, and potentially senses the thermal regime early in the uplift history because of the event depth being at c. 15 km and the reset ages being more than 3 Ma for this model. Average uplift rates are considerably lower than present-day near-surface values and the temperature gradient is within about 10% of the assumed, pre-uplift gradient of 20°C/km. These values apply to a zone 20–40 km from the surface trace of the Alpine fault (and at 15 km depth).

Discussion

The results of the numerical modeling presented in the preceding sections satisfactorily agree with both the surface heat flow measurements, reset K-Ar ages and fission track ages, as well as geotherms at greater depth implied by the earthquake focal depth distribution. Our modeling, although preliminary in nature, provides a good first-order approach to the tectonic and thermal evolution of the region, and demonstrates where new data may be able to distinguish between different uplift histories. Some additional comments on the implications of the modeling are warranted.

Thermal Regime

Different models have been tested to investigate the stability of the solution and to evaluate the effects of boundary conditions. For the deformation computation, the choice of viscosity ratio and the rheological stratification can produce slight differences, but the general thermal features do not change and the resultant changes in the values of surface heat flow generally do not exceed 20%. The thermal results are more sensitive to boundary conditions, especially the present-day uplift and erosion rates. The differences in the models for the Franz Josef and Haast profiles are mainly produced by different uplift and erosion profiles. The choice of the average thermal diffusivity of the crust has a small effect, with likely variations of up to $\pm 20\%$ causing inverse temperature variations attaining the same magnitude.

Two competing factors, crustal thickening, and uplift and erosion, control the regional thermal evolution. Our calculations for the Franz Josef and Haast profiles evidence that surface heat flows are higher than the background value if uplift rates exceed about 3 mm/y, typically within a distance of 30 to 40 km of the Alpine fault. Normally, an erosion rate of 3 mm/y would produce a significant heat flow high anomaly relatively quickly (<1 Ma). The reason for the reduced surface heat flow beyond 30–40 km from the Alpine fault is a heat sink effect caused by the crustal thickening. This effect is predicted to be greatest about 60 km from the Alpine fault, where the surface heat flow as a zone is a minimum of 40 mW/m^2. Such a difference in heat flow may not be easy to detect in practice, and no heat flow data are available in this part of the South Island.

It is suggested by the NUVEL model for plate motions, that convergent rate at the Alpine fault may be significantly lower than 20 mm/yr, and, if so, this has implications for the balance between crustal shortening occurring as observed uplift and erosion, and inferred crustal thickening. We have calculated the effect of two extreme models of crustal shortening: one with 100% uplift and erosion; and the other with no uplift and erosion and 100% crustal thickening. The former produces increased surface heat flows, which exactly follow the trend of the uplift rate. The latter produces extensive reduced low heat flow zone over the entire region of crustal thickening. Therefore, the existence or absence of the predicted low heat flow zone beneath the eastern foothills of the Southern Alps may in turn constrain the balance between uplifted and eroded crust versus crustal thickening. It is possible that this balance may have changed with time, therefore the results would need to be interpreted carefully.

The modeling confirms the observations that the heat flow adjacent to the Alpine fault on the Haast profile is significantly less than that on the Franz Josef profile. It is unclear whether the differences between the respective model predictions and the observations on each profile are real because the uncertainties on the two heat flow values are poorly constrained. The relatively large uncertainties in the heat flow determinations (estimated to be at least $\pm 25\%$) raises questions about whether attempts at additional measurements are worthwhile in such mountainous terrain, especially if the holes are only several hundred meters deep. Difficult drilling conditions, unknown effects of water movement in fractures, and large terrain corrections make such holes a relatively expensive investment for constraining tectono-thermal models such as the ones discussed above. It is shown in Figures 11 and 12 that a profile of reset apatite ages provides a good proxy for the surface thermal regime in the region of high uplift and erosion. We conclude that a few strategically located heat flow holes which are at least 500 m deep are likely to yield more valuable thermal data than a greater number of shallower heat flow holes.

Thermal Regime from Fission Track Analysis

To estimate the uplift and erosion rate from fission track data, it is customary to assume a constant geothermal gradient for the studied region. Our study of synthetic fission tracks, however, reveals that the averaged geothermal gradient varies both in location and in time duration covered by the annealing events. Figure 11 demonstrates that the apatite data reflects the most recent events and produces the highest geothermal gradients and uplift rates; while K-Ar ages reflect events which occurred during much earlier times and at a greater depth, and therefore generate a lower geothermal gradient and uplift rate. It is also found that the averaged geothermal gradient from a single method varies with location. The geothermal gradient near the fault could be 100% higher than that 20 km from the fault, as shown by the apatite data in Figure 11c. Even relatively low uplift and erosion rates of ~ 1 mm/y cause significant thermal anomalies if continued for several million years. The general conclusion is, in regions with strongly varying uplift rates, a constant geothermal gradient assumption may be questionable and lead to overestimation of uplift rates.

It is clear from the modeling that the various types of temperature-sensitive dating methods, such as fission track and K-Ar ages, are capable of providing information suitable for constraining uplift histories, *particularly when used together*. That is, a suite of age data is ideally needed from the same location, in order to constrain the cooling path for the rock presently at that location. In the case of the Southern Alps, there appears to be considerable age data beyond about 20 km east of the Alpine fault where ages are either partially reset or date the Mesozoic metamorphic events, and there are some totally reset data close to the fault (0–5 km). There appears to be a lack of reset ages in the critical 5–20 km range. This coincides with the axis of the Southern Alps, where sampling is more difficult, but this is also where the uplift and erosion rates may be a maximum. Our modeling on both the Franz Josef and the Haast profiles indicates that the outcrop interval from the Alpine fault to about 20 km east of the fault is probably the most important for distinguishing both the early and late uplift histories based on the different dating methods. Because of the two-dimensional nature of the uplift path (in fact 3-D because of the presence of shear parallel to the Alpine fault), the outcrop at, for example, 20 km east of the fault is potentially holding information about the original thermal regime from around 50 km east of the fault, and during the early part of the uplift episode. At 5 km from the fault, the rock probably holds information from up to 20 km from the surface trace of the fault, and from the more recent, rapid uplift phase. The lateral changes in the reset ages should therefore be able to constrain the dip on the Alpine fault to about 20 km depth. The satisfactory agreement with our modeling suggests that in the Franz Josef area the dip may be around 45°. Further modeling is recommended, however this should also be accompanied by additional age determinations in the areas most useful for distinguishing the tectono-thermal history. The challenge may be to achieve sufficient accuracy with the relatively young age dating that is required. Ideally, apatite fission track analyses must resolve ages of 0.1–0.3 Ma with uncertainties of 0.05 Ma, and zircon ages of around 1.0 Ma require uncertainties of less than ± 0.2 Ma.

There is one source of data which presently appears to conflict with most other data, in particular, the fission track ages. This is the formation pressure inferred from fluid inclusion studies (e.g., CRAW *et al.*, 1994). The range of temperatures derived from the inclusions appears to be consistent with the age data, with pre-uplift (Cenozoic) formation temperatures of 350–450°C, and late vein inclusions in the region of rapid uplift near the Franz Josef region registering trapping temperatures of 300–400°C. The depth of the pre-uplift greenschist facies rock is assumed by CRAW *et al.* (1994) to be 10–15 km, or from a pressure of 3–4 kbar (0.3–0.4 MPa), based on an undisturbed temperature gradient of around 25°C/km. The depth of some of the fluid inclusions in late stage veins is inferred to be as low as 0.6 to 0.8 kbar, based on CO_2 and salinity properties, which implies depths as low as 2 km if lithostatic pressures are assumed. If hydrostatic pressures are assumed, then substantially greater depths are possible (6–8 km).

The presence of zircon fission ages of around 1 Ma in the region of maximum uplift, the thermal implications that this represents the time when rock temperatures cooled below about 250°C ($\pm$ 20°C), and the geodynamic implications that in 1 Ma the rock has probably come from 8–10 km depth, mean that fluid inclusions with even higher trapping temperatures probably formed at greater depth on the uplift path. The discrepancy in inferred depths could mean that the zircon fission track annealing temperature is considerably higher than assumed, or that the fluid inclusion trapping pressure is much higher than inferred, or that the inclusions are the result of brief episodes of fluid upflow which are atypical of pressure and temperature conditions in the uplift zone. It is beyond the scope of this paper to resolve this apparent conflict.

Frictional Heating

We have made computations of models with and without frictional heating for both the Franz Josef and the Haast profiles. The results indicate that, for the Franz Josef profile, the model without frictional heating fits the fission track ages, K-Ar ages and T-t history, but possibly not the surface heat flow, marginally better than results of the model with frictional heating. For the Haast profile, the model without frictional heating appears to fit the surface heat flow value and the zircon fission track ages (not shown) better than the model with frictional heating. The differences are close to the measurement errors and, therefore, any conclusions based on the comparisons must be made cautiously. However, it indicates that frictional heating, if it exists, may not play a significant role.

The role of frictional heating on the Alpine fault has been controversial. Although SCHOLZ et al. (1979) and SCHOLZ (1982) suggested high shear strength on the Alpine fault, and ADAMS (1981) and ADAMS and GABITES (1985) proposed frictional heating over a 15 km zone to explain the observed metamorphism, ALLIS (1981, 1982) believed that shear stress on the Alpine fault is small and differential uplift rates alone are sufficient to explain the metamorphism and recent Ar age data (KAMP et al., 1989). JOHNSTON and WHITE (1983) believe shear heating may have occurred, but limited to a very short period of 1 y– 0.1 m.y. and the effects were limited within about 1 km of the fault zone, based on Fe/Mg geothermometry. Frictional heating on the San Andreas Fault, California, is also suggested to be low, as indicated by low heat flows and principal stress orientations (e.g., HICKMAN, 1991), and one explanation for the low stress is high pore pressure (RICE, 1993). On balance, our modeling of the thermal regime beneath the Southern Alps indicates that shear adjacent to the Alpine fault may be similar to that on the San Andreas, and consistent with low frictional heating, however, the evidence is not yet conclusive.

Brittle-ductile Transition

Earlier thermal modeling by ALLIS *et al.* (1979), and KOONS (1987) indicated that the rapid uplift and erosion in parts of the Southern Alps has caused thermal weakening of the crust, with temperatures of around 300°C possible at 2 km depth, and 400°C at less than 5 km depth. The brittle-ductile transition was suggested to be as shallow as 3 km depth if uplift rates of 10 mm/y have persisted for more than 1.5 Ma. HOLM *et al.* (1989) suggested that the depth of the transition zone could be at 6–8 km depth and at 300–350°C, based on fluid inclusion interpretations. This was shown to be consistent with likely strain rates and theoretical rheological relationships for the crust (after SIBSON, 1983). In the present modeling, the temperature at mid-crustal depths within the uplift zone is cooler than indicated in the earlier thermal models. Inspection of the isotherms in Figure 4a shows the 350°C isotherm to be generally at *greater* depth beneath the uplift zone than it is towards the sides of the model in the undeformed crust, where it occurs at 18 km depth. If the 350°C isotherm is regarded as an indicator of the base of the brittle upper crust, then based on thermal criteria alone, the uplift zone is not necessarily weakened by high temperatures. In the Franz Josef model with frictional heating, a 10 km wide zone coinciding with the fault zone and extending from around 8 km to 20 km depth, has anomalously high temperatures in the range of 350–400°C. This particular model is consistent with a relatively narrow zone of thermal weakening centered on the fault zone. In both the Haast models (Figures 6 and 7) the 350°C isotherm is, in general, depressed beneath the uplift zone.

The amount of depression of isotherms at mid-crustal depths along both the Franz Josef and the Haast profiles is dependent on the amount of crustal shortening, in particular the fraction which is accommodated as crustal thickening. Notwithstanding these uncertainties, we conclude that thermal weakening of the crust beneath the Southern Alps may not be the main factor contributing to the relative aseismicity of the region. In fact, assuming quartz-dominant lithologies for the entire crust, the depression of the 300–350°C isotherms due to crustal thickening beneath the Southern Alps might indicate that the crust here is actually stronger than that adjacent to the collision zone. This conclusion is similar to that of HOLM *et al.* (1989) who suggested the effects of pressure reduction may be more important in controlling the depth of the brittle-ductile transition zone than temperature.

The modeling also predicts a relatively cool upper mantle beneath the Southern Alps. Seismic wave velocities would be higher and attenuations would be lower, if other conditions are the same and temperature is the dominant factor. The inversion of upper mantle velocities (P_n, S_n) beneath New Zealand by HAINES (1979) did not reveal an anomaly coinciding with the Southern Alps, but the size of the inferred anomaly is probably too small to have been resolved by that inversion. Seismic velocity profiles may be estimated if data are available on the rock type and their velocity dependence on temperature and pressure is based on laboratory experiments. An integrated study of seismic profiling and inversion, rock experiments in high temperature and pressure, gravity and thermo-tectonic modeling, are needed to resolve uncertainties in the extent of crustal thickening occurring beneath the Southern Alps.

Conclusions

The Alpine fault in South Island of New Zealand is the boundary of oblique continental collision. The Pacific continental crust adjacent to the boundary has shortened by some 70 km and thickened to 40 km. Uplift occurs along the Southern Alps, with the maximum erosion rate sufficiently high to remove all newly uplifted rocks. Metamorphic rocks have been exhumed all along the fault, with the highest metamorphic grade schist closest to the fault. The late Cenozoic uplift and erosion has raised isotherms towards the surface, and exhumed rocks from below the depths where various types of age signatures are locked into minerals. Previously published reset ages derived from apatite and zircon fission tracks, and from K-Ar dating, therefore provide constraints for models of the thermal regime at the time the ages are set. Additional constraints are heat flow determinations from two drillholes about 5 km from the Alpine fault. The heat flow values, when corrected for topographic effects, are 190 ± 50 mW/m^2 at Franz Josef (near Mt. Cook) and 90 ± 25 mW/m^2 near Haast, and are consistent with the higher uplift rates at Franz Josef. These values contrast with a background heat flow of 60 ± 5 mW/m^2 in areas undisturbed by the late Cenozoic tectonism. Hot springs occur in the Southern Alps, but they are restricted to the north of the Mt. Cook area, and their occurrence cannot be explained solely by high upper crustal temperatures caused by the uplift and erosion history. The presence of anomalously high permeability in the uppermost 3–5 km of crust, due to open fractures, is considered to be a major factor contributing to the hot spring occurrences.

Our finite-element tectono-thermal modeling of the Southern Alps of New Zealand deciphers the geodynamic process and its thermal consequences for the last 10 Ma. Two cross sections, Franz Josef in the north and Haast in the south, have been modelled. The models use the present-day uplift and erosion rates as an upper boundary condition, and assume that the remainder of the compression occurs as crustal thickening. For the Franz Josef profile this means about 70% of the compressed crust is uplifted and eroded, whereas for the Haast profile this figure is around 50%.

Our models indicate that there are two competing factors affecting the thermal structure: crustal thickening tends to reduce surface heat flow, and uplift and erosion tend to increase the surface heat flow. This produces a maximum heat flow of around 150–190 mW/m² in the zone of maximum uplift and erosion, with the higher value resulting from the inclusion of a frictional heating component on the Alpine fault. On the Haast profile the maximum predicted heat flow is in the range of 100–140 mW/m², with the predicted heat flow at the drillhole site 10 mW/m² less than the maximum values. When the effects of thermal conductivity assumptions in the model are considered, the observed heat flow at Franz Josef is midway between the range of predicted values, and at Haast the heat flow is slightly below the predicted value for the model without frictional heating. Small adjustments to the rate of uplift at the heat flow site, or possibly to the thermal diffusivity of the upper crust are capable of improving the match of the model results to heat flow data.

The modeling predicts a low heat flow anomaly of 40 mW/m² in the eastern foothills of the Southern Alps, about 60 km east of the Alpine fault. This is due to the dominance of the effects of crustal thickening beneath this zone. There is no data to confirm this anomaly. Depressed temperatures are also predicted in the lower crust and upper mantle beneath the Southern Alps due to the crustal and lithospheric thickening. Modelled isotherms in the range 500–800°C slope downwards from around 30 km depth, 100 km east of the Alpine fault, to around 70 km depth beneath the Alpine fault. This coincides with a band of seismicity and is consistent with a westward-dipping zone of brittle mantle beneath the Southern Alps. With the possible exception of a relatively narrow zone coinciding with the Alpine fault, the depth to the brittle-ductile transition is also greater beneath the uplift zone than in the undeformed crust on either side. The Southern Alps are therefore not necessarily thermally weakened by the rise of typical lower crustal temperatures to shallow depth. The effects of decompression along the uplift path and relatively high pore pressures at shallow depth may be the main causes of a weak crust, and the apparent aseismicity of the central Southern Alps.

The modeling has achieved reasonable matches to the reset age data close to the Alpine fault in the vicinity of the Franz Josef profile. Predicted apatite fission track ages are less than 0.5 Ma within 20 km of the surface position of the Alpine fault; zircon fission track ages increase from around 1 Ma close to the fault, to over 5 Ma beyond 15 km from the fault; and reset K-Ar (muscovite) ages increase from 3 Ma within 5 km of the fault, to 6 Ma 8 km from the fault. The inferred pore pressures from fluid inclusion in late stage veins appear to be the only data which is in conflict with the modelled thermal regime. There is weak evidence that the model without a frictional heat source provides a better fit to the age data. Rapid cooling of rocks close to the Alpine fault in the last 1 Ma is not due to a recent increase in the rate of compression—it is a consequence of the strongly two-dimensional uplift path close to the fault, and the change from predominantly horizontal movement to vertical movement as the fault zone is approached. The modeling highlights the importance of using multiple types of reset age data in the zone 0–25 km from the Alpine fault, in order to resolve the mode of deformation, the dip of the Alpine fault, and the resultant thermal regime. Our modeling suggests outcrop within

25 km of the fault has probably originated from depths of 10–20 km, and from over 50 km from the surface position of the Alpine fault. The modeling also shows that the reset apatite fission track ages are a good proxy for the surface heat flow in the area of maximum uplift and erosion. Further numerical modeling and new age data are recommended.

Acknowledgements

YS thanks the National Natural Science Foundation of China (NSFC) and the Institute of Geological and Nuclear Sciences, New Zealand for their support for this research. The research was supported by Contract C05315 from the New Zealand Foundation for Research Science and Technology. The thermal conductivity measurements on core from the Franz Josef hole were made by R. Munroe and J. Sass of the U.S.G.S. We thank F. Evision, R. Funnell, J. Latter, P. Whiteford, and M. Reyners for stimulating discussions during the course of the work, and Kelin Wang, Peter Kamp and an anonymous reviewer for helpful suggestions regarding an earlier version of the manuscript.

REFERENCES

ADAMS, J. (1979) *Vertical drag on the Alpine fault, New Zealand*. In *The Origin of the Southern Alps* (eds. Walcott, R. L., and Cresswell, M. M), The Roy. Soc. of New Zealand Bull. *18*, 47–54.

ADAMS, J. (1980), *Contemporary Uplift and Erosion of the Southern Alps, New Zealand*, Geol. Soc. Am. Bull. Part II, *91*, 1–114.

ADAMS, C.J. (1979), *Age and origin of the Southern Alps*. In *The Origin of the Southern Alps* (eds. Walcott, R. L., and Cresswell, M. M), The Roy. Soc. of New Zealand Bull. *18*, 73–78.

ADAMS, C. J. (1979), *Uplift rates and thermal structure in the Alpine Fault Zone and Alpine Schists, Southern Alps, New Zealand*. In *Thrust and Nappe Tectonics*. (eds. McClay, K. R., and Price, N. J.). (Backwell Sci. Pub., Oxford 1981) pp. 211–222.

ADAMS, C. J., and GABITES, J. E. (1985), *Age of Metamorphism and Uplift in the Haast Schist Group at Haast Pass, Lake Wanaka and Lake Hawea, South Island, New Zealand*, N. Z. J. Geol. Geophys. *28*, 85–96.

ALLIS, R. G. (1981), *Continental Underthrusting beneath the Southern Alps of New Zealand*, Geology *9*, 303–307.

ALLIS, R. G. (1982), *Reply on 'Continental Underthrusting beneath the Southern Alps of New Zealand'*, Geology *10*, 485–491.

ALLIS, R. G. (1986), *Mode of Crustal Shortening Adjacent to the Alpine Fault, New Zealand*, Tectonics *5*, 15–32.

ALLIS, R. G., HENLEY, R. W., and CARMAN, A. F. (1979), *The thermal regime beneath the Southern Alps*. In *The Origin of the Southern Alps* (eds. Walcott, R. L., and Cresswell, M. M.), The Roy. Soc. of New Zealand Bull. *18*, 79–85.

ANDERSON, H., and WEBB, T. (1994), *New Zealand Seismicity: Patterns Revealed by the Upgraded National Seismic Network*, New Zealand J. Geol. Geophys. *37*, 477–494.

BYERLEE, J. D. (1978), *Friction of Rocks*, Pure and Appl. Geophys. *116*, 615–626.

CHEN, W.-P., and MOLNAR, P. (1983), *Focal Depths of Intracontinental and Intraplate Earthquakes and their Implications for the Thermal and Mechanical Properties of the Lithosphere*, J. Geophys. Res. *88*, 4183–4214.

CLIFF, R. A. (1985), *Isotopic Dating in Metamorphic Belts*, J. Geol. Soc. London *142*, 97–110.

CRAW, D. (1988), *Shallow-level Metamorphic Flids in a High Rate Metamorphic Belt; Alpine Schist, New Zealand*, J. Metamorphic Geology *6*, 1–16.

CRAW, D., RATTENBURY, M. S., and JOHNSTONE, R. D. (1994), *Structures within Greenschist Facies Alpine Schist, Central Southern Alps*, New Zealand J. Geol. Geophys. *37*, 101–111.

DEMETS, C., GORDAN, R. G., ARGUS, D. F., and STEIN, S. (1990), *Current Plate Motions*, Geophys. J. Int. *101*, 425–478.

EVISON, F. F. (1971), *Seismicity of the Alpine Fault, New Zealand*. In *Recent Crustal Movements*, Roy. Soc. New Zealand Bull. *9*, 161–165.

FUNNELL, R. H., and ALLIS, R. G., in preparation. *Heat Flow Variations in the South Island of New Zealand*.

FUNNELL, R. H., CHAPMAN, D. S., ALLIS, R. G., and ARMSTRONG, P. A. (1995), *Thermal State of the Taranaki Basin, New Zealand*, J. Geophys. Res.

HAINES, J. (1979), *Seismic Wave Velocities in the Uppermost Mantle Beneath New Zealand*, N.Z. J. Geol. Geophys. *22*, 245–257.

HATHERTON, T. (1980). *Shallow Seismicity in New Zealand 1956–75*, J. Royal Soc. N.Z. 10, 19–25.

HICKMAN, S. H. (1991), *Stress in the Lithosphere and the Strength of Active Faults*, Contributions in Tectonophysics, U.S. National Report 1987–1985, 759–775.

HOLM, D. K., NORRIS, R. J., and CRAW, D. (1989), *Brittle and Ductile Deformation in a Zone of Rapid Uplift: Central Southern Alps, New Zealand*, Tectonics 8, 153–168.

HURFORD, A. J., (1986), *Cooling and Uplift Patterns in the Lepontine Alps, South-central Switzerland and an Age of Vertical Movement of the Insubric Fault Line*, Contrib., Mineral. Petrol. *92*, 413–427.

HYNDMAN, R. D., and WANG, K. (1993), *Thermal Constraints on the Zone of Major Thrust Earthquake Failure: The Cascadia Subduction Zone*, J. Geophys. Res. 2039–2060.

JOHNSTON, D. C., and WHITE, S. H. (1983), *Shear Heating Associated with Movement along the Alpine Fault, New Zealand*, Tectonophysics *92*, 241–252.

KAMP, P. J. J. (1992), *Tectonic Architecture of the Mountain Front-foreland Basin Transition, South Island, New Zealand, Assessed by Fission Track Analysis*, Tectonics 11, 98–113.

KAMP, P. J. J., GREEN, P. F., and WHITE, S. H. (1989), *Fission Track Analysis Reveals Character of Collisional Tectonics in New Zealand*, Tectonics 8, 169–195.

KAMP, P. J. J., and TIPPETT, J. M. (1993), *Dynamics of Pacific Plate Crust in the South Island (New Zealand) Zone of Oblique Continent-continent Convergence*, J. Geophys. Res. *98*, 16105–16118.

KAPPELMEYER, O., and HAENEL, R. (1974), *Geothermics, with Special Reference to Application*, Geoexploration Monograph series 1, 4, Gebrüder Borntraeger, Berlin.

KOONS, P. O. (1987), *Some Thermal and Mechanical Consequences of Rapid Uplift: An Example from the Southern Alps, New Zealand*, Earth Planet. Sci. Lett. *86*, 307–319.

LEES, C. H. (1910), *On the Isotherms in Mountain Ranges in Radioactive Districts*, Proc. Roy. Soc. *A 83*, 339–346.

PANDEY, O. P. (1981), *Terrestrial Heat Flow in New Zealand*, Unpublished Ph.D. Thesis, Victoria University of Wellington.

POWELL, W. G., CHAPMAN, D. S., BALLING, N., and BECK, A. E., *Continental heat flow density*. In *Handbook of Terrestrial Heat Flow Density* (eds. Haenel, R., Rybach, L., and Stegena, L.) (Kluwer Academic Publishers, Dordrecht 1988) pp. 167–218.

RANALLI, G., and MURPHY, D. C. (1987), *Rheological Stratification of the Lithosphere*, Tectonophysics *132*, 281–295.

REYNERS, M. (1987), *Subcrustal Earthquakes in the Central South Island, New Zealand, and the Root of Southern Alps*, Geology *15*, 1168–1171.

REYNERS, M., and COWAN, H. (1993), *The Transition from Subduction to Continental Collision: Crustal Structure in the North Canterbury Region, New Zealand*, Geophys. J. Int. *115*, 1124–1136.

RICE, J. R. (1993), *Spatio-temporal Complexity of Slip on a Fault*, J. Geophy. Res. *98*, 9885–9907.

SCHOLZ, C. H., BEAVAN, J., and HANKS, T. C. (1979), *Frictional Metamorphism, Argon Depletion, and Tectonic Stress on the Alpine Fault, New Zealand*, J. Geophys. Res. 84, 6770–6782.

SCHOLZ, C. H. (1982), *Comment on 'Continental Underthrusting beneath the Southern Alps of New Zealand'*, Geology 10, 479–485.

SHI, Y., and WANG, C.-Y. (1987), *Two-dimensional Modeling of the P−T−t Paths of Regional Metamorphism in Simple Overthrust Terrains*, Geology *15*, 1048–1051.

SIBSON, R. H., WHITE, S. H., and ATKINSON, B. K. (1979), *Fault rock distribution and structure within the Alpine Fault Zone: A preliminary account*. In *The Origin of the Southern Alps* (eds. Walcott, R. L., and Cresswell, M. M.), The Roy. Soc. of New Zealand Bull. *18*, 55–65.

TIPPETT, J. M., and KAMP, P. J. J. (1993), *Fission Track Analysis of the Late Cenozoic Vertical Kinematics of Continental Pacific Crust, South Island, New Zealand*, J. Geophys. Res. *98*, 16119–16148.

WALCOTT, R. I. (1978), *Present Tectonics and Late Cenozoic Evolution of New Zealand*, Geophys. J. R. Astr. Soc. *52*, 137–164.

WALCOTT, R. I. (1979), *Plate motion and shear strain rates in the vicinity of the Southern Alps*. In *The Origin of the Southern Alps* (eds. Walcott, R. L., and Cresswell, M. M.), The Roy. Soc. New Zealand Bull. *18*, 5–12.

WELLMAN, H. W. (1979), *An uplift map for the South Island of New Zealand, and a model for uplift of the Southern Alps*. In *The Origin of the Southern Alps* (eds. Walcott, R. L., and Cresswell, M. M.), The Roy. Soc. New Zealand Bull. *18*, 13–20.

WILLETT, S. D., *Modelling thermal annealing of fission tracks in apatite*. In *Short Course on Low Temperature Thermochronology: Mineralo.* (eds. Zentilli, M., and Reynolds, P. H. eds.), (Assoc. Canada Short Course Handbook 1992) pp. 42–47.

WOOD, R. A. (1991), *Structure and Seismic Stratigraphy of the Western Challenger Plateau*, N. Z. J. Geol. Geophys. *34*, 1–9.

WOODWARD, D. J. (1979), *The crustal structure of the Southern Alps, New Zealand, as determined by gravity*. In *The Origin of the Southern Alps* (eds. Walcott, R. L., and Cresswell, M. M.), The Roy. Soc. New Zealand Bull. *18*, 95–98.

(Received March 2, 1995, accepted May 20, 1995)

PAGEOPH, Vol. 146, Nos. 3/4 (1996)

0033–4553/96/040503–29$1.50 + 0.20/0

Stress Field and Seismicity in the Indian Shield: Effects of the Collision between India and Eurasia

T. N. GOWD,[1] S. V. SRIRAMA RAO[1] AND K. B. CHARY[1]

Abstract—Intraplate stresses and intraplate seismicity in the Indian subcontinent are strongly affected by the continued convergence between India and Eurasia. The mean orientation of the maximum horizontal compression in the Indian subcontinent is subparallel to the direction of the ridge push at the plate boundary as well as to the direction of compression expected to arise from the net resistive forces at the Himalayan collision zone, indicating that the intraplate stresses in the subcontinent, including the shield area, are caused by plate tectonic processes. Spatial distribution of historic and instrumentally recorded earthquakes indicate that the seismic activity is mostly confined to linear belts while the remaining large area of the shield is stable. The available conventional heat flow data and other indicators of heat flow suggest hotter geotherms in the linear belts, leading to amplification of stresses in the upper brittle crust. Many of the faults in these linear belts, which happen to be 200–80 m.y. old, are being reactivated either in a strike-slip or thrust-faulting mode. The reactivation mechanisms have been analyzed by taking into consideration the amplification of stresses, pore pressures, geological history of the faults and their orientation with respect to the contemporaneous stress field. The seismicity of the Indian shield is explained in terms of these reactivation mechanisms.

Key words: Intraplate stresses, intraplate seismicity, linear belts, stable region, pre-existing faults, pore pressure, stress amplification, reactivation mechanisms.

Introduction

Until recently the Indian shield was considered as a stable continental region relatively free from earthquakes. However, the Koyna earthquake of December 10, 1967 with a magnitude of 6.3, following the impoundment of water in the Shivaji Sagar Lake in 1962 (GUPTA *et al.*, 1969) and a number of other earthquakes of $M > 5.0$ that occurred within a short period of time in the Indian shield (CHANDRA, 1977), have given rise to doubt regarding its aseismic nature. Following the occurrence of the globally deadliest earthquake of M 6.4 at Killari near Latur in Maharashtra State on September 30, 1993 (GUPTA, 1993), Indian shield is no longer considered to be seismically inactive. This anomalous seismicity of the Indian shield is considered to be due to high intraplate stresses and the associated tectonic

---

[1] National Geophysical Research Institute, Hyderabad–500 007, (A.P.), India.

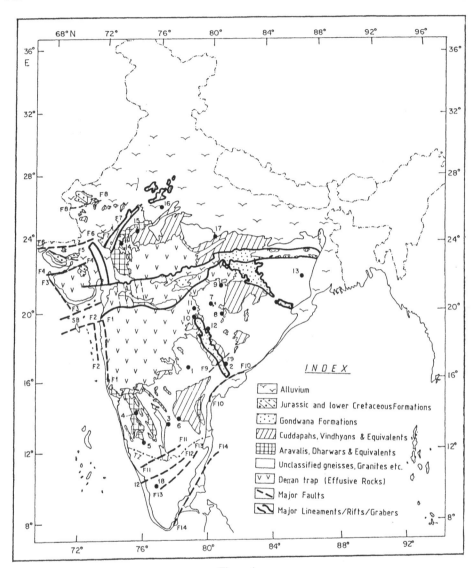

Figure 1

Geological map of India. This map has been prepared by modifying the geological map of the Indian shield published by VALDIYA (1973). The faults F1, F2, F3, F4, F5 and F6 are after BISWAS (1987); F7 and F8 after VALDIYA (1973); F10 after SASTRI *et al.* (1973); F11, F12, F13, F14 after KATZ (1978); Southern boundary of the NTS lineament zone delineated on the basis of the tectonic map of India (after ERMENKO *et al.*, 1969); F9 is delineated in the present study. F1: West coast fault; F2: Off-shore fault; F3: Saurashtra fault; F4: North Kathiawar fault; F5: Kutch fault; F6: Nagar-Parkar fault; F7: Marginal fault of the Delhi Mobile belt; F8: A fault in the Jaisalmer area in southern Rajasthan; F9: Bhadrachalam fault; F10: Marginal fault of the Krishna-Godavari coastal basin; F11: Kabbani fault; F12: Bhavani fault; F13: Attur fault; F14: Boundary fault of the Cauvery basin; A: Khandwa region where the Aravallis trend across the Narmada Tapti lineament zone in SSE direction. Source references for heat-flow values of different sites in the stable region: 1 (GUPTA *et al.*, 1987); 2 (RAO *et al.*, 1970); 3, 4 (GUPTA *et al.*, 1991); 5 (GUPTA and RAO, 1970); 6 (VERMA *et al.*, 1969); 7, 8, 9 (GUPTA *et al.*, 1993); 10, 11, 12, 15, 16, 17 (RAVISHANKER, 1988); 13, 18 (RAO *et al.*, 1976); 14 (GUPTA and GAUR, 1984).

movements caused by the continental collision between India and Eurasia (CHAN-
DRA, 1977; KAILASAM, 1979; VALDIYA, 1989, 1993; GAUR, 1993; RAVAL, 1993;
KHATTRI, 1994), particularly after the Himalayan orogeny was initiated in mid-
Miocene about 20 Ma. Though CHANDRA (1977) suggested some two decades ago
that the orientation of the zones of weakness (pre-existing faults), with respect to
the ambient stress field, is an important factor in determining the faults along which
earthquakes are likely to occur in the Indian shield, the reactivation mechanisms
have not been analyzed to date due to the paucity of stress data.

Recently GOWD et al. (1992b) evaluated tectonic stress field in the Indian
subcontinent, and prepared a map of maximum horizontal compressive stress
orientations in the Indian subcontinent using orientations derived from borehole
breakouts, hydraulic fracturing stress measurements, and earthquake focal mechanisms. The mean orientation of $S_{H_{max}}$ in the mid-continent stress province, which
includes most of the Indian subcontinent excepting the Bengal basin and the
southern shield, is N23°E, subparallel to the direction of compression expected to
arise from the net resistive forces at the Himalayan collision zone, suggesting that
it is largely determined by the tectonic collision processes. Though the sediments in
the Bengal basin, which extend from the northeastern margin of the Indian shield
to the southern Indo-Burman ranges, are being compressed in an E–W direction
due to the convergence of the Indian and Burmese plates, the stress field in the
basement and crust beneath the Bengal basin and in the subducted slab beneath the
Indo-Burman ranges is similar to the one prevailing in the mid-continent stress
province. The stress pattern evaluated by GOWD et al. (1992b) thus reveals that the
Indian subcontinent, including the Indian shield, is being compressionally stressed
in a NNE direction due to the resistive forces imposed along the collision boundary,
causing higher stresses in the intraplate region to the south of the collision zone
also.

Although there are numerous faults in the Indian shield (Fig. 1), only few of
them are being reactivated to cause earthquake activity. We have endeavored to
understand the mechanisms responsible for the reactivation of such faults, taking
into consideration the magnitude and direction of the contemporaneous stresses,
amplification of stresses as an effect of hotter geotherms, pore pressures, and
geological history of faults. The details are presented and discussed in this paper.

Seismicity

We have prepared an up-to-date catalogue of instrumentally recorded intraplate
earthquakes ($M \geq 4.5$) that occurred in the Indian shield (Table 1), based on
epicentral estimates reported by various agencies. All the historical earthquakes
listed by CHANDRA (1977), based on the descriptions available in the catalogues of
earthquakes published by OLDHAM (1883), MILNE (1911), TURNER et al. (1911,

Table 1

Catalogue of Earthquakes (M > 4.5) of the Indian Shield

DATE	LAT (°N)	LONG (°E)	LOCATION	MAG/ INTEN	REF
West Coast Seismic Belt (I)					
1618 MAY 26	18.90	72.10	Bombay	IX	1
1678	19.10	73.20	Vasai and Agashi, N. Konkan	VI	1
1751 DEC 09	19.10	73.20	Vasai, Shasthi, Bombay	VI	1
1764 AUG	17.90	73.70	Mahabaleshwar and vicinity	VII	1
1826 MAR 20	16.10	73.60	Moje Morvade	VI	1
1832 OCT 04	15.80	73.70	Majkur in Moje Ugate District	VI	1
1856 DEC 25	20.00	73.00	Bombay, Surat	VII	1
1935 JUL 20	20.00	73.00	Bombay, Surat, Kapadvanj	VI	2
1965 JUN 04	17.00	73.00	Rathnagiri	5.4	3
1967 APR 25	18.20	73.40	Mahad	5.6	3
1967 SEP 13	17.40	73.70		5.5	4
1967 SEP 13	17.40	73.70	Koyna Region	5.8	3
1967 SEP 13	17.40	73.70	Koyna Region	5.6	3
1967 DEC 10	17.40	73.70	Koyna Region	5.0	5
1967 DEC 10	17.40	73.90	Koyna Region	6.0	5
1967 DEC 11	17.28	73.67		4.7	4
1967 DEC 11	17.30	73.70	Koyna region	5.2	5
1967 DEC 11	17.48	73.83		4.5	4
1967 DEC 12	17.33	73.91		5.1	4
1967 DEC 12	17.34	73.79		4.6	4
1967 DEC 12	17.40	73.90	Koyna region	5.0	5
1967 DEC 12	17.60	73.90	Koyna region	5.4	5
1967 DEC 13	17.30	73.70	Koyna region	5.5	3
1967 DEC 13	17.20	73.66		4.6	3
1967 DEC 13	17.55	73.60	Koyna region	5.6	3
1967 DEC 24	17.50	73.90	Koyna region	5.5	5
1967 DEC 25	17.20	73.90	Koyna region	5.1	5
1968 FEB 07	17.40	73.70	Koyna region	VI	3
1968 MAR 04	17.40	73.70	Koyna region	VI	3
1968 JUL 31	17.30	74.00		5.7	3
1968 AUG 31	17.30	74.00	Koyna region	5.7	3
1968 OCT 29	17.30	73.90	Pophali	5.4	5
1969 NOV 03	17.30	73.90	Koyna region	5.7	5
1970 MAY 27	17.40	73.70	Koyna region	4.5	3
1970 SEP 25	17.40	73.60	Koyna region	5.0	3
1970 SEP 26	18.00	74.00	Koyna region	5.5	3
1973 OCT 17	17.40	73.70	Koyna region	VI	6
1974 FEB 17	17.50	73.10	Near west coast	5.0	3
1974 FEB 17	17.30	73.70	Koyna region	4.6 M_L	3
1974 APR 17	17.50	73.10	Near west coast	5.0 M_b	3
1976 MAR 14	17.30	73.70	Koyna region	4.8 M_s	7
1977 SEP 19	17.30	73.64	Koyna region	5.0 M_l	3
1980 SEP 02	17.27	73.71		4.7b	4
1980 SEP 20	17.24	73.87		4.8b	4
1980 SEP 20	17.26	73.64		5.2b	4
1980 SEP 25	17.40	74.20	Koyna	4.9	3
1980 OCT 04	17.27	73.82		4.8b	4

Table 1 (Contd.)

DATE	LAT (°N)	LONG (°E)	LOCATION	MAG/ INTEN	REF
1983 SEP 25	17.29	73.97		4.6b	4
1984 NOV 14	17.28	73.96		4.5b	4
Kutch Seismic Belt (II)					
1819 JUN 16	23.60	69.60	Kutch	XI	1
1828 JUL 20	23.20	69.90	Bhuj	VI	1
1845 APR 19	23.80	68.90	Lakhpat	VIII	1
1845 JUN 19	23.80	68.90	Delta of Indus, Lakhpat	VII	1
1903 JAN 14	24.00	70.00	Kutch	VII	3
1940 OCT 31	24.50	70.20		5.6	8
1956 JUL 21	23.00	70.00	Anjar	6.1	5
1965 MAR 26	24.40	70.00	N. of Kutch	5.3	3
1982 JAN 31	24.10	69.60	Pak border	4.8	3
1982 JUL 18	23.40	70.66		4.8b	4
1985 APR 07	24.30	69.90	Indo-Pak border	5.0	3
1989 DEC 10	24.67	70.99		4.7	8
1991 JAN 20	23.08	69.50		4.9	4
1991 SEP 10	24.18	68.64		4.7	8
1991 SEP 10	24.26	68.81		4.8	8
Transcontinental Lineament Zone (III)					
1846 MAY 27	23.00	80.00	Narmada	VI	1
1863 NOV 18	21.80	75.30	Nimar and Burwani country	VI	1
1864 APR 29	22.30	72.80	Ahmedabad, Surat	VII	1
1868 JUL 31	24.00	85.40	Hazaribagh	VI	1
1868 SEP 30	24.00	85.00	Manbhum vicinity	VII	1
1903 MAY 17	23.00	80.00	Jabalpur	VI	1
1919 APR 21	22.00	72.00	Bhavnagar Para, vicinity	VIII	2
1927 JUN 02	23.50	81.00	Son Valley	6.5	9
1938 MAR 14	21.50	75.70	Satpura	6.3	9
1938 JUN	22.30	71.60	Paliyad	VI	2
1938 JUL 14	22.40	71.80	Paliyad	VI	2
1938 JUL 19	22.40	71.80	Paliyad	VI	2
1938 JUL 23	22.40	71.80	Paliyad, Wadhwan, Limbdi	VII	2
1940 OCT 31	22.50	71.40	Dwaraka and vicinity	VI	2
1967 JAN 06	21.57	74.27		4.5	4
1970 MAR 23	21.70	73.00	Broach	5.4	5
1971 JUN 18	21.70	73.00	Broach	–	3
1974 OCT 20	21.70	74.20	Near Taloda	4.6	10
1986 APR 27	20.56	73.34	Valsad	4.6	11
1986 APR 18	22.34	79.25		4.9	8
Delhi Mobile Belt (IV)					
1848 APR 26	24.40	72.70	Mt. Abu	VII	1
1882 DEC 15	24.90	72.70	Mt. Abu	VI	10
1962 SEP 01	24.00	73.00	N. Gujarat	5.0	3
1966 JUN 20	28.70	76.90		5.1	3
1969 OCT 24	24.80	72.40	Mt. Abu	5.3	5
Bhadrachalam Seismic Belt (V)					
1954 JAN 05	18.00	81.30	Kothagudem	–	12
1963 DEC 05	17.30	80.10	N. of Guntur	–	5

Table 1 (Contd.)

DATE	LAT (°N)	LONG (°E)	LOCATION	MAG/ INTEN	REF
1968 JUL 29	17.60	80.80	Bhadrachalam and K'Gudem	4.5	12
1968 JUL 29	17.60	80.80	Bhadrachalam and K'Gudem	4.5	12
1969 APR 13	17.90	80.60	Bhadrachalam	5.3	5
1969 APR 14	18.00	80.50	Bhadrachalam	5.2	3
1969 APR 15	18.00	80.70	Bhadrachalam	4.6 M_s	7
1969 APR 30	17.90	80.60	Bhadrachalam	4.5 M_s	7
Ongole Seismic Belt (VI)					
1800 OCT 19	15.60	80.10	Ongole	VI	1
1859 JUL 21	16.30	80.50	Guntur	VI	1
1959 OCT 12	15.70	80.10	Ongole	VI	13
1959 OCT 13	15.60	80.10	Ongole	VI	14
1967 MAR 19	15.62	80.16		5.2	4
1967 MAR 27	15.60	80.10	Ongole	5.4	5
1987 DEC 03	15.50	80.21		4.5b	4
Southern Granulitic Terrain (VII)					
1807 DEC 10	13.10	80.30	Madras	VI	1
1816 SEP 16	13.10	80.30	Madras	VI	1
1822 JAN 29	12.50	79.70	Madras, Chittore, Nellore	VI	1
1823 MAR 02	13.00	80.00	Madras, Srilanka	VI	1
1858 AUG 23	11.40	76.00	Malabar	VI	1
1859 JAN 03	12.50	79.00	North Arcot	VI	1
1889 AUG 12	13.10	80.30	Madras	VI	1
1900 FEB 08	10.80	76.80	Coimbatore	VII	15
1901 APR 27	12.00	75.00	N. of Calicut	VI	1
1953 JUL 26	9.90	76.30	Cochin	VI	12
1972 JUL 29	11.00	77.00	Coimbatore	VI	3
1984 NOV 27	12.87	78.60	Tirupattur	4.5	12
1988 JUN 07	9.82	77.22	Idukki	4.5	12
Jaisalmer Region					
1907 JUL 12	26.00	72.00	Mallani	VI	1
1971 JAN 05	26.00	71.40		4.8	4
1971 MAY 18	26.99	71.80		4.9	4
1974 MAY 18	26.90	71.70	N. India	5.0	3
1991 NOV 08	26.32	70.60		5.6	8
1991 NOV 20	26.33	70.86		4.5	8
Stable Region					
1764 JUN 04	24.00	88.00	Banks of Ganga	VIII	1
1828 AUG 22	13.00	75.00	Malabar Coast	VII	1
1837 JUN 15	19.50	85.10	Ganjam	VI	1
1839 MAY 11	25.30	86.50	Jamalpur	VI	1
1843 APR 01	15.20	76.90	Bellary, Deccan	VII	1
1858 OCT 02	18.30	84.00	Chicacole	VI	1
1861 JUL 31	25.40	83.00	Varanasi	VI	1
1866 MAY 23	25.00	87.00	Bengal	VIII	1
1869 JUL 04	20.20	74.20	Nasik, Chandore	V	1
1869 SEP 01	14.50	80.00	Nellore	VI	1
1916 JAN 07	13.00	77.50	Bangalore	VI	3

Table 1 (Contd.)

DATE	LAT (°N)	LONG (°E)	LOCATION	MAG/ INTEN	REF
1917 APR 17	18.00	84.00	Vizianagaram	VII	3
1920 JUL 10	25.00	83.80		VII	13
1926 DEC 21	25.00	77.50	Lukwasa	VII	10
1929 APR 10	25.00	77.50	Near Lukwasa	VII	13
1930 JUN 25	25.00	77.50	Near Lukwasa	VII	13
1957 AUG 25	22.00	80.00	Balaghat	5.5	16
1963 MAY 08	21.70	84.90	SE. of Bihar	5.2	5
1964 APR 15	21.70	88.00	Midnapore	5.5	5
1966 APR 10	14.70	80.00	Tambaram	VI	10
1969 APR	16.60	79.30	Nagarjunasagar	VII	10
1969 MAY 03	23.00	86.60	Bankura	5.7	3
1972 MAY 16	12.40	77.00	Mandya, Mysore	4.6	17
1972 MAY 17	12.40	77.00	Mandya, Mysore	4.5	17
1978 SEP 27	23.60	75.40		4.5b	4
1979 AUG 05	22.10	86.00		4.7b	4
1980 MAR 30	17.50	81.84		4.4b	4
1982 OCT 14	20.39	84.42		4.7b	4
1983 JUN 30	17.90	78.50	Medchal/Hyderabad	4.9b	4
1984 MAR 20	12.70	77.80	SE. of Banglore	4.5	10
1985 FEB 17	24.60	85.50	N. India	4.8	3
1987 APR 18	25.50	79.24		4.8b	4
1993 SEP 30	18.00	76.54	Latur/Killari	6.4	18

1. Abbreviations: LAT-latitude; LONG-longitude; MAG-magnitude; INTEN-Intensity on Modified Mercalli Scale; REF-reference.
2. The following are the references listed in the Ref. column:
 1: CHANDRA (1977)
 2: TANDON (1959)
 3: India Meteorological Department, New Delhi, India
 4: ISC
 5: Coast and Geodetic Survey
 6: CHAUDHURY and SRIVASTAVA (1974)
 7: INDRA MOHAN et al. (1981)
 8: HYPO CENTER LISTING [SOURCE: NOAA]
 9: GUTENBERG and RICHTER (1954)
 10: RAMALINGESWARA RAO and SITAPATI RAO (1984)
 11: RASTOGI (1992)
 12: GUBIN (1968)
 13: INTERNATIONAL SEISMOLOGICAL SUMMARY, Kew, England
 14: RAO (1966)
 15: BASU (1964)
 16: Publications of Indian J. Meteorology and Geopyhysics
 17: ARORA et al. (1973)
 18: GUPTA (1993)

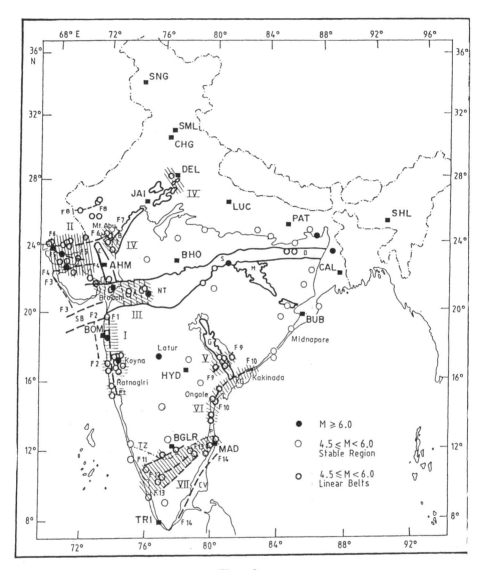

Figure 2

A map of the Indian shield showing epicenters of historical as well as instrumentally recorded earthquakes of magnitude ≥4.5 of of intensity VI on MM scale. I through VII are the linear seismic belts as indicated in Table 1. B: Delhi Mobile belt; C: Cambay basin; NTSD: Narmada-Tapti-Son-Damodar lineament; A: Khandwa region where the Aravallis trend across the NTSD in SSE direction; M: Mahanadi graben; G: Godavari graben; KG: Krishna-Godavari basin; CV: Cauvery basin; TZ: Transition zone.

1912, 1913) and KELKAR (1968), are also included in the catalogue. A map of the Indian shield, showing epicenters of all the earthquakes listed in Table 1, has been prepared (Fig. 2). The map reveals that many intraplate earthquakes are confined to some linear belts while the remaining earthquakes are scattered over the rest of the Indian shield, implying that the Indian shield, excluding the linear belts, is a seismically stable region. Accordingly, earthquakes in their chronological order have been listed in Table 1, separately for each of the linear belts and for the stable region. Table 1 reveals that about 80% of the earthquakes occurred in the linear belts while the rest took place in the stable region.

The west coast seismic belt is the most seismically active linear belt in the Indian shield. This seismic belt extends over a distance exceeding 400 km throughout the west coast fault (F1 in Fig. 2). The Shivaji Sagar Lake at Koyna, which became world famous for reservoir-induced seismicity, is located in this seismic belt. Table 1 reveals that a number of earthquakes of moderate intensity occurred in this belt, even before the impoundment of water in the lake. After impoundment of the reservoir in 1962, a major earthquake of magnitude 6.3 occurred on December 10, 1967 at the Koyna dam (GUPTA et al., 1969). Reservoir-induced seismicity at Koyna is unique in the world in the sense that earthquakes of $M > 4$ have continued to occur for more than 25 years after the initial impoundment in 1962 (GUPTA, 1992) and to date several reservoir-induced earthquakes of $M \geq 5.0$ have occurred in the Koyna region.

The Kutch seismic belt extends over a distance of approximately 250 km in an E-W direction and 150 km in a N–S direction and is associated with the Nagar-Parkar (F6), the Kutch (F5), and the North Kathiawar (F4) faults of the Kutch basin (Fig. 2). The Kutch earthquake of 1819, with a magnitude of 7.8 (JOHNSTON and KANTER, 1990), is the largest earthquake not only in this seismic belt but also in the entire Indian shield. Three earthquakes of magnitudes exceeding 6.0 have been experienced in this belt, indicating that this belt is seismically more potential.

The trans-continental lineament zone extends across the Indian shield over a distance of about 1600 km from the west coast up to the northeastern margin of the Indian shield and is confined to the Narmada-Tapti-Son-Damodar lineament (NTSD, Fig. 2). Within this belt, earthquake activity is clustered in the Khandwa region ('A' in Fig. 2). The Satpura earthquake of 1938 with a magnitude of 6.5 is the largest earthquake experienced in this region. As many as 500 mild tremors of magnitudes equal to or less than 2.3 have occurred in this region during the months of November and December, 1993. At the southwestern extremity of this zone (Lower Narmada region) two earthquakes of magnitude 5.4 have occurred in quick succession near Broach, one in 1970 and another in 1971.

The Delhi mobile belt ('B', Fig. 2) extends in the northeastern direction from Mt. Abu to New Delhi spanning a distance of about 500 km. Seismic activity is concentrated near the southwestern end of this belt at Mt. Abu, and to some extent at the northeastern end near Delhi. The Bhadrachalam seismic belt (F9), about 170 km long, trends almost perpendicular to the Godavari graben. The rifted margins of the graben are free from earthquake activity. The Ongole seismic belt is associated with a boundary fault (F10) along the western margin of the Krishna-

Table 2

*Favorable mode of reactivation (FMR) and the K-factor values of the faults associated with the linear belts. L: Approximate length of the fault; $\mathscr{L}$: Average strike of the fault; ψ: $S_{H_{max}}$ orientation in the vicinity of the faults; θ_s: Reactivation angle for strike-slip mode with respect to local $S_{H_{max}}$ orientation; θ_r: Reactivation angle for thrust faulting mode (dip of the fault); *: Value estimated on the basis of average $S_{H_{max}}$ orientation of N23°E for the mid-continent stress province; SS: Strike slip mode of reactivation; TF: Thrust faulting mode of reactivation; S_3: least principal stress*

Area	Associated faults	L	$\mathscr{L}$	ψ	θ_s	θ_r	FMR	K
West Coast seismic belt	West Coast fault zone (F1)	500	N 7W	N10E	17	>60	SS	1.4
	Offshore fault (F2)	210	N10W		20	>60	SS	1.3
Kutch seismic belt	Saurashtra fault (F3)	250	N45W	N19E	64	>60	None	>S_3
	N. Kathiawar fault (F4)	275	N69E		50	45	TF	2.5
	Kutch fault (F5)	200	N78W		83	45	TF	2.5
	Nagar-Parkar fault (F6)	30	N79E		60	45	TF	2.5
Transcontinental lineament Zone	Boundary faults along the NTSD lineament zone	900	N78E	N-S	78	>60	None	>S_3
	Lower Narmada fault near Broach	–	N85E		85	45	TF	2.5
	Aravalli trend (A) across NTSD (Khandwa region)	160	N15W		15	>60	SS	1.5
Delhi mobile belt	Marginal fault including F7	600	N39E	N31E	8	>60	SS	1.7 (1.4)*
Bhadrachalam seismic belt	Bhadrachalam fault (F9)	170	N50E	N23E	27	>60	SS	1.3
Ongole seismic belt	F10 along the western margin of the Krishna subbasin	50	N10E		12	>60	SS	1.5
	F10 along the western margin of the Godavari subbasin	150	N55E	N 2W	57	>60	None SS	>S_3 (1.4)*
Cambay basin	Marginal faults	310	N25W	N19E	44	>60	SS	1.9
Godavari rift	Marginal faults	420	N30W	N23E	53	>60	None	>S_3

Godavari coastal basin of Cretaceous and Tertiary sediments, and extends over a distance of about 200 km along the east coast (Fig. 2). The earthquake activity in this belt is mainly concentrated near its southern end at Ongole town (Krishna subbasin). The remaining part of this belt (Godavari subbasin) is characterized solely by low seismic activity. The Delhi mobile belt, the Bhadrachalam and the Ongole seismic belts are characterized by the strongest earthquakes producing magnitudes not more than 5.5.

Seismogenic Faults

The average strike of the seismogenic and nonseismogenic faults of the various seismic belts described above, is shown in Table 2. All these faults, excepting those in the Kutch basin, are characterized by steep dips exceeding 60°.

The northern margin of the Kutch basin was faulted along the Nagar Parkar fault F6 (Fig. 2), while the southern margin was faulted along the North Kathiawar fault F4 (BISWAS, 1987). The northern side of the Nagar Parkar fault is known to have risen by 10 to 26 feet (Allah Bund) during the Kutch earthquake of 1819 (KRISHNAN, 1966), implying that this earthquake was induced by reactivation of the Nagar Parkar fault due to thrust-faulting. According to KHATTRI (1994), the seismicity of the Kutch basin is related to a N-S compressional stress regime originating in the collision of the Indian plate and is brought about by reactivation of the Kutch basin faults in thrust faulting mode. It may be understood from the fault plane solutions of the Broach earthquake of March 23, 1970 (CHANDRA, 1977) that the earthquake was produced by reactivation of one of the nodal planes striking almost E-W like NTS and dipping to the north by 45°, by predominently thrust-faulting with a component of left-lateral strike-slip motion. Since Early Jurassic or even earlier the Kutch basin and the lower Narmada-Tapti valleys formed a single tectonothermal province and the other faults in the Kutch basin, i.e., the Kutch fault (F5) and the North Kathiawar fault (F4) must have developed under the same tectonothermal environment as the Nagar Parkar fault and the lower Narmada fault. Hence the Kutch basin faults (F4, F5, F6) can be understood to have a moderate dip of 45°, corresponding to that of the lower Narmada fault at Broach.

Reactivation Mechanisms

It can be shown, according to the Coulomb-Navier criterion, that a pre-existing fault with no cohesive strength can be reactivated when the following condition is satisfied.

$$S_1'/S_3' = (S_1 - P)/(S_3 - P) = (1 + \mu \cot \theta)/(1 - \mu \tan \theta) = \beta \qquad (1)$$

$$P/P_0 = K,$$

where S_1' and S_3' are the effective maximum and minimum principal stresses and S_1 and S_3 are the total maximum and minimum principal stresses respectively, P is the pore pressure in the fault zone and P_0 is the hydrostatic pressure. For $\theta < 45°$ the equation (1) can be solved for K as shown below

$$K = S_3(\beta - m)/[P_0(\beta - 1)] \qquad (2)$$

$$m = (S_1/S_3).$$

Equation (2) can be rewritten in the case of strike-slip faulting as:

$$K = S_h(\beta - m)/[P_0(\beta - 1)] \qquad (3)$$

$$m = (S_H/S_h)$$

and in the case of thrust faulting as:

$$K = S_v(\beta - m)/[P_0(\beta - 1)] \qquad (4)$$

$$m = (S_H/S_V),$$

where S_h and S_H are the total minimum and maximum horizontal principal stresses and S_V is the total vertical stress. For $\theta > 45°$, it can be shown that

$$K = (S_3/2P_0)\{[m(\sin \theta + \cos \theta) + (\sin \theta - \cos \theta)]$$

$$- (m - 1)(\sin \theta + \cos \theta)(\sin \theta - \cos \theta)/\mu\}. \qquad (5)$$

Equations (3) and (4) are useful for evaluating pore pressures needed for the reactivation of strike-slip faults and thrust faults respectively, for a given set of values of m and θ.

We have estimated m values for the linear belts by taking into consideration the amplification of stresses due to hotter geotherms and also we have determined the global average of hydrofrac stress data as detailed below, in order to apply the above equations for understanding the reactivation mechanisms in the Indian shield.

Global Hydrofrac Data Set

The global hydrofrac data compiled by us includes about 160 data sets, each containing magnitude of $S_{H_{max}}$ and $S_{h_{min}}$ at a given depth. Most of the data comes from the USA (HAIMSON, 1977; HAIMSON and DOE, 1983; VERNIK and ZOBACK, 1992; ZOBACK *et al.*, 1980) and Europe (BAUMGARTNER, 1987; BJARNASON *et al.*, 1986; CORNET and VALETTE, 1984; CORNET and JULIEN, 1989; CORNET and BURLET, 1992; HAIMSON *et al.*, 1989; KLEE and RUMMEL, 1993; PINE *et al.*, 1983; RUMMEL and BAUMGARTNER, 1985), and also some data originate from India

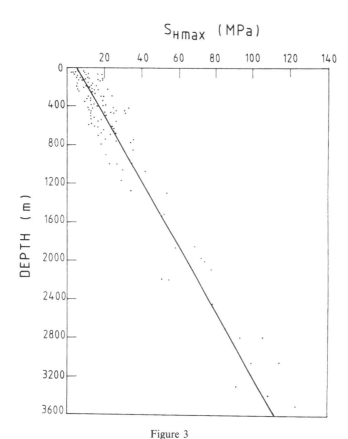

Figure 3

$S_{H\,max}$ vs. depth plot based on the global hydrofrac stress data set compiled by the authors. The best straight line fit to the data is given by $S_{H\,max}$ (MPa) $= 5.6 + 29\ Z$ (km).

(GOWD *et al.*, 1986, 1987, 1992a; SRIRAMA RAO and GOWD, 1990a,b), China (LI, 1986), Japan (TSUKHARA and IKEDA, 1987) and Australia (ENEVER and CHOPRA, 1986). The data covers a depth range reaching 3500 m and includes deep hole stress measurements at Cajon Pass extending 3500 m depth in the U.S.A. The stress data from crystalline terrains and hard and compact sedimentary rocks such as lime-stones only are included in this compilation. $S_{H\,max}$ vs. Z (depth) and $S_{h\,min}$ vs. Z plots have been prepared using these global hydrofrac data and the same are presented in Figures 3 and 4, respectively. The figures show that much of the data come from shallow depths reaching 500 m while there is a good number of data points in the depth range of 500–3500 m.

    The best straight line fits to the data indicate that $S_{H\,max}$ increases with depth with a gradient of about 29 MPa/km and $S_{h\,min}$ with a gradient of about 13.5 MPa/km. We consider that these gradients represent an average stress field of the globe.

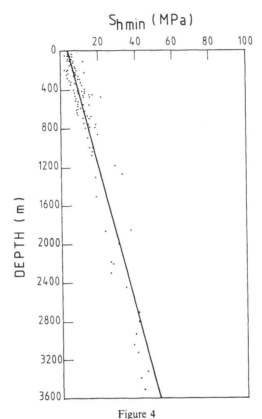

Figure 4

$S_{h_{min}}$ vs. depth plot based on the global hydrofrac stress data set compiled by the authors. The best straight line fit to the data is given by $S_{h_{min}}$ (MPa) $= 3.9 + 13.5\,Z$ (km).

Amplification of Stresses in the Linear Belts

Along the West Coast (F1) and in the Narmada-Tapti-Son lineament zone (NTS) there are a multitude of hot springs with surface temperatures in the range of 32–70 °C (RAVISHANKAR, 1987). Temperature vs. depth plots for a number of 300–600 m deep boreholes in the Narmada-Tapti lineament zone (RAVISHANKAR, 1991) indicate a steady and linear increase in temperature with gradients ranging from about 36°C/km to 60°C/km. Ground temperature data in the West Coast fault zone yielded a temperature gradient of 47°C/km (RAVISHANKAR, 1987). RAVISHANKAR (1988) used a silica heat-flow technique developed by SWANBERG and MORGAN (1979) and determined heat-flow values for a number of sites along the West Coast and in the Narmada-Tapti-Son lineament zone. The silica method heat-flow values vary from about 110 mW m^{-2} to 130 mW m^{-2}, with a mean of 120 mW m^{-2} in the West Coast fault zone and Narmada-Tapti region. The conventional heat-flow data reveal that the oil fields near Bombay to the west of the West

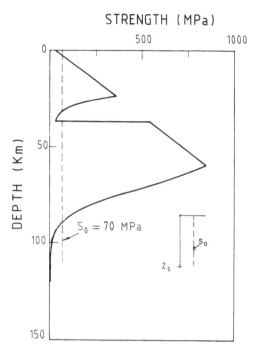

Figure 5

Mechanical strength of the crust and upper mantle as a function of depth in the stable region; Frictional strength (strike-slip fault) is calculated as per Eq. (8) using a value of 0.85 for the coefficient of sliding friction, $\theta = 30°$ and $k = 1$, ductile strength is calculated using Eq. (9) for the middle and lower crust and Eq. (10) for the upper mantle.

Coast fault zone (F1) are characterized by mean heat flow of some $80\,\text{mW m}^{-2}$ (PANDA, 1985). The conventional heat-flow data for the Sonhat basin of the Son valley and the Gondwana basins of the Damodar valley are characterized by mean heat flow of $107\,\text{mW m}^{-2}$ and $74\,\text{mW m}^{-2}$ respectively (RAO and RAO, 1983). The conventional and the silica method heat-flow values thus indicate that the West Coast fault zone and the transcontinental lineament zone (NTSD, Fig. 2) are characterized by heat flow of nearly $80\,\text{mW m}^{-2}$ and more.

On the basis of more than 600 observations, PANDA (1985) concluded that the heat flow in the Cambay basin has a mean of $75 \pm 18.4\,\text{mW m}^{-2}$. The central part (Khetri copper belt) and the northeastern part (Tosham area) of the Delhi mobile belt (B, Fig. 2) are characterized by a high heat flow of $74.0\,\text{mW m}^{-2}$ (GUPTA et al., 1967) and $90\,\text{mW m}^{-2}$ (SUNDAR et al., 1990), respectively. We presume that the entire belt, including the Mt. Abu area, is characterized by a high heat flow of at least $75\,\text{mW m}^{-2}$. The areas around the Kutch basin, including the Lower Narmada-Tapti valley to the south, the Cambay basin to the east, and the Jaisalmer and Bikaner basins in the north, are characterized by a high surface heat flow of

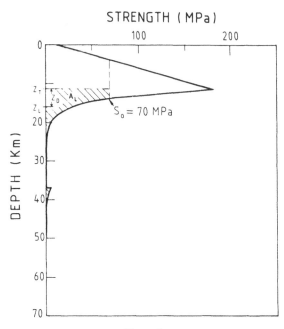

Figure 6
Mechanical strength of the crust and upper mantle as a function of depth in the seismic belts; frictional strength (strike-slip fault) of the upper brittle crust is calculated as per Eq. (8) using a value of 0.85 for the coefficient of sliding friction, $\theta = 30°$ and $k = 1$; ductile strength is calculated using Eq. (9) for the middle and lower crust and Eq. (10) for the upper mantle.

75 mW m^{-2} and more. The geodynamics of the region (RAVAL, 1994) strongly suggest that the heat flow in this basin also should be high. Hence we presume that the Kutch basin is also characterized by a high surface heat flow of 80 mW m^{-2} comparable to the other linear belts. The surface heat-flow values on either side of the Bhadrachalam fault (F9) are in the range of 84–104 mW m^{-2} (RAO *et al.*, 1970; RAO and RAO, 1983), implying that the fault zone is characterized by a high heat flow of more than 80 mW m^{-2}. The Krishna-Godavari coastal basin has a mean value of 100 ± 38 mW m^{-2} (PANDA, 1985), implying that the heat flow along its marginal fault (F10) could be more than 80 mW m^{-2}. Heat-flow values reported for 18 sites (Fig. 1) reveal that the stable region of the Indian shield is characterized by an average heat flow of 45 mW m^{-2}, which is slightly higher than the normal heat flow in the Precambrian shields.

In light of the above discussion, we have calculated geotherms for the stable region (G_S) as well as for the linear belts (G_L) by assuming surface heat flow of 45 mW m^{-2} and 80 mW m^{-2}, respectively. According to these geotherms, the brittle to ductile transition temperature of 300°C (CHEN and MOLNAR, 1983; SIBSON, 1982) is attained at a depth of about 27 km in the stable region and at a depth of about 12.5 km in the linear belts.

The Byerlee's frictional law (BYERLEE, 1968) for normal stresses < 200 MPa can be expressed in terms of the principal stresses (S_1, S_3) as below

$$\frac{(S_1 - S_3)}{2} \sin 2\theta = 0.85 \left[\frac{(S_1 + S_3)}{2} - \frac{(S_1 - S_3)}{2} \cos 2\theta - KP_0 \right] \tag{6}$$

$$K = P/P_0$$

where θ is the angle between the S_1 direction and the fault plane, P is the pore pressure in the fault zone and P_0 is the hydrostatic pressure. For a specific case of $\theta = 30°$, the above equation can be expressed as below

$$S_1 = 4.85\, S_3 - 3.85\, KP_0. \tag{7}$$

In the case of the global average stress field where $S_{h\,min}$ (MPa) $= 13.5\, Z$ (km), the frictional strength (S_f) of strike-slip faults $(S_3 = S_{h\,min})$ subjected to hydrostatic pore pressures $(K = 1)$ can be shown to be

$$S_f \,(\text{MPa}) = (S_1 - S_3) = 15.42 + 14.2\, Z \,(\text{km}). \tag{8}$$

Using the above expression we have calculated the frictional strength of the upper brittle crust as a function of depth in the stable region (Fig. 5) and in the linear belts (Fig. 6). We have computed ductile creep strength of the middle and lower crust using the following flow laws of steady-state creep of quartz (BRACE and KOHLSTEDT, 1980) and that of Olivine (GOETZE, 1978) for the upper mantle.

Quartz

$$S_D = (\varepsilon/A)^{1/n} \exp(Q/nR\phi). \tag{9}$$

Olivine

$$S_D = 8500\{1 - [R\phi \, \ln(A/\varepsilon)/Q]^{1/2}\} \tag{10}$$

where ε is the strain rate, S_D is the ductile strength (differential stress) in MPa, $\phi = 273 + T$ (T being ambient temperature in °C), R is the gas constant. In these computations, we have applied the values of temperature (Q) as per the geotherms G_S and G_L. Also a strain rate of $10^{-16}\, S^{-1}$ has been assumed as it is considered by ZOBACK et al. (1993) to be a reasonable upper limit for plate interiors. Ductile strength vs. depth curve thus computed for the stable region and the linear belts is presented in Figures 5 and 6, respectively.

The strength profile for the stable region (Fig. 5) reveals that the crust has the maximum frictional strength at about 30 km depth, and the upper mantle at about 60 km depth. The profile is similar to the one calculated by MANGLIK and SINGH (1991) for the Kolar region in southern India. Whereas the strength profile for the linear belts (Fig. 6) indicated that the crust has the maximum strength at about 12.5 km depth only after which it decreases rapidly as the Moho is approached.

Interestingly, the upper mantle beneath the linear belts has nominal strength. The strength profile of the linear belts is similar to the one determined by ZOBACK *et al.* (1993) for the mildly seismic KTB borehole region in Germany.

Ridge push plate boundary force, and the isostatically compensated loads associated with the Tibetan plateau uplift appear to be the main sources of intraplate stresses in the Indian subcontinent. The tectonic forces arising from ridge push (F_r) and isostatically compensated loads associated with plateau uplift (F_i) have been estimated to be 2–$3 \times 10^{12} \, \text{Nm}^{-1}$ and 0–$4 \times 10^{12} \, \text{Nm}^{-1}$, respectively (KUSZNIR, 1991). Hence, these two forces together $(F = F_r + F_i)$ may have a maximum value of $7 \times 10^{12} \, \text{Nm}^{-1}$ in respect of the Indian subcontinent.

The strength profile in Figure 5 indicates that the load-bearing section of the lithosphere has a thickness (Z_s) of about 100 km in the stable region. The combined plate boundary force (F) of $7 \times 10^{12} \, \text{Nm}^{-1}$, acting over this section, can induce $S_{H_{max}}$ (F/Z_s) resulting in an average depth-independent magnitude of 70 MPa and is termed here as S_0. Whereas in the linear belts, stresses $> S_0$ can be transmitted by a relatively small section as indicated in Figure 6 while the rest of the lithosphere remains passive. It may be understood from the figure that such stresses can be transmitted not only by the upper brittle crust but also significantly by the ductile crust lying immediately below the brittle to ductile transition depth (Z_T). The effective thickness (Z_L) of the upper lithospheric section capable of transmitting stresses $> S_0$ can, therefore, be expressed as

$$Z_L = Z_T + Z_D. \tag{11a}$$

Z_D is the effective thickness of the ductile crust and upper mantle capable of transmitting S_0, and is given by

$$Z_D = A_L / S_0, \tag{11b}$$

where A_L is the area shown as a hatchured portion in Figure 6. The average depth-independent magnitude of $S_{H_{max}}$, induced beneath the linear belts due to the combined plate boundary force of $7 \times 10^{12} \, \text{Nm}^{-1}$, can be shown to be as below:

$$S_L = (Z_S / Z_L) S_0. \tag{12}$$

On substituting $Z_L = 16$ km as determined from Eq. (11), $Z_S = 100$ km, and $S_0 = 70$ MPa, the above equation yields a value of about 440 MPa for S_L. The vertical gradient of $S_{H_{max}}$ in the upper brittle crust of the linear belts can be expressed, as a first approximation, as

$$\dot{S}_{H_{max}} = 2S_L / Z_L \quad (\text{MPa/km}). \tag{13}$$

The above equation yields a value of about 55 MPa/km for the vertical gradient of $S_{H_{max}}$, indicating that $S_{H_{max}}$ is amplified almost two times in the linear belts, when compared to the global average of 29 MPa/km.

ZOBACK *et al.* (1993) experimentally evaluated the stress field in the KTB borehole up to 6 km depth and published stress vs. depth plots. The plots reveal

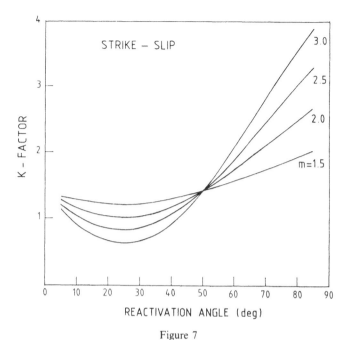

Figure 7

K factor vs. reactivation angle plots for strike-slip faulting; the relations $S_{h_{min}}$ (MPa) = 13.5 Z (km) and pore pressure P_0 (MPa) = 10 Z (km) have been used in computing these plots as per Eqs. (3) and (5).

that the vertical gradient of $S_{H_{max}}$ and $S_{h_{min}}$ in the region is about 47 MPa/km and 20 MPa/km, respectively. This demonstrates that the vertical gradient of $S_{H_{max}}$ estimated by us for the upper brittle crust beneath the linear belts of the Indian shield, is slightly higher than the one in the KTB borehole site. This higher value is justified in view of higher heat flow in the linear belts as compared to the one in the KTB borehole site. The results thus reveal that our estimate of the vertical gradient of $S_{H_{max}}$ in the linear belts of the Indian shield is in agreement with the experimental data of the KTB borehole.

Because of the Poisson's effect, $S_{h_{min}}$ is also amplified in the linear belts following the amplification of $S_{H_{max}}$. For plain strain conditions, it can be shown that the Poisson's ratio, $v = R/(1 + R)$ where, $R = \dot{S}_{h_{min}}/\dot{S}_{H_{max}}$. The above equation yields a value of 0.30 for the Poisson's ratio of the upper brittle crust of the KTB borehole region, where $R = 0.426$. Assuming that the same value is also valid for the upper brittle crust beneath the linear belts, we find that

$$\dot{S}_{h_{min}} = 23.5 \text{ MPa/km}.$$

Since the Poisson's ratio of the upper brittle crust of the shield areas is expected to be slightly lower, because of their geological history, we have constrained $\dot{S}_{h_{min}}$ to 22 MPa/km such that

$$m = (\dot{S}_{H_{max}}/\dot{S}_{h_{min}}) = 2.5.$$

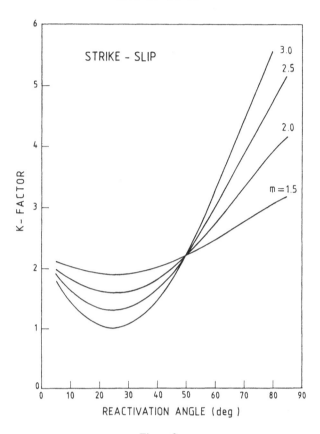

Figure 8

K factor vs. reactivation angle plots for strike-slip faulting due to amplified stress field in the linear belts; the relations $S_{h\,min}$ (MPa) $= 22\,Z$ (km) and pore pressure P_0 (MPa) $= 10\,Z$ (km) have been used in computing these plots as per Eqs. (3) and (5).

This estimate of $\dot{S}_{h\,min}$ indicates that the linear belts are characterized by a strike-slip stress regime with $m = 2.5$. However the Kutch region, including the Broach area in the lower Narmada valley, seems to be exceptional. This region has been geodynamically far more active than the other linear belts and therefore its elastic and rheological character is expected to be somewhat different, perhaps with a slightly higher Poisson's ratio such that $S_{h\,min}$ is equal to or greater than S_V (thrust faulting stress regime, with $m = 2.0$).

Pore Pressures

We have computed K-factor values for the strike-slip mode of reactivation for different values of m and reactivation angle (θ), using equations (3) and (5). K vs. θ plots thus computed for the global average stress field $[S_{h\,min}$(MPa) $= 13.5\,Z$ (km)] and for the amplified stress field $[S_{h\,min}$(MPa) $= 22\,Z$ (km)] are presented in

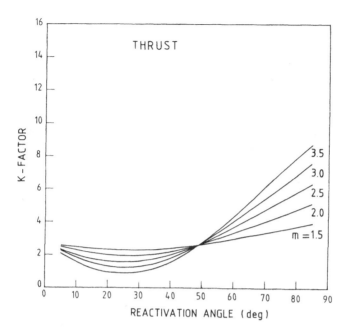

Figure 9

K factor vs. reactivation angle plots for thrust faulting; the relations S_V (MPa) $= 27\,Z$ (km) and pore pressure P_0 (MPa) $= 10\,Z$ (km) have been used in computing these plots as per Eqs. (3) and (5).

Figures 7 and 8, respectively. The plots demonstrate that a minimum pore pressure is needed to reactivate strike-slip faults of $\theta = \theta^* = 25°$ for a given value of m. These plots indicate that strike-slip faults with $\theta = 2\theta^* = 50°$ can be reactivated at pore pressures equivalent to average and amplified $S_{h\,\text{min}}$, i.e., 1.35 and 2.2 times hydrostatic pressure, respectively, irrespective of the value of m. These limiting angles are governed by the condition that $2\theta^* = \tan^{-1}(1/\mu)$, where $\mu = 0.85$ is the coefficient of sliding friction as per the Byerlee's law. Figures 7 and 8 reveal that pore pressures exceeding $S_{h\,\text{min}}$ are required for the reactivation of strike-slip faults with $\theta > 2\theta^*$. Since the upper bound to pore pressure levels is limited by the formation of hydraulic fractures which would allow fluids to escape from the overpressured zones (SIBSON, 1990), build-up of pore pressures greater than $S_{h\,\text{min}}$ and tensile strength together is impossible. Hence, pre-existing faults with $\theta > 2\theta^*$ can never be reactivated and they remain locked until the stress field is favorably rotated.

In the case of thrust fault stress regime, thrust faults dipping at $50°$ ($\theta = 50°$) can be reactivated by lithostatic pore pressures (Fig. 9) since vertical stress (S_V) happens to be the minimum principal stress, whereas those with dips exceeding $50°$ remain locked up as supralithostatic pore pressures are needed. Figure 9 reveals that thrust faults dipping at $10°-40°$ can be reactivated at pore pressures of about 2.0 to 2.2 times the hydrostatic pressure for $m = 2.0$, and at about 1.0 to 1.6 times the hydrostatic pressure for $m = 3.5$. Also the results reveal that near lithostatic pore pressures are needed in the case of $m \leq 1.25$. In view of the fact that extremely low strain rates prevail in the intraplate regions, possibilities for the build-up of lithostatic pore pressures in such regions are very remote. Hence, reactivation of thrust faults in an intraplate region such as the Indian shield is possible only under the action of amplified stress fields.

The reactivation angles (θ) of all the above discussed seismogenic faults have been determined with respect to the $S_{H_{max}}$ orientation in and around the respective linear belts for the strike-slip mode (θ_S), while the dip of these faults is taken as θ_T for the thrust-faulting mode. However, due to nonavailability of local stress data, reactivation angles of the Bhadrachalam fault and the marginal faults of the Godavari graben have been obtained with respect to the average $S_{H_{max}}$ orientation for the midcontinent stress province, i.e., N23°E evaluated by GOWD *et al.* (1992b). In accordance with our above assessment of the nature of the stress regime of the linear belts, we have evaluated K-factor for the reactivation of the faults in the Kutch seismic belt due to the thrust faulting stress regime, and reactivation of those in all the other linear belts due to the strike-slip stress regime. In this evaluation, we have used the above estimated vlaues of $m = 2.0$ and $m = 2.5$, respectively. The values of θ and K-factor thus obtained are presented in Table 2.

The equations (3), (4) and (5) are derived based on the condition that the fault plane is vertical in the case of strike-slip faults, and strikes at right angles to the orientation of $S_{H_{max}}$ in the case of thrust faults such that the respective intermediate principal stresses S_V and $S_{h_{min}}$ are contained in it (fault plane) and play no role in the dynamics of faulting. Table 2 shows that these conditions are not strictly satisfied. However, the values of θ_T in the case of the strike-slip mode and θ_S in the case of the thrust faulting mode are generally greater than 60° (Table 2), and under these circumstances, K-factor values do not differ significantly from those presented in Figures 7, 8 and 9. The reactivation mechanisms (mode of reactivation and K factor) of the various faults of the linear belts, analyzed on the basis of the above equations, may therefore be considered as valid. The table reveals that a majority of the seismogenic faults of the linear belts can be reactivated in the strike-slip mode at pore pressures of 1.3 to 1.7 times the hydrostatic pressure, wheras the faults in the Kutch seismic belt can be reactivated in the thrust-faulting mode at pore pressures of about 2.5 times the hydrostatic pressure. The details are discussed below.

Table 2 reveals that the West Coast fault can be reactivated in the strike-slip mode when the pore pressures in the fault zone obtain a value of 1.4 times the hydrostatic pressure. The moderate seismic activity in the West Coast seismic belt, prior to the impoundment of water in the Shivaji Sagar Lake at Koyna in 1962, may be attributed to the high strain energy induced by the amplified stress field. However, seismic activity in the Koyna region has radically changed after the impoundment of water. The reservoir induced seismicity in the region is understood to be due to the reactivation of the N-S trending Koyna fault (GUPTA *et al.*, 1980), by a strike-slip mode as can reliably be inferred from the focal mechanism solutions of the strongest earthquake of the region, determined by SINGH *et al.* (1975). The Koyna fault is parallel to the West Coast fault which passes through the western neighborhood of the former (KAILASAM, 1979; KAILA and KRISHNA, 1992). Though the West Coast fault and the Koyna fault are characterized by the same values of m and θ, the latter is seismically more active than the former. This discrepancy is due to the hydraulic head of about 100 m created at the surface by the impoundment of water. This high hydraulic head is causing a build-up of pore pressure in the Koyna fault zone at a considerably faster rate. Hence the uniqueness of the reservoir-induced seismicity at Koyna, as described earlier, should be attributed to the high strain energy density as well as to the high pore pressure build-up rate in the area.

Similar seismic activity in the other seismic belts i.e., the Bhadrachalam, the Ongole (Krishna subbasin), the Delhi mobile belt, and the Khandwa region (NTS) is caused by reactivation of the respective faults in the strike-slip mode due to the high strain energy density. However, the frequency of earthquakes in these belts is considerably less than in the Koyna region because surface water from the river beds is only recharged into the seismogenic faults of the former at much slower rate than in the latter (Koyna region).

The seismogenic faults of the Kutch seismic belt can be reactivated in the thrust-faulting mode only when pore pressures within the fault zones attain a critical value of about 2.5 times the hydrostatic pressure (Table 2). The occurrence of less frequent but strong earthquakes is a characteristic feature of this belt (Table 1) and this is in agreement with the above discussed reactivation mechanism involving nearly lithostatic pore pressures of obviously longer recurrence period.

Table 2 reveals that the marginal faults of the Godavari rift, Damodar valley coal basins and the boundary faults of the NTS lineament zone are locked up and cannot be reactivated. The fact that there is very nominal seismic activity along these marginal faults corresponds with the above observation.

The above analysis is based on the consideration that the seismogenic faults are being reactivated due to the stress field prevailing in the vicinity of the linear belts. On the presumption that the entire shield is subjected to a homogeneous compressional stress field at the seismogenic depths, with $S_{H_{max}}$ orientation equal to the average orientation of $S_{H_{max}}$ for the midcontinent stress province (N23°E), K values have also been worked out and are shown in brackets in Table 2 only when they differ significantly from the above reported K values. Table 2 reveals that the marginal fault F10 of the Godavari subbasin and the North Kathiawar fault (F4) come under this category. The mode of reactivation of the other seismogenic faults remains the same in either case.

Eleven of the thirteen earthquakes of the southern granulitic terrain occurred in 19th Century and their epicenters have been located by CHANDRA (1977), based on historical records. If we assume that these epicenters are truthfully located, the three prominent faults (F11, F12 and F13) in the granulitic terrain appear to be seismically active and hence the earthquake activity in this region can be considered as occurring in a linear belt formed by these faults. Our present understanding of the NW-SE orientation for $S_{H_{max}}$ in this region is tentative (GOWD *et al.*, 1992b). Heat-flow data are not available for this region excepting for one site (site 18, Fig. 1). It is not possible to analyze the reactivation mechanisms in the southern granulitic terrain due to uncertainty in our assessment of $S_{H_{max}}$ orientation and because of the nonavailability of heat-flow data in the region. Also we could not analyze the reactivation mechanisms causing seismic activity in the Jaisalmer region (F8) due to the paucity of data.

Only 33 of the 156 earthquakes of the Indian shield have occurred in the stable region extending over a vast area. Some of these earthquakes could even be regarded as occurring in a linear belt extending along the northern margin of the shield. Although there are some favorably oriented older deep-seated faults in the stable region, none of them seems to be seismically active and the region appears to be relatively free of earthquake activity. These older faults are characterized by cold geotherms, indicating that amplification of stresses in these fault zones is not possible, and we may therefore consider that the global average stress field possibly represents the state of stress in the stable region. The density of strain energy we computed reveals that the volumetric as well as distortional strain energy induced in these older faults is several folds less than in the above discussed younger seismogenic faults of the linear belts. These younger faults with larger strain energy have more potential to undergo unstable slip (ruptures), resulting in earthquake activity. Equally important is the fact that the older faults are less permeable because of their partial healing, implying that their pore pressure build-up rate is likely to be low, whereas in the younger seismogenic fault zones, characterized by high permeability due to hotter geotherms, the required critical pore pressures can be attained at a much faster rate. We may conclude that the favorably oriented faults in the stable region are not being reactivated like those in the linear belts due to these complexities, even though pore pressure needed for reactivating the former in strike-slip mode are slightly less (Fig. 7) than for reactivating the latter (Fig. 8).

However, strong earthquakes can also occur in the stable region due to reactivation of blind thrusts, such as the one at Latur (SEEBER, 1994). Reactivation mechanisms of such blind thrusts can be analyzed in light of what has been discussed above. The possible presence of a few more such thrusts under the cover of Deccan traps cannot be ruled out in view of our ignorance of the tectono-thermal environment under which Latur-like thrust fault can develop in an intraplate region. Hence Latur-like earthquakes can also possibly occur in other parts of the Deccan volcanic province.

Geological History of the Faults

Development of the faults in the Kutch basin and in its eastern and southern proximity was related to the break-up of the eastern Gondwana land from the western Gondwana land in the Late Triassic/Early Jurassic (BISWAS, 1987). Rifting of the Kutch region was initiated in the Late Triassic as evidenced by continental Rhaetic sediments in the northern part of the basin. The Kutch basin formed during the Early Jurassic time by the subsidence of a block between the Nagar-Parkar fault and the North Kathiawar fault, and the basin became fully marine during the Middle Jurassic. In Early Cretaceous time the entire region lying to the west of the East Cambay fault and to the north of the lower Narmada fault subsided to form an extensive platform on which large volumes of deltaic sediments were deposited as it subsided (BISWAS, 1983). The Lower Narmada valley at the westernmost end of the NTS lineament zone was rifted in the Early Cretaceous as indicated by the deltaic sediments of the same age in the region. India was rifted from Madagascar in the Late Cretaceous (KRISHNAN, 1966) about 80 Ma (AGRAWAL et al., 1992). The West Coast fault and the Koyna fault were derived from the same thermo-mechanical forces that were responsible for the rifting of India from Madagascar. The Cambay graben came into existence in the Late Cretaceous as a rift valley and the lower Narmada rift basin further subsided as indicated by the marine sediments (the Bagh beds) of the Late Cretaceous age (BISWAS, 1987). The West Coast fault was reactivated during the Early Tertiary times as the present western continental shelf subsided along it. The Cambay graben subsided during the Tertiary time, accumulating thick Tertiary sediments over the Deccan Trap floor (BISWAS, 1987). The Kutch basin also subsided marginally during this period. Rifting of the marginal basins, including the Krishna-Godavari basin on the east coast of India, was related to the continental break-up and commencement of sea-floor spreading between India and Australia-Antarctica, about 130 Ma (JOHNSON et al., 1976). The K-G basin subsided along the Eastern Ghat trends (NE-SW) and along the marginal fault (F10) which delimits the basin on the west. The upper Gondwana sediments of Middle Jurassic to Early Cretaceous are the oldest rocks in the basin (SASTRI et al., 1973). The Bhadrachalam fault (F9) is subparallel and close to the K-G basin marginal fault (F10) (Fig. 1). This might originally have been a cross fault within the Godavari graben after which it must have been extended in northeast and southwest directions into the basement crystalline complex by the same forces that were responsible for rifting of the K-G basin.

Collectively these facts thus reveal that the seismogenic faults of the seismic belts (Table 2) have developed in an extensional stress regime 200–80 Ma, with some possibly even later than 80 Ma. These younger faults have come under a compressional stress regime following the collision between India and Eurasia about 40 Ma (POWELL, 1979; PATRIAT and ACACHE, 1984), and are now being reactivated due to the various mechanisms previously discussed. JOHNSTON and KANTER (1990) studied more than 800 earthquakes ($M \geq 4.5$) of the stable continental crust and found that areas that have undergone extension (250–25 Ma) are more likely to experience an earthquake of any size than in the ancient shields (unaffected areas). The seismogenic character of the younger faults of the Indian shield discussed above is in agreement with the above finding of JOHNSTON and KANTER.

Acknowledgements

Dr. T. N. Gowd expresses his grateful thanks to the Organizing Committee of the ISMPG and its Chairman Prof. Ren Wang for inviting him and for generously providing financial assistance to him to participate in the symposium. The authors profoundly thank Dr. H. K. Gupta, Director, National Geophysical Research Institute for his encouragement and for according permission to present this paper at the ISMPG.

REFERENCES

AGRAWAL, P. K., PANDEY, O. P., and NEGI, J. G. (1992), *Madagascar: A Continental Fragment of the Paleosuper Dharwar Craton of India*, Geology *20*, 543–546.

ARORA, S. K., SUBBARAMU, K. R., MURTHY, S. V. N., and BHAT, M. K. (1973), *Recent Mandya Earthquakes and the Aftershock Microseismic Activity*, Indian J. Meteorol. Geophys. *24*, 375–382.

BASU, K. L. (1964), *A Note on the Coimbatore Earthquake of 8 February, 1900*, Indian J. Meteorol. Geophys. *15*, 281–286.

BAUMGARTNER, J. (1987), *Application of Hydraulic Fracturing Technique for Stress Measurements in Fractured Rocks (in German)*, Ph. D. Thesis, 223 pp., Ruhr University Bochum, Bochum, Germany.

BISWAS, S. K., *Cretaceous of Kutch-Kathiawar region*. In Proc. Seminar. *Cretaceous Stratigraphy of India* (Indian Assoc. Palynostratigraphers, 1983), pp. 41–65.

BISWAS, S. K. (1987), *Regional Tectonic Framework, Structure and Evolution of the Western Marginal Basins of India*, Tectonophysics *135*, 307–327.

BJARNASON, B., STEPHANSSON, O., TORIKKA, A., AND BERGSTROM, K. (1986), *Four years of hydrofracturing rock stress measurements in Sweden*. In *Rock Stress*, Proc. Int. Symp. on Rock Stress Measurements, Stockholm, Sept. 1–3, 1986, 421–427.

BRACE, W. F., and KOHLSTEDT, D. L. (1980), *Limits of Lithospheric Stress Imposed by Laboratory Experiments*, J. Geophys. Res. *85*, 6248–6252.

BYERLEE, J. D. (1968), *Brittle Ductile Transition in Rock*, J. Geophys. Res. *73*, 4741–4750.

CHANDRA, U. (1977), *Earthquake of Peninsular India—A Seismotectonic Study*, Bull. Seismol. Soc. Am. *67*, 1387–1413.

CHAUDHURY, H. M., and SRIVASTAVA, H. N., *Earthquake activity in India during* 1970–1973. In Proc. Symp. *Earthquake Eng.*, 5th, University of Rookee, India, 1974, pp. 427–434.

CHEN, W. P., and MOLNAR, P. (1983), *Focal Depth of Intracontinental and Intraplate Earthquakes and their Implications for the Thermal and Mechanical Properties of the Lithosphere*, J. Geophys. Res. *88*, 4183–4214.

CORNET, F. H., and VALETTE, B. (1984), *In situ Stress Determination from Hydraulic Injection Test Data*, J. Geophys. Res. 89, *11527–11537*.

CORNET, F. H., and JULIEN, P. (1989), *Stress Determination from Hydraulic Test Data and Focal Mechanisms of Induced Seismicity*, Int. J. Rock Mech. and Geomech. Abstr. *26*, 235–248.

CORNET, F. H., and BURLET, D. (1992), *Stress Field Determinations in France by Hydraulic Tests in Boreholes*, J. Geophys. Res. *97*, 11829–11849.

ENEVER, J. R., and CHOPRA, P. N. (1986), *Experience with hydraulic fracture stress measurements in granite*. In *Rock Stress*, Proc. Int. Symp. on Rock Stress Measurements, Stockholm, Sept. 1–3, 1986, 411–420.

ERMENKO, N. A., NEGI, B. S., KASIANOV, M. V. et al. (1969), Tectonic Map of India, Bull. Oil Nat. Gas Comm., India 6, 1–11.

GAUR, V. K. (1993), Mitigating Earthquake Hazards, Current Science 65, 509.

GOETZE, C. (1978), The Mechanism of Creep in Olivine, Phil. Trans. R. Soc. Lond, A288, 99–119.

GOWD, T. N., SRIRAMA RAO, S. V., CHARY, K. B., and RUMMEL, F. (1986), In situ Stress Measurements Carried out for the First Time in India Using Hydraulic Fracturing Method. Proc. Indian Acad. Sci. (Earth Planet. Sci.) 95, 311–319.

GOWD, T. N., SRIRAMA RAO, S. V., and CHARY, K. B., Hydraulic fracturing technique for the determination of in situ stresses at great depth-results of stress measurements at Malanjkhand, M.P. In Proc. Crustal Dynamics (Indian Geophys. Union, Hyderabad, India 1987) pp. 125–133.

GOWD, T. N., SRIRAMA RAO, S. V. and CHARY, K. B. (1992a), In situ Stress Measurements by Hydraulic Fracturing at the Underground River Bed Power House Site, Sardar Sarovar Project, Kevadia, Gujarat State, Tech. Rep. No. NGRI-92-ENVIRON-121, 41 pp., National Geophysical Research Inst., Hyderabad, India.

GOWD, T. N., SRIRAMA RAO, S. V. and GAUR, V. K. (1992b), Tectonic Stress Field in the Indian Subcontinent, J. Geophys. Res. 97, 11789–11888.

GUBIN, I. E. (1968), Seismic Zoning of Indian Peninsula, Bull. Int. Inst. Seismology Earthquake Eng. 5, 109–139.

GUPTA, H. K. (1992), Reservoir-induced Earthquakes, Current Sci. 62, 183–197.

GUPTA, H. K. (1993), The deadly Latur Earthquake, Science 262, 1666–1667.

GUPTA, H. K., NARAIN, H., RASTOGI, B. K., and MOHAN, I. (1969), A Study of the Koyna Earthquake of December 10, 1967, Bull. Seismol. Soc. Am. 59, 1149–1162.

GUPTA, H. K., RAO, C. V. R. K., RASTOGI, B. K., and BHATIA, S. C. (1980), An Investigation of Earthquakes in Koyna Region, Maharashtra for the Period October 1973 through December 1976, Bull. Seismol. Soc. Am. 70, 1833–1847.

GUPTA, M. L., and GAUR, V. K. (1984), Surface Heat Flow and Probable Evolution of Deccan Volcanism, Tectonophysics 105, 309–318.

GUPTA, M. L., and RAO, G. V. (1970), Heat Flow Studies under Upper Mantle Project, Bull. N. G. R. I. 8, 87–112.

GUPTA, M. L., VERMA, R. K., RAO, R. U. M., HAMAZA, V. M., and RAO, G. V. (1967), Terrestrial Heat Flow in Khetri Copper Belt, Rajasthan, India, J. Geophys. Res. 21, 4215–4220.

GUPTA, M. L., SHARMA, S. R., SUNDAR, A., and SINGH, S. B. (1987), Geothermal Studies in the Hyderabad Granite Region and the Crustal Thermal Structure of the Southern Indian Shield, Tectonophysics 140, 257–264.

GUPTA, M. L., SUNDAR, A., and SHARMA, S. R. (1991), Heat Flow and Heat Generation in the Archaen Dharwar Cratons and Implications for Southern Indian Shield Geotherm and Lithospheric Thickness, Tectonophysics 194, 107–122.

GUPTA, M. L., SUNDAR, A., SHARMA, S. R., and SINGH, S. B. (1993), Heat flow in the Bastar Craton, Central Indian Shield: Implications for Thermal Characteristics of Proterozoic Cratons, Phys. Earth Planet. Int. 78, 23–31.

GUTENBERG, B., and RICHTER, C. F., Seismicity of the Earth, 2nd ed. (Princeton University Press, Princeton, N.J. 1954).

HAIMSON, B. C. (1977), Crustal Stress in the Continental United States as Derived from Hydrofracturing Tests, Geophys. Monograph (Am. Geophys. Un.) 20, 576–592.

HAIMSON, B. C., and DOE, T. W. (1983), State of Stress, Permeability and Fractures in the Precambrian Granite of Northern Illinois, J. Geophys. Res. 88, 7355–7372.

HAIMSON, B. C., TUNBRIDGE, L. W., LEE, M. Y., and COOLING, C. M. (1989), Measurement of Rock Stress Using the Hydraulic Fracturing Method in Cornwall, U.K.—Part II. Data Reduction and Stress Calculation, Int. J. Rock Mech. Geomech. Abstr. 26, 361–372.

JOHNSON, B. D., POWELL, C. MCA., and VEEVERS, J. J. (1976), Spreading History of the Eastern Indian Ocean and Greater India's Northward Flight from Antarctica and Australia, Geol. Soc. Am. Bull. 87, 1560–1566.

JOHNSTON, A. C., and KANTER, L. R. (1990), Earthquakes in Stable Continental Crust, VIGYAN Scientific American, 54–61.

KAILA, K. L., and KRISHNA, V. G. (1992), *Deep Seismic Sounding Studies in India and Major Discoveries*, Current Science *62*, 117–154.

KAILASAM, L. N. (1979), *Plateau Uplift in Peninsular India*, Tectonophysics *61*, 243–269.

KATZ, M. B. (1978), *Tectonic Evolution of the Archean Granulite Belt of Sri Lanka-South India*, J. Geol. Soc. India *19*, 185–205.

KELKAR, Y. N. (1968), *Earthquakes in Maharashtra in Last 300 Years*, Kesari Daily (A Marathi language newspaper) of January 7, 1968, Poona, India.

KHATTRI, K. N. (1994), *A Hypothesis for the Origin of Peninsular India*, Current Sci. *67*, 590–597.

KLEE, G., and RUMMEL, F. (1993), *Hydrofrac Stress Data for the European HDR Research Project Test Site Soultz-Sous-Forests*, Int. J. Rock Mech. and Geomech. Abstr. *30*, 973–976.

KRISHNAN, M. S. (1966), *Tectonics of India*, Bull. Indian Geophys. Union *3*, 1–35.

KUSZNIR, N. J. (1991), *The Distribution of Stress with Depth in the Lithosphere: Thermo-rheological and Geodynamic Constraints*, Phil. Trans. Roy. Soc. Lond. *A337*, 95–110.

LI, F. (1986), *In situ stress measurements, stress state in the upper crust and their application to rock engineering*. In *Rock Stress*, Proc. Int. Symp. on Rock Stress Measurements, Stockholm, Sept. 1–3, 1986, 69–77.

MANGLIK, A., and SINGH, R. N. (1991), *Rheology of Indian Continental Crust and Upper Mantle*, Proc. Indian Acad. Sci. (Earth Planet. Sci.) *100*, 389–398.

MILNE, J. (1911), *A Catalogue of Earthquakes-A.D. 7 to A.D. 1899*, Brit. Assoc. Advan. Sci. Rep. Portsmouth Meeting, 1911 Appendix No. *1*, 649–740.

MOHAN, I., SITARAM, M. V. D., and GUPTA, H. K. (1981), *Some Recent Earthquakes in Peninsular India*, J. Geol. Soc. India *22*, 292–298.

OLDHAM, T. A. (1883), *Catalogue of Indian Earthquakes*, Mem. Geol. Surv. India *19*, 163–215.

PANDA, P. K. (1985), *Geothermal maps of India and their significance in Resources Assessment*. In *Petroliferous Basins of India*, Vol. III, 202–210, Petroleum Asia Journal, Dehradun, India.

PATRIAT, P., and ACHACHE, J. (1984), *India-Eurasia Collision Chronology has Implications for Crustal Shortening and Driving Mechanism of Plates*, Nature *311*, 615–620.

PINE, R. J., LEDINGHAM, P., and MERRIFIELD, C. M. (1983), *In situ Stress Measurements in the Carnmenellis Granite-II. Hydrofracture Tests at Rosemanowes Quarry to Depths of 2000 m*, Int. J. Rock Mech. and Geomech, Abstr. *20*, 63–72.

POWELL, C. M. (1979), *A speculative tectonic history of Pakistan and surroundings: Some constraints from the Indian Ocean*. In *Geodynamics of Pakistan* (ed. Farah, A. and DeJong, K. A.) (Geol. Survey of Pakistan, Quetta 1979) pp. 5–24.

RAO, B. R., and RAO, P. S. (1984), *Historical Seismicity of Peninsular India*, Bull. Seismol. Soc. Am. *74*, 2519–2533.

RAO, G. V., and RAO, R. U. M. (1983), *Heat Flow in Indian Gondwana Basins and Heat Production of their Basement Rocks*, Tectonophysics *91*, 105–117.

RAO, R. U. M., VERMA, R. K., RAO, G. V., HAMZA, V. M., PANDA, P. K., and GUPTA, M. L. (1970), *Heat Flow Studies in Godavari Valley (India)*, Tectonophysics *10*, 165–181.

RAO, R. U. M., RAO, G. V., and HARINARAIN (1976), *Radioactive Heat Generation and Heat Flow in the Indian Shield*, Earth Planet. Sci. Lett. *30*, 57–64.

RAO, T. M. (1966), *Geological studies of tectonic features influencing the occurrence of earthquakes in the Eastern Ghats in Peninsular India*. In Proc. Symp. Earthquake Eng., 3rd, University of Roorkee, Rorkee, India, 443–454.

RASTOGI, B. K. (1992), *Seismotectonics Inferred from Earthquakes and Earthquake Sequences in India during the 1980s*, Current Science *62*, 101–108. ·

RAVAL, U. (1993), *Collision boundary, stress build-up and role of basement heterogeneities: a peg effect*. In *Proc. Indian Geophys. Union*, Hyderabad, India 129–140.

RAVAL, U. (1994), *On Certain Large Scale Gravity Field Patterns over the Indian Subcontinent*. In *Proc. Space Applications in Earth System Science*, 153–168. Indian Geophys. Union, Hyderabad, India.

RAVISHANKER (1988), *Heat Flow Map of India and its Geological and Economic Significance*, Indian Mineral *42*, 89–110.

RAVISHANKER (1987), *Status of Geothermal Exploration in Maharashtra and Madhya Pradesh (Central Region)*, Rec. Geol. Surv. India *15*, 7–29.

RAVISHANKER (1991), *Thermal and Crustal Structure of SONATA. A zone of Mid-Continental Rifting in Indian Shield*, J. Geol. Soc. of India *37*, 211–220.

RUMMEL, F., and BAUMGARTNER, J. (1985), *Hydraulic Fracturing in situ Stress and Permeability Measurements in the Research Bore-hole Konzen, Hohes Venn (West Germany)*, N. Jb. Geol. Palaont. Abh. *171*, 183–193.

SASTRI, V. V., SINHA, R. N., SINGH, G., and MURTI, V. S. (1973), *Stratigraphy and Tectonics of Sedimentary Basins on East Coast of Peninsular India*, Am. Assoc. Petroleum Geologists Bull. *57*, 655–678.

SEEBER, L. (1994), *The earthquake that Shook the World*, New Scientist, 25–29.

SIBSON, R. H. (1982), *Fault Zone Models, Heat Flow, and the Depth Distribution of Earthquakes in the Continental Crust of the United States*, Bull. Seismol. Soc. Am. *72*, 151–163.

SIBSON, R. H. (1990), *Condition for fault-valve behaviour*. In *Deformation Mechanisms, Rheology, and Tectonics* (eds., Knipe, R. J. and Rutter, E. H.), Geol. Soc. Sp. Pub. No. *54*, 15–28.

SINGH, D. D., RASTOGI, B. K., and GUPTA, H. K. (1975), *Surface Wave Radiation Pattern and the Source Parameters of Koyna Earthquake of December 10, 1967*, Bull. Seismol. Soc. Am. *65*, 711–731.

SRIRAMA RAO, S. V., and GOWD, T. N. (1990a), *In situ Stress Measurements at Tatanagar*, Unpublished Tech. Rep., 5 pp., National Geophysical Research Institute, Hyderabad, India.

SRIRAMA RAO, S. V., and GOWD, T. N. (1990b), *In situ Permeability and Tectonic Stresses at Bamnia-Kalan Village, near Rajpura-Dariba Mines (Hindustan Zinc Ltd.), Rajasthan*, Technical Report, 12 pp., National Geophysical Research Institute, Hyderabad, India.

SUNDAR, A., GUPTA, M. L., and SHARMA, S. R. (1990), *Heat Flow in the Trans-Aravalli Igneous Suite, Tusham, India*, J. Geodynamics *12*, 89–100.

SWANBERG, C. A., and MORGAN, P. (1979), *The Linear Relation between Temperatures Based on the Silica Content of Groundwater and Regional Heat Flow: A New Heat Flow Map of the United States*, Pure Appl. Geophys. *117* (1/2), 227–241.

TANDON, A. N. (1959), *The Rann of Kutch Earthquake of 21st July 1956*, Indian J. Meteorol. Geophys. *10*, 137–146.

TSUKAHARA, H., and IKEDA, R. (1987), *Hydraulic Fracturing Stress Measurements and in situ Stress Field in the Kanto-Tokai Area, Japan*, Tectonophysics *135*, 329–345.

TURNER, H. H., and others (1911), *Seismological Investigations, XII. Seismic Activity, 1899–1903 Inclusive*, Brit. Assoc. Advan. Sci., 16th Rep., 57–65.

TURNER, H. H., and others (1912), *Seismological Investigations, II. Seismic Activity, 1904–1909 Inclusive*, Brit. Assoc. Advan. Sci., 17th Rep., 70–88.

TURNER, H. H., and others (1913), *Seismological Investigations, II. Seismic Activity in 1910*, Brit. Assoc. Advan. Sci., 18th Rep., 46–51.

VALDIYA, K. S. (1973), *Tectonic Framework of India: A Review and Interpretation of Recent Structural and Tectonic Studies*, Geophys. Res. Bull. *11*, 79–114.

VALDIYA, K. S. (1989), *Neotectonic Implication of Collision of Indian and Asian Plates*, Indian Journal of Geology *61*, 1–13.

VALDIYA, K. S. (1993), *Latur Earthquake of 30 September 1993: Implications and Planning for Hazard-preparedness*, Current Science *65*, 515–517.

VERMA, R. K., RAO, R. U. M., GUPTA, M. L., RAO, G. V., and HAMAZA, V. M. (1969), *Terrestrial Heat Flow in Various Parts of India*, Bull. Volcanologique *32*, 66–68.

VERNIK, L., and ZOBACK, M. D. (1992), *Estimation of Maximum Horizontal Principal Stress Magnitude from Stress Induced Well Bore Breakouts in the Cajon Pass Scientific Research Borehole*, J. Geophys. Res. *97*, 5109–5119.

ZOBACK, M. D., TSUKAHARA, H., and HICKMAN, S. (1980), *Measurements at Depth in the Vicinity of the San Andreas Fault: Implications for the Magnitudes of Shear Stress at Depth*, J. Geophys. Res. *85*, 6157–6173.

ZOBACK, M. D., and others (1993), *Upper-crustal Strength Inferred from Stress Measurements to 6 km Depth in the KTB Borehole*, Nature *365*, 633–635.

(Received January 28, 1995, revised July 24, 1995 and August 10, 1995)

PAGEOPH, Vol. 146, Nos. 3/4 (1996)

0033–4553/96/040533–17$1.50 + 0.20/0

$(\partial\mu/\partial T)_P$ of the Lower Mantle

YANBIN WANG[1] and DONALD J. WEIDNER[1]

Abstract —We estimate $(\partial\mu/\partial T)_P$ of the lower mantle at seismic frequencies using two distinct approaches by combining ambient laboratory measurements on lower mantle minerals with seismic data. In the first approach, an upper bound is estimated for $|(\partial\mu/\partial T)_P|$ by comparing the shear modulus (μ) profile of PREM with laboratory room-temperature data of μ extrapolated to high pressures. The second approach employs a seismic tomography constraint $(\partial \ln V_S/\partial \ln V_P)_P = 1.8-2$, which directly relates $(\partial\mu/\partial T)_P$ with $(\partial K_S/\partial T)_P$. An average $(\partial K_S/\partial T)_P$ can be obtained by comparing the well-established room-temperature compression data for lower mantle minerals with the K_S profile of PREM along several possible adiabats. Both $(\partial K_S/\partial T)$ and $(\partial\mu/\partial T)$ depend on silicon content [or (Mg + Fe)/Si] of the model. For various compositions, the two approaches predict rather distinct $(\partial\mu/\partial T)_P$ vs. $(\partial K_S/\partial T)_P$ curves, which intersect at a composition similar to pyrolite with $(\partial\mu/\partial T)_P = -0.02$ to -0.035 and $(\partial K_S/\partial T)_P = -0.015$ to -0.020 GPa/K. The pure perovskite model, on the other hand, yields grossly inconsistent results using the two approaches. We conclude that both vertical and lateral variations in seismic velocities are consistent with variation due to pressure, temperature, and phase transformations of a uniform composition. Additional physical properties of a pyrolite lower mantle are further predicted. Lateral temperature variations are predicted to be about 100–250 K, and the ratio of $(\partial \ln \rho/\partial \ln V_S)_P$ around 0.13 and 0.26. All of these parameters increase slightly with depth if the ratio of $(\partial \ln V_S/\partial \ln V_P)_P$ remains a constant throughout the lower mantle. These predicted values are in excellent agreement with geodynamic analyses, in which the ratios $(\partial \ln \rho/\partial \ln V_S)_P$ and $(\partial \ln \rho/\partial \ln V_S)_P$ are free parameters arbitrarily adjusted to fit the tomography and geoid data.

Key words: Lower mantle, seismic tomography, thermoelasticity, composition models, geodynamics, shear modulus.

Introduction

Lateral temperature and density variations in the earth's lower mantle can provide insights into the dynamic processes of the earth's deep interior. While lateral variations in P- and S-wave velocities are becoming well established through tomographic studies (e.g., WOODHOUSE et al., 1987; WOODHOUSE and DZIEWONSKI, 1989; SU and DZIEWONSKI, 1994; DZIEWONSKI et al., 1993; FORTE et al., 1993; DZIEWONSKI, 1993), linking these values to temperature and density requires

---
[1] Center for High Pressure Research (A National Science Foundation Science and Technology Center) and Department of Earth and Space Sciences, University at Stony Brook, Stony Brook, NY 11794–2100, U.S.A.

additional information about the properties of the materials of the lower mantle, including thermal expansion and the temperature derivative of the seismically determined shear velocity. Considerable laboratory data have been recently gleaned that helps to constrain the bulk modulus and thermal expansion of mantle minerals at mantle conditions (e.g., recent results on (Mg, Fe)SiO$_3$ perovskite by MAO et al., 1991; YAGI et al., 1993; WANG, et al., 1994). In contrast, the shear velocities of the lower mantle phases are known only at room pressure and temperature (YEGANEH-HAERI et al., 1989a; YEGANEH-HAERI, 1994).

In this paper, we utilize seismic results in conjunction with the laboratory data to constrain the temperature derivative of the shear modulus at mantle conditons and seismic frequencies. Two methods are used to reach the conclusions. The first employs the radial variation of shear modulus from seismic earth models to extract the pressure dependence of the shear modulus of the lower mantle. By comparison with the ambient laboratory shear moduli, $\partial\mu/\partial T$ of the lower mantle is deduced as a function of the (Mg + Fe)/Si ratio. The second method relies on the observed $v = \partial \ln V_S/\partial \ln V_P$ of the lower mantle. $\partial\mu/\partial T$ is estimated from v and $\partial K_S/\partial T$, which is in turn calculated from the combined radial variation of velocities and mineral properties. The $\partial\mu/\partial T$ thus defined depends again on the (Mg + Fe)/Si ratio. The two estimates of $\partial\mu/\partial T$ are consistent for a mantle with a pyrolite composition and strongly disagree for a mantle with a pyroxene stoichiometry.

By requiring consistency, we deduce both the lower mantle composition and the value of $\partial\mu/\partial T$. With these values, we deduce the magnitude of the lateral variations in temperature and density within the lower mantle. Within this framework, we conclude that radial variations in mantle seismic velocities and lateral variations within the lower mantle are the result of the intrinsic variations of a uniform chemical composition and due to phase transformations and the effects of pressure and temperature on the elastic properties and density. In this analysis, we cannot conclude whether the temperature variation of the shear modulus results from anharmonic properties or from anelastic processes. However, compositional effects need not be invoked.

1. Laboratory Data on Candidate Minerals for the Lower Mantle

The lower mantle is most likely composed of primarily (Mg, Fe)SiO$_3$ perovskite with certain amounts of magnesiowüstite (Mg, Fe)O and CaSiO$_3$ perovskite (e.g., RINGWOOD and MAJOR, 1971; LIU, 1975; LIU and RINGWOOD, 1975; ITO and MATSUI, 1978; KNITTLE and JEANLOZ, 1987; MAO et al., 1989). These three components probably account for 95% of the volume of the mantle (ANDERSON, 1989). Possible phases enriched in aluminum remain elusive at this stage (e.g., MADON et al., 1989; AHMED-ZAÏD and MADON 1991; IRIFUNE et al., 1991; FITZ GERALD and RINGWOOD, 1991; YUSA et al., 1993). MgSiO$_3$ perovskite itself may

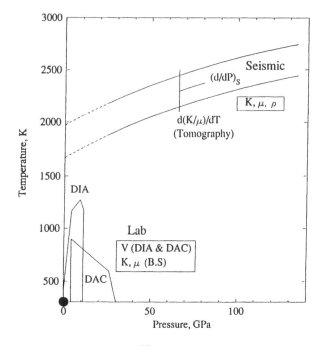

Figure 1

Pressure and temperature conditions of the available laboratory data for (Mg, Fe)SiO$_3$ perovskite as compared with those in the lower mantle. Solid circle represents the condition (ambient) where elasticity data for MgSiO$_3$ perovskite were obtained using Brillouin spectroscopy; pressure and temperature conditions, where P-V-T data for the perovskite are obtained using large-volume, high-pressure apparatus (DIA) and the diamond-anvil cells (DAC), are outlined. Seismic earth models (e.g., PREM) give K_S, μ, and ρ at each depth as well as their pressure derivatives; seismic tomography data provide information on $[\partial(\mu/K_S)/\partial T]_P$. The temperature conditions of the seismic data are plotted according to estimated lower-mantle geotherm discussed in the text.

also be able to accommodate Al in the structure (e.g., GASPARIK, 1990; WENG *et al.*, 1982; IRIFUNE, 1994). As the MgO-FeO-SiO$_2$ system accounts for about 90% of the volume of the mantle, we shall consider a mantle of only two components, (Mg, Fe)SiO$_3$ perovskite and magnesiowüstite.

High-pressure and high-temperature data have just become available for (Mg, Fe)SiO$_3$ perovskite through pressure-volume-temperature (P-V-T) measurements (e.g., MAO *et al.*, 1991; YAGI *et al.*, 1993; WANG *et al.*, 1994). Figure 1 summarizes the pressure and temperature conditions of these studies, compared with the conditions of the lower mantle. In these P-V-T studies, information on bulk modulus and its temperature derivative is obtained by differentiating the volume with respect to pressure and temperature to the first and second order, therefore resulting in progressively larger errors. Since most of the high-temperature measurements were obtained at conditions outside the perovskite stability field, issues such as structural stability and chemical stability (especially mobility of iron)

Table 1

Room-temperature Laboratory Data for Lower Mantle Minerals[a]

Parameters	$Mg_{1-x}Fe_xSiO_3$	$Mg_{1-x}Fe_xO$
V_0, cc/mol	$24.46 + 1.03x$	$11.25 + 1.00x$
ρ_0, g/cm^3	$4.108 + 1.07x$	$3.583 + 2.28x$
K_{T0}, GPa	261	$161 - 7.5x$
μ_0, GPa	177[b]	$131 - 77.x$
K'_{T0}	4.0	4.0

[a] See HEMLEY *et al.* (1992) and WANG *et al.* (1994) for discussion of sources of data.
[b] YEGANEH-HAERI (1994).

must be carefully addressed (WANG *et al.*, 1994). On the other hand, room-temperature static compression studies have yielded consistent equation-of-state results to over 100 GPa (KNITTLE and JEANLOZ, 1987). No dependence on iron content could be detected for the bulk modulus of perovskite from static compression studies (MAO *et al.*, 1991). The ambient bulk modulus K_{T0} deduced from these P-V measurements agrees quite well with Brillouin spectroscopy data (YEGANEH-HAERI, 1994).

The most recent Brillouin scattering data, obtained on high-purity MgSiO$_3$ samples that suffered much less from structural instability problems under the laser beam, give $K_{S0} = 264$ GPa and $\mu_0 = 177$ GPa (YEGANEH-HAERI, 1994). This μ_0 is the only information we have on shear properties of the perovskite. There are no shear modulus data at high temperature or high pressure and nothing is known about its iron dependence. However, silicates which have iron substitution for magnesium in similar sites, such as olivine and garnet, show little or no dependence of shear modulus on iron content (ANDERSON *et al.*, 1992; GOTO *et al.*, 1976). In addition, the total amount of iron in perovskite will be relatively small, since iron tends to partition to other phases such as magnesiowüstite. Thus, to the first order, we assume that there is no iron dependence on μ for perovskite. The uncertainty introduced by this assumption is rather minor, for example, if we take the iron dependence as observed in olivine (ANDERSON *et al.*, 1992) and apply it for a perovskite with a maximum iron content of 10%, we get a decrease in μ by only 3 GPa.

Considerably more data are available for magnesiowüstite; a summary on the elasticity data of which has been given by Hemley *et al.* (1992). Data on the bulk and shear moduli and their pressure and temperature derivatives have been obtained with ultrasonic techniques to a few GPa (e.g., SPETZLER, 1970; JACKSON and NESLER, 1982; SUMINO and ANDERSON, 1984). Recent P-V-T measurements to much higher pressures give K_T and $\partial K_T/\partial T$ data that are consistent with the lower pressure ultrasonic data (FEI *et al.*, 1992).

In what follows, we shall use only the room-temperature data that are well established through various measurements and in different laboratories as summarized in Table 1. Recent studies have reached consistent conclusions that the lower mantle iron content, Fe/(Mg + Fe), is about 0.12(1) regardless of the silicon content of the model (STIXRUDE et al., 1992; HEMLEY et al., 1992; WANG et al., 1994). Therefore, the principal uncertainty in lower mantle composition is the silica content or the (Mg + Fe)/Si ratio. In the following, we calculate the ambient moduli K_{T0} and μ_0 for different (Mg + Fe)/Si ratios, using the Hashin-Shtrikman's bounds which differ only by about 1 GPa.

In order to combine these mineral physics data with seismically measured properties of the mantle, we convert the isothermal bulk moduli to adiabatic by using $K_{S0} = K_{T0}(1 + \alpha_0\gamma_0 T)$, where α_0 is the volumetric thermal expansion and γ_0 the Grüneisen parameter of the material; subscript 0 indicates ambient values. The correction is very small because of the low temperature (300 K); thus uncertainties in α_0 and γ_0 contribute insignificantly to the uncertainty in K_{S0}. Adiabatic and isothermal shear moduli, on the other hand, should be identical to the first order (e.g., POIRIER, 1991) and thus will be denoted as μ.

2. Estimating $(\partial\mu/\partial T)_P$ and $(\partial K_S/\partial T)_P$ Using Mineral Physics Data

Following previous workers, we assume that the lower mantle is chemically homogeneous and that the temperature distribution is close to adiabatic (e.g., DAVIES and DZIEWONSKI, 1975; BUTLER and ANDERSON, 1978; STIXRUDE et al., 1992; HEMLEY et al., 1992; WANG et al., 1994). The foot temperature (T_0) of the lower mantle adiabat is estimated to be 1800 (±200) K from previous studies (e.g., BROWN and SHANKLAND, 1981; STIXRUDE et al., 1992; HEMLEY et al., 1992; AKAOGI and ITO, 1993; WANG et al., 1994).

In order to obtain constraints on $(\partial K_S/\partial T)_P$ and $(\partial\mu/\partial T)$ for the lower mantle, we start from a model composition and compress it to the lower mantle pressure using previously determined room-temperature equations of state. We then compare the adiabatic bulk modulus of the model composition with that of PREM (DZIEWONSKI and ANDERSON, 1981) to estimate the temperature derivative $(\partial K_S/\partial T)_P$ of the lower mantle

$$\langle(\partial K_S/\partial T)_P\rangle = [K_S(300, P) - K_S(T, P)]/(T - 300) \qquad (1)$$

where $\langle\cdots\rangle$ indicates average over temperature, $K_S(300, P)$ is the bulk modulus of the model at 300 K and pressure P, $K_S(T, P)$ is the seismic value at the same pressure (or depth), T the temperature at that depth. Uncertainties in temperature estimates are of the order of 10%; therefore this equation supplies a quite accurate estimate of average temperature derivatives (see examples in Table 2). In general, $(\partial K_S/\partial T)_P$ is quite temperature independent and we need not be concerned with its

Table 2

Estimated $|\partial K_S/\partial T|$ and $|\partial \mu/\partial T|$ for Two Lower Mantle Models at 1071 km

	(Mg + Fe)/Si = 1.0	(Mg + Fe)/Si = 1.5
K_{T0}, GPa	261[a]	237[b]
μ_0, GPa	177[a]	163[b]
$K_S(300, P)$, GPa	422[c]	397[c]
Mineral Physics		
$\quad|\partial K_S/\partial T|$,[d] GPa/K	0.032(4)	0.018(2)
$\quad|\partial \mu/\partial T|$,[d] GPa/K	<0.031(7)	<0.024(6)
Tomography		
$\quad|\partial \mu/\partial T|$,[e] GPa/K	0.059(8)	0.030(6) (ν = 1.8)
	0.095(15)	0.043(4) (ν = 2.0)

[a] YEGANEH-HAERI (1994).

[b] Hashin-Shtrikman average of properties of perovskite and magnesiowüstite, based on a total iron content Fe/(Mg + Fe) = 12% and a partitioning coefficient [{Fe/(Mg + Fe)}$_{MW}$/{Fe(Mg + Fe)}$_{PV}$] of 4.0.

[c] Adiabatic bulk modulus calculated at $P = 41.86$ GPa (corresponding to the pressure at 1071 km).

[d] *Mineral physics* estimate, see Section 2. Numbers in parentheses correspond to temperature uncertainty of ± 200 K.

[e] *Tomography* estimate, see Section 3, for ν = 1.8 and 2.0, respectively. Numbers in parentheses are temperature uncertainties propagated from estimates on $|\partial K_S/\partial T|$.

variation over the temperature interval of interest therefore we shall drop the $\langle \rangle$ sign.

Table 2 gives the estimates $(\partial K_S/\partial T)_P$ values at 1071 km depth ($P = 41.86$ GPa) for two lower-mantle composition models, pyrolite [(Mg + Fe)/Si = 1.5] and pure perovskite [(Mg + Fe)/Si = 1.0], using the procedure outlined above. The temperature at 1071 km depth is around 2160 (200) K according to the adiabatic thermal gradient given by BROWN and SHANKLAND (1981). We use a third-order Birch-Murnaghan equation of state to calculate the isothermal elastic moduli at 300 K and 41.86 GPa (Table 1). An average thermal expansion of 1.6×10^{-5} K^{-1} and a Grüneisen $\gamma_0 = 1.5$ (WANG *et al.*, 1994) are used to convert $K_T(300$ K, $P)$ to the adiabatic $K_S(300$ K, $P)$. They differ only by about 3 GPa. It can be seen that a silicon-rich lower mantle requires higher $(\partial K_S/\partial T)_P$ values and that the range [-0.031 (3) GPa/K for the perovskite model and -0.018 (2) GPa/K for pyrolite] is consistent with previous analyses (e.g., STIXRUDE *et al.*, 1992; WANG *et al.*, 1994; ZHAO and ANDERSON, 1994).

The high-pressure shear modulus can be predicted in a similar manner. However, as no high-pressure data are available for perovskite, we utilize seismic data to estimate the pressure dependence of the shear modulus by using $\mu(P) = \mu_0 + \mu' P$, where μ' is approximated by the pressure derivative of the seismic shear modulus at the foot of the mantle adiabat. The PREM data have been analyzed in terms of

seismic equations of state, based on the same assumptions of composition homogeneity and adiabatic temperature distribution in the lower mantle. Various analyses using either the third- or the fourth-order equation-of-state provide generally
consistent results with $K_{S0} = 213$ (10) GPa, $\mu_0 = 130$ (5) GPa, $\rho_0 = 4.00$ (2) g/cc,
$K'_0 = 3.9$ (2), and $\mu'_0 = 1.5–1.8$ at T_0 and zero pressure (e.g., SAMMIS *et al.*, 1975;
DAVIES and DZIEWONSKI, 1975; BUTLER and ANDERSON, 1978; JEANLOZ and
KNITTLE, 1986). Based on the Grüneisen-Debye theory of equation of state, the
thermodynamic Grüneisen γ has been estimated to be 1.2(1) for the lower mantle
(BROWN and SHANKLAND, 1981). From various models that relate γ to volume
dependence of the elastic moduli, ANDERSON (1987) estimated that μ' ranges from

M/Si from lab and seismic data

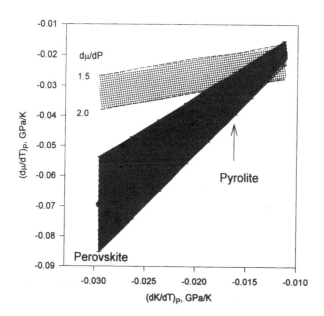

Figure 2

Dependence of $(\partial \mu / \partial T)$ on $(\partial K_S / \partial T)_P$ and on lower mantle composition. For each composition the
$(\partial K_S / \partial T)_P$ value is calculated by extrapolating the ambient K_S of the model composition to 1071 km
depth (41.86 GPa but room temperature) and by comparing with K_S of PREM at this depth [41.86 GPa
and 2160 (200) K]. Composition dependence of $(\partial K_S / \partial T)_P$ is illustrated by the solid circles in the heavily
shaded region, representing the ratio of (Mg + Fe)/Si from 1.0 (left) to 1.8 (right) at an increment of 0.1.
The lightly shaded region is the $(\partial \mu / \partial T)$ values calculated based on the mineral physics estimate
discussed in Section 2. Composition dependence of $(\partial \mu / \partial T)$ in this region is reflected by the dependence
of $(\partial K_S / \partial T)_P$ on the (Mg + Fe)/Si ratio. Vertical range represents μ' from 1.5 (top) to 1.8 (bottom). The
heavily shaded region corresponds to the required $(\partial \mu / \partial T)$ values from seismic tomography [equation (3)],
for each composition and its $(\partial K_S / \partial T)_P$. Vertical range represents ν from 1.8 (top) to 2.0 (bottom). Only
the models that fall within the area where the two shaded regions intersect are consistent with all the
data. This requires (Mg + Fe)/Si > 1.4 for the lower mantle. Iron content [Fe/(Mg + Fe)] is 0.12 for all
the compositions and satisfies the density requirement of PREM.

1.2 to 1.7 in the lower mantle. The estimate using $\mu(300, P) = \mu_0 + \mu_0'P$ (with $\mu_0' = 1.5-1.8$) gives an upper bound as μ' generally decreases with increasing pressure. The estimated room-temperature shear modulus is then compared with seismically observed μ at the same pressure (but at a high temperature) to obtain an average $(\partial\mu/\partial T)$ with an equation similar to equation (1). The estimates are listed in Table 2 for two composition models, pyrolite and perovskite, at 1071 km depth. Since all high-temperature shear modulus data used in this calculation are derived from seismic observations at seismic frequencies, the deduced temperature derivative of the shear modulus is for seismic frequencies. Anelastic processes, that can give rise to a frequency dependence of the shear modulus, become expressed as a frequency dependence of the temperature derivative of the shear modulus. Thus, the results obtained here are appropriate for the lower mantle. The biggest assumption that we must make, is that $\langle(\partial\mu/\partial T)_P\rangle$ is relatively temperature-independent. Even if there are strong contributions to $(\partial\mu/\partial T)_P$ from anelastic processes, we need only assume that they are broad-band enough that the temperature derivative does not vary significantly over the temperature interval.

In Table 2, we summarize these "mineral physics" estimates for two lower-mantle composition models with $(Mg + Fe)/Si = 1.0$ and 1.5, respectively, at 1071 km depth. Similar estimates of the "mineral physics" $(\partial K_S/\partial T)$ and $(\partial\mu/\partial T)$ for various model compositions with $(Mg + Fe)/Si$ ranging from 1.0 to 1.8 are graphically presented in Figure 2 by the lightly shaded band, the width of which represents the estimated uncertainty of the analysis. Clearly, $(\partial\mu/\partial T)$ is rather insensitive to the $(Mg + Fe)/Si$ ratio from these estimates, because the shear modulus of the perovskite dominates the Hashin-Shtrikman average.

3. Seismic Tomography Constraints on $(\partial\mu/\partial T)_P$ of the Lower Mantle

Seismic tomography studies have shown that $v = (\partial \ln V_S/\partial \ln V_P)_P = 1.8-2.0$ throughout the lower mantle (WOODHOUSE, et al., 1987; WOODHOUSE and DZIEWONSKI, 1989; SU and DZIEWONSKI, 1994; DZIEWONSKI et al., 1993; FORTE et al., 1993; DZIEWONSKI, 1993). Assuming that the lower mantle is chemically homogeneous and that variations in V_P and V_S are due to lateral temperature fluctuation, the ratio of $(\partial \ln V_S/\partial \ln V_P)_P$ can be written in terms of temperature derivatives of K_S and μ:

$$v = (\partial \ln V_S/\partial \ln V_P)_P = (\Gamma - 1)[1 + (4/3)\mu/K_S]/[\delta_s - 1 + (\Gamma - 1)(4/3)\mu/K_s] \quad (2)$$

where $\delta_s = -[1/(\alpha K_s)](\partial K_S/\partial T)_P = (\partial \ln K_S/\partial \ln \rho)_P$, $\Gamma = -[1/(\alpha\mu)](\partial\mu/\partial T)_P = (\partial \ln \mu/\partial \ln \rho)_P$, and α the volumetric thermal expansion. Equation (2) leads to the following in situ relationship between $(\partial\mu/\partial T)_P$ and $(\partial K_S/\partial T)_P$

$$(\partial\mu/\partial T)_P = \{v(\partial K_S/\partial T)_P + \alpha(v - 1)\mu(K_S/\mu + 4/3)\}/\{K_S/\mu - (v - 1)(43)\}. \quad (3)$$

$(\partial\mu/\partial T)_P$ varies linearly with $(\partial K_S/\partial T)_P$ with a slope of $v/\{K_S/\mu - (v-1)(4/3)\}$, which is about 2–3.5 in the lower mantle; varying $(\partial K_S/\partial T)_P$ from -0.018 to -0.031 GPa/K (or changing the composition from pyrolite to perovskite) would cause $|(\partial\mu/\partial T)_P|$ to increase by about 0.04 GPa/K. On the other hand, $(\partial\mu/\partial T)_P$ is proportional to α with a slope of $(v-1)\mu(K_S/\mu + 4/3)/\{K_S/\mu - (v-1)(4/3)\}$; doubling α from $1 \times 10^{-5}\,\mathrm{K}^{-1}$ to $2 \times 10^{-5}\,\mathrm{K}^{-1}$ would only cause $|(\partial\mu/\partial T)_P|$ to change by 0.01 GPa/K. Therefore, $(\partial\mu/\partial T)_P$ is most sensitive to $(\partial K_S/\partial T)_P$.

Equation (3) provides a "tomography" estimate for $(\partial\mu/\partial T)$ of the lower mantle *in situ*. In Table 2, we compare the $(\partial\mu/\partial T)$ values for the pure perovskite and the pyrolite models based on both the tomography and the mineral physics estimates of $(\partial\mu/\partial T)$. Note that the "tomography" $(\partial\mu/\partial T)$ values for the perovskite lower mantle (-0.059 GPa/K for $v = 1.8$ and -0.095 for $v = 2.0$) are about two or three times the "mineral physics" estimate $[-0.031\,(7)\,\mathrm{GPa/K}]$. The pyrolite mantle, on the other hand, requires an average $(\partial\mu/\partial T)$ about -0.030 and -0.043 GPa/K, for $v = 1.8$ and 2.0, respectively, agreeing quite well with the "mineral physics" $(\partial\mu/\partial T)$ of $-0.24\,(6)$ GPa/K.

4. Constraints on Composition of the Lower Mantle

For various composition models of the lower mantle, we calculate the $(\partial K_S/\partial T)$ values required by ambient mineral physics data in order to match the seismically determined K_S at 1071 km depth, as outlined in Section 2. These $(\partial K_S/\partial T)_P$ values are then used to estimate $(\partial\mu/\partial T)$, using equation (3) in order to be consistent with seismic tomography data. The heavily shaded area represents the range of $(\partial\mu/\partial T)$ with $v = 1.8$–2.0, with solid circles representing the mid-points of the two situations (again an average $\alpha = 1.6 \times 10^{-5}\,\mathrm{K}^{-1}$ is used in the calculation). The lightly shaded, nearly horizontal area is the range of "mineral physics" $(\partial\mu/\partial T)$ discussed in section 2. The two shaded areas intercept only at $|(\partial K_S/\partial T)| < 0.02$ GPa/K and $|(\partial\mu/\partial T)| < 0.035$ GPa/K, where the laboratory and seismic data are consistent. Therefore, we conclude that models with $(\mathrm{Mg + Fe})/\mathrm{Si} > 1.4$ are consistent with both laboratory and seismic observations. The pyrolite model, $(\mathrm{MgFe})/\mathrm{Si} \approx 1.5$, lies within this range. In all of the models we have assumed the same iron content of 0.12, which has been shown in previous studies to be consistent with the density of PREM, independent from the silicon content of the models (e.g., STIXRUDE et al., 1992; WANG et al., 1994).

The range of $|(\partial K_S/\partial T)| \leq 0.02$ GPa/K thus obtained is consistent with the laboratory P-V-T measurements on MgSiO_3 perovskite of WANG et al. (1994), who reported $(\partial K_T/\partial T) = -0.023(11)$ GPa/K and therefore $|(\partial K_S/\partial T)| \leq 0.02$ GPa/K. No laboaotry data exist on $|(\partial\mu/\partial T)|$ for the perovskite. However, systematics on $|(\partial\mu/\partial T)|$ for silicates and oxides indicates that 0.035 GPa/K is an abnormally large value. YEGANEH-HAERI et al. (1989a) suggested that this large $|(\partial\mu/\partial T)|$ may be due to the ferroelastic nature of the orthorhombic structured perovskite. Supporting evidence has been found in other orthorhombic perovskites such as $SrAlO_3$ (YEGANEH-HAERI et al., 1989b), which exhibits large $|(\partial C_{ij}(\partial T)|$.

The pure perovskite lower mantle model $[(Mg + Fe)/Si = 1.0]$ requires an unreasonably high $|(\partial\mu/\partial T)] > 0.055$ GPa/K, according to equation (3); on the other hand, laboratory estimates indicate that $|(\partial\mu/\partial T)|$ must be smaller than 0.035 GPa/K. Such inconsistency renders a perovskite model unacceptable within the framework of a chemically homogeneous lower mantle.

Throughout this analysis, we have assumed that the seismic tomography observation $(\partial \ln V_S/\partial \ln V_P)_P = 1.8$–$2.0$ is due to lateral temperature fluctuations in a chemically uniform, solid-state lower mantle. There are several other proposals to explain this observation. DUFFY and AHRENS (1992) suggested that the tomography results may be explained by partial melting. Their model assumes that a certain amount of water is present in the lower mantle producing partial melting, causing a decrease in either shear or bulk modulus. Theoretical calculations on simple oxide MgO indicate a gradual decrease in $\delta_S = -(\partial K_S/\partial T)/(\alpha T)$ with pressure; both AGNON and BUKOWINSKI (1990) and ISAAK et al. (1992) suggest that if the perovskite behaves the same way as does MgO, then v would increase as is evident from equation (2). KARATO (1993), on the other hand, suggested that anelasticity may be responsible for the high v values in the lower mantle. All of these issues are important but are difficult to evaluate quantitatively for the lower mantle at this stage. The similar effects of anharmonic and anelastic processes on the temperature derivative of the shear modulus make it difficult to distinguish between these in the absence of laboratory-based high frequency acoustic data. Our analysis indicates that it is not necessary to invoke any of these assumptions to explain the tomography results.

All above models assume that lateral temperature fluctuations are responsible for the tomography observations. Another possibility may be chemical variation. YUEN et al. (1993) suggested that large chemical inhomogeneities are responsible, based on their exceptionally large temperature fluctuations (in excess of 1000 K) estimated for a chemically homogeneous lower mantle. However, these estimates were based on the room temperature $(\partial \ln V_{P,S}/\partial \ln \rho)_T$ data on MgO (CHOPELAS, 1992), with the assumption that under lower mantle pressures $(\partial \ln V_{P,S}/\partial \ln \rho)_P = (\partial \ln V_{P,S}/\partial \ln \rho)_T$. Our analysis, based on laboratory data on the silicate perovskite, gives estimates of lateral temperature variations only about 200 K (see Section 5). Therefore, there is no need to assume chemical inhomogeneities to explain the seismic tomography data.

Another possibility is lateral variation in iron content in an otherwise homogeneous lower mantle. This may be evaluated to the zeroth order with a pure perovskite lower mantle model. Since there is no detectable dependence in iron content (x_{Fe}) on the bulk modulus K_S of the perovskite, $(\partial \ln V_P/\partial x_{Fe})_P = (1/2)[(4/3)(\partial\mu/\partial x_{Fe})/(K_S + (4/3)\mu) - \partial \ln \rho/\partial x_{Fe}]$, and $(\partial \ln V_S/\partial x_{Fe})_P = (1/2)[(\partial \ln \mu/\partial x_{Fe}) - \partial \ln \rho/\partial x_{Fe}]$.

Thus, we have

$$(\partial\mu/\partial x_{Fe})[4(v-1)\mu - 3K_S]/[\mu(3K_S + 4\mu)] = (v-1)(\partial \ln \rho/\partial x_{Fe}). \qquad (4)$$

There are three possibilities, depending on $[4(v-1)\mu - 3K_S]/\mu(3K_S + 4\mu)$ to be (a) $= 0$, (b) > 0, and (c) < 0. In case (a), we have $v = (3/4)(K_S/\mu) + 1$. For typical lower-mantle values of $K_S/\mu \approx 1.9$, $v = 2.425$. But this implies that $(\partial \ln \rho/\partial x_{Fe}) = 0$, inconsistent with laboratory findings of $(\partial \ln \rho/\partial x_{Fe}) > 0$. Case (b) $(v > 2.425)$ requires that $(\partial\mu/\partial x_{Fe}) > 0$, i.e., μ increases with increasing iron content. This is inconsistent with the notion that generally μ decreases with iron content. Case (c) $(v < 2.425)$ is consistent with $(\partial\mu/\partial x_{Fe}) < 0$, and the magnitude of v is also compatible with the tomography results. However, the required $|\partial\mu/\partial x_{Fe}|$ is too large to be acceptable. To demonstrate this point, we take a tomographic ratio of $v = 2.0$ and $(\partial \ln \rho/\partial x_{Fe}) = 0.26/(mol.Fe)$ for perovskite (Table 1), $(\partial\mu/\partial x_{Fe})$ is estimated from (4) to be $-1.5 \mu/(mol.Fe)$. As μ is in the order of 200 GPa for perovskite under lower mantle conditions, the $|\partial\mu/\partial x_{Fe}|$ thus required is in the order of 300 GPa/(mol.Fe)! Perhaps an even more serious problem is the sign of the ratio $(\partial \ln \rho/\partial \ln V_S)$. Whereas temperature causes both velocity and density to decrease, iron content would lower μ and increase ρ, thus resulting $(\partial \ln \rho/\partial \ln V_{P,S}) < 0$. Current geodynamic analyses based on both seismic tomography and dynamic geoid data all indicate positive $(\partial \ln \rho/\partial \ln V_{P,S})$ (e.g., HAGER and CLAYTON, 1989; FORTE et al., 1993; DZIEWONSKI et al., 1993). Therefore, we conclude that variation in iron content alone cannot explain the tomography observations.

5. Geodynamic Properties of a Pyrolite Lower Mantle

We have concluded that the acceptable composition models for the lower mantle are those with $(Mg + Fe)/Si > 1.4$. Pyrolite, $(Mg + Fe)/Si = 1.5$, falls within this range. Experimental uncertainties in the existing P-V-T data do not allow any discrimination between $(Mg + Fe)/Si = 1.4$ and 1.7. In the following we choose pyrolite as our model composition and calculate its physical properties and, when possible, compare the results with observations.

Figure 3 shows both $(\partial\mu/\partial T)$ and $(\partial K_S/\partial T)$ as a function of pressure (depth) of a pyrolite lower mantle. The two curves of $(\partial\mu/\partial T)$ represent two v values of 1.8 (upper curve) and 2.0 (lower curve). Volumetric thermal expansion (α) is chosen to be 1.6×10^{-5} K^{-1} at 1071 km depth and a dependence of $\partial\alpha/\partial P = (\partial K_T/\partial T)/K_T^2$ is used with a constant $(\partial K_T/\partial T)_P = -0.023$ GPa/K (WANG et al., 1994). This pressure dependence term becomes a very small correction in the calculation. It is interesting that both $(\partial\mu/\partial T)$ and $(\partial K_S/\partial T)$ are relatively constant below 40 GPa or 1000 km depth. This constancy over a wide pressure range again suggests that effects of anharmonicity and anelasticity play insignificant roles for both perovskite and magnesiowüstite under lower mantle conditions.

PREM and Pyrolite

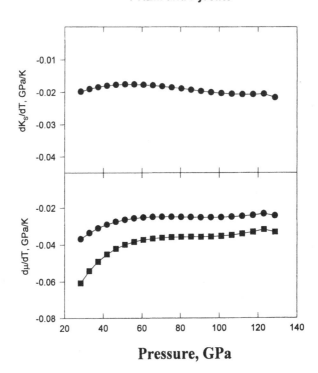

Pressure, GPa

Figure 3

Estimated variations in $(\partial K_S/\partial T)$ and $(\partial\mu/\partial T)$ as a function of pressure (or depth) for a pyrolite lower mantle. The values of $(\partial K_S/\partial T)$ are estimated by equation (1), and $(\partial\mu/\partial T)$ from tomography constraints [equation (3)]. Two curves for $(\partial\mu/\partial T)$ are for $v = (\partial \ln V_S/\partial \ln V_P)_P = 1.8$ (upper curve) and 2.0 (lower curve), respectively. K_S and μ profiles of PREM are used in the calculation. Note that $(\partial K_S/\partial T)$ remains constant throughout the lower mantle. $(\partial\mu/\partial T)$ is also relatively constant, especially above 40 GPa (1000 km depth).

Table 3

Lateral Variations in Properties at 1071 km Depth

Parameter	Pyrolite (This study)	Geodynamic model FORTE et al. (1993)	Geodynamic model DZIEWONSKI et al. (1993)
α, K^{-1}	$1.6(1) \times 10^{-5}$[a]	–	–
ΔT, K	110–260	–	
$\partial \ln \rho/\partial \ln V_P$	0.25–0.46	0.27–0.36[b]	0.15–0.36[b]
$\partial \ln \rho/\partial \ln V_S$	0.13–0.26	0.08–0.29[b]	0.15–0.36[b]
$\partial \ln V_P/\partial T$, $10^{-4}\,K^{-1}$	$-0.24--0.46$	–	–
$\partial \ln V_S/\partial T$, $10^{-4}\,K^{-1}$	$-0.39-0.82$	–	–

[a] WANG et al. (1994).
[b] Range of values over the entire lower mantle.

The constraints on $(\partial \mu / \partial T)$ and $(\partial K_S / \partial T)$ obtained in this study also allow estimates of several other parameters that are important for geodynamic analyses (Table 3). Since

$$(\partial \ln V_S / \partial T)_P = 1/2[\partial \ln \mu / \partial T)_P + \alpha] \qquad (5a)$$

$$(\partial \ln V_P / \partial T)_P = 1/2[(\partial \ln[K_S + (4/3)\mu] / \partial T)_P + \alpha], \qquad (5b)$$

lateral temperature fluctuation can be estimated from seismic tomography results. Assuming a relatively constant lateral variation in seismic velocities, viz $\delta \ln V_S = 1\%$ and $\delta \ln V_P = 0.5\%$, equations (5a and b) can be used to calculate lateral temperature variations in the lower mantle. For a pyrolite lower mantle, taking $|\partial K_S / \partial T| = 0.019$ and $|\partial \mu / \partial T| = 0.024$ GPa, the calculated lateral temperature variations are plotted in Figure 4 for the entire lower mantle. Figure 4 also reveals that using equations (5a) and (5b) produces somewhat different predictions in the temperature variations because the values of $\delta \ln V_P$ and $\delta \ln V_S$ were chosen somewhat arbitrarily; however, the results are generally consistent within the magnitude, which is about 100–250 K. These temperature variations are one order of magnitude smaller than previous estimates of YUEN et al. (1993), based on isothermal measurements of $(\partial \ln V / \partial \ln \rho)_T$ on MgO.

Lateral Temp. Variation

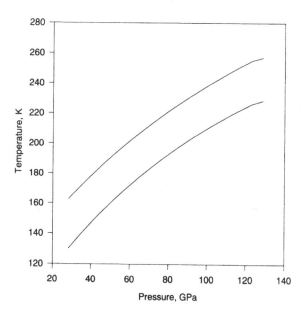

Figure 4
Estimated lateral temperature variations in the lower mantle. The upper curve is from equation (5a) and the lower one from (5b). See text for details.

One can also write relations for $(\partial \ln \rho / \partial \ln V_{P,S})_P$ by dividing equations (5a) and (5b) with $(\partial \ln \rho / \partial T)_P = -\alpha$ and obtain bounds for these ratios for the lower mantle (Table 3). These ratios should remain relatively constant since K_S and μ increase whereas α decreases with depth. It is remarkable that the absolute values can be directly compared with the results from geodynamic flow models that were employed to explain tomography and geoid data. For example, FORTE et al. (1993) concluded that for the lower mantle, the ratios $(\partial \ln \rho / \partial \ln V_P)_P$ and $(\partial \ln \rho / \partial \ln V_S)_P$ fall in the range 0.2–0.4 and 0.1–0.2, respectively, agreeing very well with 0.26–0.34 and 0.6–0.11 for the pyrolite model. In the geodynamic analyses, these ratios are free adjustable parameters and model dependent. For example, they are sensitive to the assumed viscosity profiles and convection styles in geodynamic models (e.g., HAGER and RICHARDS, 1989; FORTE et al., 1993). We are now able to explain these ratios in terms of thermoelastic properties of the mineral assembly of the lower mantle, without any additional assumptions such as partial melting, chemical inhomogeneity, or anelasticity.

6. Conclusions

We have combined mineral physics and seismology data to examine constraints on properties of the lower mantle. Assuming that the seismic tomographic observations of $v = 1.8$–2.0 reflect lateral temperature variations in the P- and S-wave velocities as well as density, a combination of seismic earth models and tomography data with laboratory mineral physics data places vigorous constraints on mineralogical models of the lower mantle. Our analysis indicates the following

(1) The PREM K_S, μ, and ρ profiles require that the mineral assemblage of the lower mantle have $|(\partial K_S / \partial T)_P| < 0.02$ and $|(\partial \mu / \partial T)_P| < 0.035$ GPa/K to satisfy the tomographic results of $v = 1.8$–2.0.

(2) For a lower mantle with $(Mg + Fe)/Si$ ratio ≈ 1, the required $|(\partial K / \partial T)_P|$ must be greater than 0.03 GPa, which requires unacceptably large $|(\partial \mu / \partial T)_P| > 0.055$ GPa if seismic tomography data are used to relate $(\partial K_S / \partial T)_P$ and $(\partial \mu / \partial T)_P$. Such large $|(\partial \mu / \partial T)_P|$ values are inconsistent with laboratory measurements.

(3) Acceptable composition models must have $(Mg + Fe)/Si$ ratios greater than 1.4. For the pyrolite model, $(Mg + Fe)/Si \approx 1.5$, a $|(\partial \mu / \partial T)_P| \sim 0.035$ GPa/K is required to satisfy both laboratory measurements on perovskite and seismic density, K_S, and μ profiles. Such large $|(\partial \mu / \partial T)_P|$ value is abnormal within silicates but is consistent with the ferroelastic nature of the perovskite (YEGANEH-HAERI et al., 1989a).

(4) Our estimates of $(\partial \mu / \partial T)_P$ and $(\partial K_S / \partial T)_P$ for the pyrolite lower mantle can be used to estimate several important parameters for geodynamic analyses. The lateral temperature fluctuation is estimated to be in the order of 100–250 K from

both compressional and shear velocities. The lateral density fluctuations relative to the velocities $[(\partial \ln \rho / \partial \ln V_P)_P$ and $(\partial \ln \rho / \partial \ln V_S)_P]$ are estimated to be about 0.1–0.2 and 0.2–0.4, respectively, in excellent agreement with previous analyses of dynamic tomography and geoid, independent from mineral physics constraints.

Acknowledgments

We thank I. Jackson and F. Guyot for critical reading of the manuscript, an anonymous reviewer for helpful comments, and W. J. Su for helpful discussions. This work is supported by the NSF grant EAR 8920239. MPI contribution No. 143.

REFERENCES

AHMED-ZAÏD, I., and MADON, M. (1991), *A High-pressure Form of Al_2SiO_5 as a Possible Host of Aluminium in the Lower Mantle*, Nature *353*, 426–428.

ANDERSON, D. L. (1987), *A Seismic Equation of State II. Shear Properties and Thermodynamics of the Lower Mantle*, Phys. Earth Planet. Int. *45*, 307–323.

ANDERSON, D. L. (1989), *Composition of the Earth*, Science *243*, 367–370.

ANDERSON, O. L., ISAAK, D., and ODA, H. (1992), *High-temperature Elastic Constant Data on Minerals Relevant to Geophysics*, Rev. Geophys. *30*, 57–90.

AGNON, A., and BUKOWINSKI, M. S. T. (1990), δ_S *at High Pressure and d* $\ln V_S/d \ln V_P$ *in the Lower Mantle*, Geophys. Res. Lett. *17*, 1149–1152.

AKAOGI, M., and ITO, E. (1993), *Heat Capacity of $MgSiO_3$ Perovskite*, Geophys. Res. Lett. *20*, 105–108.

BROWN, J. M., and SHANKLAND, T. J. (1981), *Thermodynamic Parameters in the Earth's Mantle as Determined from Seismic Profiles*, Geophys. J. Roy. Astr. Soc. *66*, 579–596.

BUTLER, R., and ANDERSON, D. L. (1978), *Equation of State Fits to the Lower Mantle and Outer Core*, Phys. Earth Planet. Int. *17*, 147–162.

CHOPELAS, A. (1992), *Sound Velocities of MgO to very High Compression*, Earth Planet. Sci. Lett. *114*, 185–192.

DAVIES, G. F., and DZIEWONSKI, A. M. (1975), *Homogeneity and Constitution of the Earth's Lower Mantle and Core*, Phys. Earth Planet. Int. *10*, 336–343.

DUFFY, T. S., and AHRENS, T. J., *Lateral variations in lower mantle seismic velocity*. In *High-pressure Research: Application to Earth and Planetary Sciences* (Syono, Y., and Manghnani, M. H., eds.) (AGU, Washington, D.C. 1992) pp. 197–205.

DZIEWONSKI, A. M., and ANDERSON, D. L. (1981), *Preliminary Reference Earth Model*, Phys. Earth Planet. Int. *25*, 297–356.

DZIEWONSKI, A. M. (1993), *Tomographic Image of the Earth's Interior: A Review*, EOS, Trans. AGU *74*, 572.

DZIEWONSKI, A. M., FORTE, A. M., SU, W. J., and WOODWARD, R. L., *Seismic tomography and geodynamics*. In *Relating Geophysical Structures and Processes*. The Jeffreys Volume (Aki, K., and Dmowska, R., eds. IUGG, vol. 16) (AGU, Washington, D.C. 1993) pp. 67–105.

FEI, Y., MAO, H. K., SHU, J., and HU, J. (1992), *P-V-T Equation of State of Magnesiowüstite $(Mg_{0.6}Fe_{0.4})O$*, Phys. Chem. Minerals *18*, 416–422.

FITZ GERALD, J. D., and RINGWOOD, A. E. (1991), *High Pressure Rhombohedral Perovskite Phase $Ca_2AlSiO_{5.5}$*. Phys. Chem. Minerals *18*, 40–46.

FORTE, A. M., DZIEWONSKI, A. M., and WOODWARD, R. J., *Aspherical structure of the mantle, tectonic plate motions, nonhydrostatic geoid, and tomography of the core-mantle boundary*. In *Dynamics of Earth's Deep Interior and Earth Rotation*, IUGG vol. 12 (LeMouel, J. L., Smylie, D. E., and Herring, T., eds.) (AGU, Washington, D.C. 1993) pp. 135–166.

GASPARIK, T. (1990), *Phase Relation in the Transition Zone*, J. Geophys. Res. *95*, 15751–15769.

GOTO, T., OHNO, I., and SUMINO, Y. (1976), *The Determination of the Elastic Constants of Natural Almandine-pyrope Garnet by Rectangular Parallelepiped Resonance Method*, J. Phys. Earth *24*, 149–156.

HAGER, B. H., and CLAYTON, R. W., *Constraints on the structure of mantle convection using seismic observations, flow models, and the geoid*. In *Mantle Convection* (Peltier, W. R., ed.) (Gordon and Breach Science Publishers 1989) pp. 657–763.

HAGER, B. H., and RICHARDS, M. A. (1989), *Long-wavelength Variations in Earth's Geoid: Physical Models and Dynamics Implications*, Phil. Trans. Roy. Soc. Lond. *A328*, 309–327.

HEMLEY, R. J., STIXRUDE, L., FEI, Y., and MAO, H. K., *Constraints on lower mantle composition from P-V-T measurements of (Fe, Mg)SiO$_3$ perovskite and (Fe, Mg)O*. In *High-pressure Research: Application to Earth and Planetary Sciences* (Syono, Y., and Manghnani, M. H., eds.) (AGU, Washington, D.C. 1992) pp. 183–189.

IRIFUNE, T., FUJINO, K., and OHTANI, E. (1991), *A New High-pressure Form of MgAl$_2$O$_4$*, Nature *349*, 409–411.

IRIFUNE, T. (1994), *Absence of an Aluminous Phase in the Upper Part of the Earth's Lower Mantle*, Nature *370*, 131–133.

ISAAK, D. G., ANDERSON, O. L., and COHEN, R. E. (1992), *The Relationship between Shear and Compressional Velocities at High Pressures: Reconciliation of Seismic Tomography and Mineral Physics*, Geophys. Res. Lett. *19*, 741–744.

ITO, E., and MATSUI, Y. (1978), *Synthesis and Crystal Chemical Characterization of MgSiO$_3$ Perovskite*, Earth Planet. Sci. Lett. *38*, 443–450.

JACKSON, I., and NESLER, H., *The elasticity of periclase to 3 GPa and some geophysical implications*. In *High Pressure Research in Geophysics* (Akimoto, S., and Manghnani, M. H., eds.) (Center for Academic Publications, Tokyo 1982) pp. 93–113.

JEANLOZ, R., and KNITTLE, E., *Reduction of mantle and core properties to a standard state by adiabatic decompression*. In *Chemistry and Physics of Terrestrial Planets* (Saxena, S. K., ed.) (Springer-Verlag, New York 1986) pp. 275–309.

KARATO, S. (1993), *Importance of Anelasticity in the Interpretation of Seismic Tomography*, Geophys. Res. Lett. *20*, 1623–1626.

KNITTLE, E., and JEANLOZ, R. (1987), *Synthesis and Equation of State of (Mg, Fe)SiO$_3$ Perovskite to over 100 Gigapascals*, Science *255*, 1238–1240.

LIU, L. G. (1975), *Post-oxide Phases of Forsterite and Enstatite*, Geophys. Res. Lett. *2*, 417–419.

LIU, L. G., and RINGWOOD, A. E. (1975), *Synthesis of a Perovskite-type Polymorph of CaSiO$_3$*, Earth Planet. Sci. Lett. *14*, 1079–1082.

MADON, M., CASTEX, J., and PEYRONNEAU, J. (1989), *A New Aluminocalcic High-pressure Phase as a Possible Host of Calcium and Aluminium in the Lower Mantle*, Nature *342*, 422–425.

MAO, H. K., CHEN, L., HEMLEY, R. J., JEPHCOAT, A. P., and WU, Y. (1989), *Stability and Equation of State of CaSiO$_3$ Perovskite to 134 GPa*, J. Geophys. Res. *94*, 17889–17894.

MAO, H. K., HEMLEY, R. J., FEI, Y., SHU, J. F., CHEN, C., JEPHCOAT, A. P., WU, Y., and BASSETT, W. A. (1991), *Effect of Pressure, Temperature and Composition on Lattice Parameters and Density of (Fe, Mg)SiO$_3$ Perovskites to 30 GPa*, J. Geophys. Res. *96*, 8069–8079.

POIRIER, J. P. *Introduction to the Physics of the Earth's Interior* (Cambridge University, Cambridge 1991).

RINGWOOD, A. E., and MAJOR, A. (1917), *Synthesis of Majorite and Other High Pressure Garnets and Perovskites*, Earth Planet. Sci. Lett. *12*, 411–418.

SAMMIS, C., ANDERSON, D. L., and JORDAN, T. (1975), *Application of Isotropic Finite Strain Theory to Ultrasonic and Seismological Data*, J. Geophys. Res. *75*, 4478–4480.

SPETZLER, H. (1970), *Equation of State of Polycrystalline and Single-crystal MgO to 8 Kilobars and 800 K*, J. Geophys. Res. *75*, 2073–2087.

STIXRUDE, L., HEMLEY, R. J., FEI, Y., and MAO, H. K. (1992), *Thermoelasticity of Silicate Perovskite and Magnesiowüstite and Stratification of the Earth's Mantle*, Science *251*, 410–413.

SU, W. J., and DZIEWONSKI, A. M. (1994), *Joint 3-D conversion for P and S Velocity in the Mantle*, EOS, Trans. AGU *74*, 557.

SUMINO, Y., and ANDERSON, O. L., *Elastic constants of minerals*. In *Handbook of Physical Properties of Rocks*, vol III (Carmichael, R. S., ed.) (CRC Press, Florida 1984) pp. 39–138.

WANG, Y., WEIDNER, D. J., LIEBERMANN, R. C., and ZHAO, Y. (1994), *P-V-T Equation of State of (Mg, Fe)SiO₃ Perovskite: Constraints on Composition of the Lower Mantle*, Phys. Earth Planet. Int. *83*, 13–40.

WEIDNER, D. J., WANG, Y., and YEGANEH-HAERI, A. (1993), *Equation of State Properties of Mantle Perovskites*, EOS, Trans. AGU *74*, 571.

WENG, K., MAO, H. K., and BELL, P. M. (1982), *Lattice Parameters of the Perovskite Phase in the System MgSiO₃-CaSiO₃-Al₂O₃*, Carnegie Institute of Washington Yearbook *81*, 273–277.

WOODHOUSE, J. H., and DZIEWONSKI, A. M. (1989), *Seismic Modeling of the Earth's Large-scale Three-dimensional Structure*, Phil. Trans. Roy. Soc. Lond. *A328*, 291–303.

WOODHOUSE, J. H., DZIEWONSKI, A. M. GIARDINI, D., LI, X. D., and MORELLI, A. (1987), *The Emerging Three-dimensional Structure of the Earth: Results from the Modeling of Wave Forms, Free Oscillation Spectra and Seismic Travel Times*, EOS, Trans. AGU *68*, 356.

YAGI, T., FUNAMORI, N., and UTSUMI, W. (1993), *Stability and Thermal Expansion of Orthorhombic Perovskite Type MgSiO₃ under Lower Mantle Condition*, EOS, Trans. AGU *74*, 571–572.

YEGANEH-HAERI, A., WEIDNER, D. J., and ITO, E. (1989a), *Elasticity of MgSiO₃ in the Perovskite Structure*, Science *243*, 787–789.

YEGANEH-HAERI, A., WEIDNER, D. J., and LIEBERMANN, R. C. (1989b), *High Temperature Brillouin Spectroscopy*, EOS, Trans. AGU *70*, 474.

YEGANAH-HAERI, A. (1994), *Synthesis and Re-investigation of the Elastic Properties of Single-crystal Magnesium Silicate Perovskite*, Phys. Earth Planet. Int. *87*, 111–121.

YUEN, D. A., CADEK, O., CHOPELAS, A., and MATYSKA, C. (1993), *Geophysical Inferences of Thermal-chemical Structures in the Lower Mantle*, Geophys. Res. Lett. *20*, 899–902.

YUSA, H., YAGI, T., YAMANAKA, M., TAKEMURA, K., and SHIMOMURA, *A new non-quenchable high-pressure polymorph of Ca₃Al₂Si₃O₁₂* (abstract in Japanese). In *The 34th Japanese High Pressure Conference, Programs and Abstracts, Special Issue of the Review of High Pressure Science and Technology*, vol. 2 (Tokyo 1993) pp. 374–375.

ZHAO, Y., and ANDERSON, D. L. (1994), *Mineral Physics Constraints on the Chemical Composition of the Earth's Lower Mantle*. Phys. Earth Planet. Int. *85*, 273–292.

(Received January 23, 1995, revised March 24, 1995, accepted May 24, 1995)

PAGEOPH, Vol. 146, Nos. 3/4 (1996)

0033–4553/96/040551–22$1.50 + 0.20/0

Thermal Convection in a Cylindrical Annulus with a non-Newtonian Outer Surface

STUART A. WEINSTEIN

Abstract — This study presents the results of numerical simulations of a model for lithosphere-mantle coupling in a terrestrial type planet. To first order, a geologically active terrestrial type planet may consist of a metallic core, silicate mantle and lithosphere, with the lithosphere being rheologically different from the mantle. Therefore we have developed a numerical model consisting of a thin non-Newtonian fluid hoop that is dynamically coupled to a thick Newtonian fluid cylindrical annulus. Thus the rheological dichotomy between mantle and lithosphere is built into the model. Time-dependent calculations show the existence of at least two regimes of behaviors. In one regime, the behavior of the hoop switches between periods characterized by low or high speeds, in response to changes in convective vigor and planform. This regime may apply to the planet Venus where the available evidence indicates that prior to 500 myr ago, the planet was resurfaced on a time scale of <100 myr. Since that time, large-scale tectonic activity on Venus has been sharply curtailed. In the other regime, which is more like plate tectonics on Earth, the hoop speeds rise and fall on short time scales.

Key words: Cylindrical annulus, thermal convection, radiogenic isotopes, fluid hoop, surface deformation.

Introduction

To first order, a model of a geologically active terrestrial type planet consists of a thin lithosphere, a thick convecting mantle and a metallic core. The style of deformation that occurs in the lithosphere depends largely on the rheological properties of the lithosphere and the structure of thermal convection in the mantle. Two terrestrial planets, Venus and Earth, are similar in several respects such as size, mean density, relative position in the solar system, and both have an abundance of radiogenic isotopes (SURKOV *et al.*, 1987). Yet, the styles of surface deformation on the two planets are quite different.

Since the mid-1960s, geophysicists have realized that the Earth's lithosphere is comprised of mobile units called plates. Studies of plate kinematic data (e.g., DEMETS *et al.*, 1990; MINSTER and JORDAN, 1978), such as transform fault and earthquake slip vector azimuths, and ridge spreading rates show that to a good

---

[1] Department of Geology and Geophysics, SOEST, University of Hawaii, Honolulu, HI 96822, U.S.A.

approximation, the plates behave as rigidly rotating spherical caps except for narrow zones of deformation at their boundaries. Plate interiors tend to show little deformation (apart from elastic responses to surface loads) even though they have ages on the order of 100 myr. Analyses of post-glacial rebound show that the sublithospheric mantle deforms over time scales of only 10,000 years (CATHLES, 1975; PELTIER, 1980). Thus the mechanical behavior of the plates is fundamentally different from the sublithospheric mantle. This rheological dichotomy gives plates the ability to act as stress guides (ELSASSER, 1969) and modulate mantle convection through tractions generated at the base of the plate.

The Venusian lithosphere on the other hand, does not appear to be broken up into plates (SOLOMON et al., 1992; SOLOMON, 1993). The surface of Venus exhibits a variety of deformational features such as ridge belts, mountain belts, pancake domes, coronae and grid terrains. Deformation of the Venusian surface rocks exists on many scales and is widespread. Another principal difference between the Venusian and terrestrial surface is age. Recent studies suggest the age of the Venusian surface is on the order of 500 myr and is fairly uniform (PHILLIPS et al., 1992; STROM et al., 1994).

One of the principal goals of geodynamics has long been to understand the manner in which plate tectonics is coupled to a global system of thermal convection in the Earth's mantle. Perhaps a more general goal is to understand the fundamental dynamics of lithosphere-mantle interaction. In other words, geodynamicists should also seek to explain why the tectonic styles of Earth and Venus are so different, despite similarities between the two planets. Any successful model for studying the coupling of a planetary lithosphere to a convecting mantle must address the dichotomy in rheological behavior between the lithosphere and sublithospheric mantle. This study examines a computational model in which this dichotomy is built in.

Background

Previous attempts to model the behavior of plates in a fully dynamic manner have centered on the use of temperature and non-Newtonian rheology. DAVIES (1988) examined two-dimensional, time-dependent thermal convection in a fluid layer with a temperature-dependent viscosity. The idea driving the use of temperature-dependent rheology is to create a stiff cold boundary layer like the Earth's lithosphere. However, as DAVIES (1988) shows, temperature-dependent rheology makes the cold boundary layer less mobile and not more plate-like because it fails to concentrate the deformation in a small region, resulting in weak zones that could mobilize the boundary layer. In other words, convection with temperature-dependent rheology is unable to manufacture realistic plates. JACOBY and SCHMELING (1982) showed that weak zones at plate boundaries is a necessary ingredient for plate-like behavior.

Model
non-Newtonian

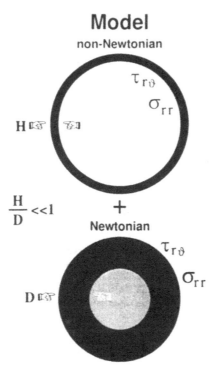

Figure 1
Schematic of the coupled non-Newtonian hoop—Newtonian annulus model.

Numerical studies by CSEREPES (1982), CHRISTENSEN (1983, 1984), KING and HAGER (1990), CHRISTENSEN and HARDER (1991), WEINSTEIN and OLSON (1992), BERCOVICI (1993, 1995), which examined the effects of non-Newtonian rheology indicate that non-Newtonian rheology may result in plate-like behavior. Non-Newtonian rheology can become very weak where stresses or strain rates are large. This property allows the cold boundary layer to be mobilized while the deformation is concentrated in narrow weak zones. Away from these zones the cold boundary layer is effectively rigid.

Model

The model examined in this study consists of a thin, incompressible, non-Newtonian fluid hoop that is dynamically coupled to the outer surface of an infinite Prandtl number, Boussinesq, Newtonian fluid cylindrical annulus (see Fig. 1). The hoop and annulus are assumed to have constant radial thicknesses of H and D respectively with $H/D \ll 1$. Physical properties other than rheology, such as thermal expansivity, thermal diffusivity, etc. (see Table 1, for symbol definitions) are the

Table 1

Symbol Definitions

Symbol	Definition
α	Thermal expansivity
ε	Internal heat generation rate
κ	Thermal diffusivity
μ_m	Annulus viscosity
μ_p	Hoop stiffness constant
ρ	Zero temperature density
$\sigma_{r\theta}$	Shear stress
$\sigma_{\theta\theta}$	Hoop stress
ΔT	Imposed temperature difference
θ	Polar angle
Ψ	Streamfunction
g	Acceleration of gravity
r	Radius
D	Annulus thickness
H	Hoop thickness
J	Jacobian
K	Thermal conductivity
N	Power-law exponent
Ra	Rayleigh number
Ra_h	Internal heat Rayleigh number
T	Temperature
U_θ	Hoop velocity

same for the two regions. The non-Newtonian hoop is dynamically coupled to the Newtonian annulus through the normal and shear tractions σ_{rr}, $\tau_{r\theta}$, generated on the base of the thin hoop by thermal convection in the Newtonian annulus. In the numerical model, the hoop is represented as a massless body with a negligibly small heat capacity. In essence, the effects of the hoop are reduced to a boundary condition on the outer boundary vorticity of the Newtonian annulus.

Model Equations

The flow in the non-Newtonian hoop is treated as though it were one-dimensional. This approximation, also known as the shallow layer approximation, is valid if $H/D \ll 1$. The one-dimensional flow is governed by the homogeneous equation of motion:

$$\frac{1}{r}\frac{\partial}{\partial r} r^2 \tau_{r\theta} + \frac{\partial \sigma_{\theta\theta}}{\partial \theta} = 0. \tag{1}$$

Averaging equation (1) over the thickness H of the thin hoop, we obtained the shallow layer equation,

$$\frac{1}{R_o}\frac{\partial \sigma_{\theta\theta}}{\partial \theta} - \frac{1}{H}\tau_{r\theta}|_{r=R_o} = 0 \tag{2}$$

where R_o is the outer radius of the annulus, $\sigma_{\theta\theta}$ is the vertically averaged hoop stress and $\tau_{r\theta}|_{r=R_o}$ is the shear traction on the base of the non-Newtonian hoop. Equation (2) is our equation of motion for the thin hoop.

The non-Newtonian hoop is assumed to have a power-law rheology with the following form:

$$\sigma_{\theta\theta} = -P + 2\mu\left[\frac{1}{r}\frac{\partial U_\theta}{\partial \theta} + \frac{U_r}{r}\right] \tag{3}$$

where

$$\mu = \mu_p(|\sigma^*_{\theta\theta}|)^{-(N-1)}. \tag{4}$$

In these expressions U_θ is the vertically averaged azimuthal velocity, P is the pressure, $\sigma^*_{\theta\theta}$ is the dimensionless hoop stress, μ_p is the hoop stiffness constant which has the dimensions of dynamic viscosity and N is the power-law index. The U_r/r term in (3) can be eliminated as the top of the Newtonian layer is assumed to be impermeable. The pressure can be related to the radial normal stress, σ_{rr}, at the outer surface of the Newtonian annulus:

$$\left[P = -\sigma_{rr} + 2\mu\frac{\partial U_r}{\partial r}\right]_{r=R_o}. \tag{5}$$

From the continuity equation

$$\frac{1}{r}\left[U_r + r\frac{\partial U_r}{\partial r}\right] + \frac{1}{r}\frac{\partial U_\theta}{\partial \theta} = 0 \tag{6}$$

we find that

$$\left[r\frac{\partial U_r}{\partial r} = -\frac{\partial U_\theta}{\partial \theta}\right]_{r=R_o} \tag{7}$$

where U_r is the radial velocity in the annulus. Thus we can now express the pressure in the following way:

$$\left[P = -\sigma_{rr} - \frac{2\mu}{r}\frac{\partial U_\theta}{\partial \theta}\right]_{r=R_o} \tag{8}$$

Our relation between the hoop velocity and hoop stress is now

$$\sigma_{\theta\theta} = \left[\sigma_{rr} + \frac{4\mu}{r}\frac{\partial U_\theta}{\partial \theta}\right]_{r=R_o}. \tag{9}$$

Adopting the following dimensionless variables,

$$U = U^*\frac{\kappa}{R_o} \quad \sigma = \sigma^*\frac{\mu_m\kappa}{R_o^2} \quad \mu = \mu^*\mu_m \quad H = R_oH^* \tag{10}$$

where κ and μ_m are the thermal diffusivity and viscosity of the annulus respectively, we can put equations (2), (9) in the dimensionless forms (with the asterisks dropped):

$$\frac{\partial \sigma_{\theta\theta}}{\partial \theta} = \frac{\tau_{r\theta}}{H}\bigg|_{r=1} \tag{11}$$

$$\sigma_{\theta\theta} = \left[\sigma_{rr}|_{r=1} + 4\mu_p(|\sigma_{\theta\theta}|)^{-(N-1)} \frac{\partial U_\theta}{\partial \theta} \right]. \tag{12}$$

Integrating equations (11), (12) we obtain

$$\sigma_{\theta\theta}(\theta) - \sigma_{\theta\theta}(0) = \frac{1}{H} \int_0^\theta \tau_{r\theta}|_{r=1}\, ds \tag{13}$$

$$U_\theta(\theta) = \frac{1}{4\mu_p} \int_0^\theta \frac{\sigma_{\theta\theta}(0) - \sigma_{rr}|_{r=1}}{(|\sigma_{\theta\theta}|)^{-(N-1)}}\, ds + C. \tag{14}$$

In the study, we will ignore the contribution from σ_{rr}, which acts as a driving force for the thin hoop. The constant $\sigma_{\theta\theta}(0)$ is determined by enforcing periodicity in (14). The equation for $\sigma_{\theta\theta}(0)$ is

$$\int_0^{2\pi} \left[\frac{\left[\sigma_{\theta\theta}(0) + \frac{1}{H} \int_0^s \tau_{r\theta}|_{r=1}\, dx \right]}{\left[|\sigma_{\theta\theta}(0) + \frac{1}{H} \int_0^s \tau_{r\theta}|_{r=1}\, dx| \right]^{-(N-1)}} \right] ds = 0. \tag{15}$$

The constant C is determined by applying the no-net torque condition which in this case requires the mean velocity of the thin hoop to vanish. The no-net torque condition prevents the hoop from imparting a spin to the Newtonian annulus. Thus our expression for the velocity is:

$$U'_\theta(\theta) = \left(U_\theta(\theta) - \frac{1}{2\pi} \int_0^{2\pi} U_\theta(s)\, ds \right) = \frac{1}{4\mu_p} \int_0^\theta \frac{\sigma_{\theta\theta}(s)}{|\sigma_{\theta\theta}(s)|^{-(N-1)}}\, ds. \tag{16}$$

The governing equations for two-dimensional time-dependent thermal convection in a Boussinesq, infinite Prandtl number, Newtonian fluid cylindrical annulus (in dimensionless form) are

$$\nabla^4 \Psi = \frac{1}{r} Ra \frac{\partial T}{\partial \theta} \tag{17}$$

$$\frac{\partial T}{\partial t} + \frac{1}{r} J(\Psi, T) = \nabla^2 T + \frac{Ra_h}{Ra} \tag{18}$$

where T is the temperature, Ψ is the streamfunction and J is the Jacobian. Ra and Ra_h are the basal and internal heating Rayleigh number with the definitions

$$Ra = \frac{g\alpha \Delta T R_o^3}{\kappa\nu}, \quad Ra_h = \frac{\rho g\varepsilon\alpha R_o^5}{\kappa\nu K} \tag{19}$$

where g is the gravity, α is the thermal expansivity, ΔT is the imposed temperature difference across the annulus, v is the kinematic viscosity of the annulus, ρ is the density, ε is the internal heat generation rate and K is the thermal conductivity. The boundary condition consists of isothermal, impermeable inner and outer boundaries with the inner boundary being stress-free and the velocity on the outer boundary given by (16). Expressed in terms of T, Ψ and U'_θ they are:

$$\frac{\partial^2 \Psi}{\partial r^2} = \frac{1}{r} \frac{\partial \Psi}{\partial r}, \quad \Psi = 0, \quad T = 1, \quad (r = .5) \tag{20}$$

$$\frac{\partial \Psi}{\partial r} = -U'_\theta, \quad \Psi = 0, \quad T = 0, \quad (r = 1.0). \tag{21}$$

The finite-difference method of WEINSTEIN and OLSON (1992) is used for obtaining solutions to (16)–(21). In their method, Ψ is split into two components Ψ_C, Ψ_B. Ψ_C satisfies the thermally driven flow subject to a rigid outer boundary and a stress-free inner boundary. Ψ_B is the flow driven by the velocity boundary condition on the outer surface. This decomposition is valid because of cylindrical annulus is Newtonian. Iteration is used to find the traction, $\tau_{r\theta}$ at the base of the non-Newtonian hoop for which the momentum fields of the non-Newtonian hoop and Newtonian annulus are in equilibrium. The solution procedure within a time-step consists of the following steps:

1. Compute Ψ_C for the current temperature distribution subject to the no-slip outer boundary condition.
2. Compute the shear stress $\tau_{r\theta}$ from Ψ_C on the rigid boundary. Now start to iterate.
3. Compute velocity in the non-Newtonian hoop using $\tau_{r\theta}$ in step 2.
4. Compute Ψ_B using the velocity obtained in step 3.
5. Add Ψ_C, Ψ_B and recalculate $\tau_{r\theta}$ on the base of the non-Newtonian hoop. Is the change from the last iteration above a specified tolerance? If YES return to step 3. If NO, equilibrium is achieved and move on to step 6.
6. Update temperature field using $\Psi_C + \Psi_B$.

The numerical resolution used for the annulus consists of 512 nodes in azimuth and 81 nodes in radius.

This model has several deficiencies with respect to the Earth or Venus. For starters, the lithosphere is nonrecyclable, i.e., it is neither manufactured nor subducted. Furthermore, the rigid push force is left out of the formulation and slab pull in this model is not quite the same as it is in a realistic geologic setting. The rigid push force arises from the elevation of the mid-ocean ridges above the sea floor which establishes a pressure head that acts to drive the lithosphere laterally away from the ridges (HAGER and O'CONNELL, 1981). However, as shown in WEINSTEIN and OLSON (1992), inclusion of the ridge push force, while it does allow the lithosphere to attain higher speeds, does not introduce substantially different dynamics into the behavior of the system. The slab pull force arises from a continuation of the stress guide effect which transmits the stress due to the pull of the slab to the lithosphere away from the subduction zone. In this model, the lithosphere (thin hoop) does not subduct and there is no stress guide effect. However, downwellings still exert a stress on the thin hoop through basal tractions. The last major deficiency is the restriction to two dimensions. As a result, this model is unable to address issues such as the generation of toroidal flow in the mantle.

Results

All of the calculations presented in this study use a dimensionless hoop thickness of 0.1, a dimensionless Newtonian annulus thickness of .5 and basal and internally heated Rayleigh numbers of 2.5×10^6 and 1×10^8, respectively. Internal

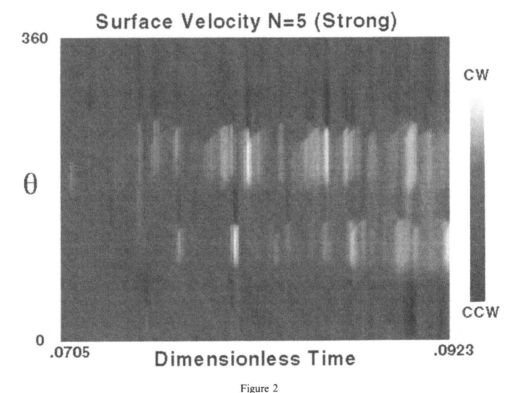

Figure 2
The evolution of the surface velocity (flow in the non-Newtonian hoop) for the calculation in which $N = 5$ and $\log \mu_p = 24.25$. The thin arrow points to 0.08 dimensionless time. CW and CCW stand for clockwise and counter-clockwise, respectively. Dimensionless time increases to the right and the polar angle increases upwards. The peak dimensionless speed is 558.

Basal Shear Stress N=5 (Strong)

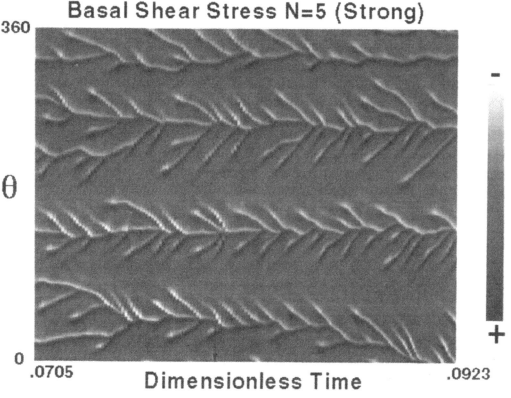

360

θ

0

.0705

.0923

Dimensionless Time

Figure 3
The evolution of the basal shear stress (drag on the base of the hoop) for the calculation in which $N = 5$
and $\log \mu_p = 24.25$.

heat generation accounts for 70% to 80% of the surface heat flow. All of the calculations use the same initial conditions. In order to avoid the high computational costs associated with spin up, thermal convection in the annulus was allowed to develop under a rigid boundary. Computation is considerably less costly when the effects of the deforming hoop are not included. After the Newtonian annulus had thermally equilibrated, the calculation was stopped and the temperature field of the last time-step is used as the initial field for all of the calculations presented below.

The governing equations and boundary conditions (16)–(21) were integrated for at least 25 transit times, and as much as 45 transit times for the longer simulations. The transit time is defined as the annulus thickness divided by the average maximum velocity of the flow. Clearly, this is not enough to fully appreciate the long-term behavior of the system. However, these simulations do show how the non-Newtonian hoop responds to changes in convection planform and vigor, and how these responses vary with hoop rheology.

Modest Power-law

The first calculation to be presented used a hoop with a power-law index of 5 and hoop stiffness constant of $\log \mu_p = 24.25$. Figures 2–4 show the evolution of the surface velocity (velocity of the non-Newtonian layer), basal shear stress ($\tau_{r\theta}$) and hoop stress ($\sigma_{\theta\theta}$) fields. The patterns in the basal shear stress field are the *footprints* of thermal convection in the annulus on the base of the hoop. In the shear stress field there are several linear structures which exist for most of, if not the entire duration of the calculation. Near a dimensionless time of 0.08 for example, four such structures are present. These linear structures represent the time history of downwelling centers which maintain a near continuous existence throughout the calculation. Intersecting these primary linear structures are smaller, short-lived tributary structures. These are due to drifting, downwelling, instabilities of the cold, outer boundary layer of the annulus. The instabilities are carried by the large-scale flow to the primary downwellings. Upwellings in these calculations are weak, due to

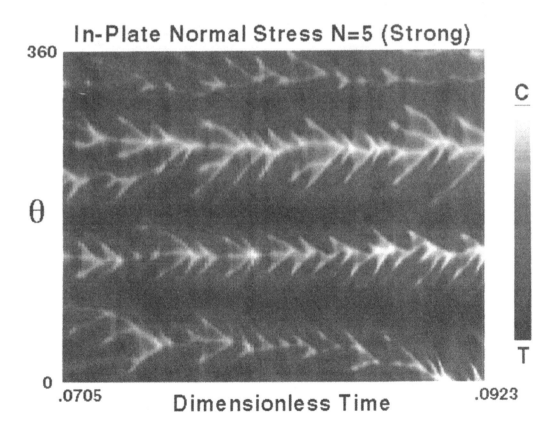

Figure 4
The evolution of the normal stress (hoop stress) for the calculation in which $N = 5$ and $\log \mu_p = 24.25$.

the large amounts of internal heat generation, and therefore their contributions to the basal shear stress are greatly overshadowed by the downwellings. The normal stress field displays a similar pattern. The primary downwellings and drifting instabilities are the loci of regions of the hoop that are in a state of deviatoric compression. Broad regions of tension exist in between the downwellings. The regions of tension lack sharp features because they are produced passively by the downwellings on either side, whereas the regions of compression are the direct result of downwellings.

The merging of drifting instabilities with the primary downwelling centers results in short-lived periods where the velocity of the non-Newtonian hoop near the merger experiences a large increase. The extra buoyancy from the merger results in greater shear and normal stresses. This is quite evident in Figure 4, where the brightest patches in the normal stress field are at the loci of the mergers. It should be noted, that a velocity increase due to the mergers is not unique to non-Newtonian rheology. An increase would occur if the hoop were Newtonian. However, the non-Newtonian rheology produces greater increases. This point will be addressed further in the next section. In this calculation the rheology of the hoop was made stiff enough (i.e., μ_p made large enough) so that at the beginning of the calculation the maximum hoop speed was only about 50. This represents only a few percent of the maximum hoop speed that would occur if the hoop had been inviscid (i.e., the outer boundary of the annulus is stress-free). In addition to the merging of drifting instabilities with primary downwellings, primary downwellings occasionally merge with each other. These types of mergers sometimes produce more spectacular velocity increases and result in a transition to a regime in which the maximum hoop speeds are consistently greater than before. This transition occurs for two reasons.

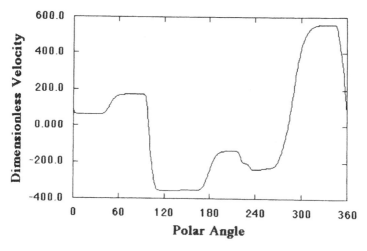

Figure 5
Surface velocity profile at a dimensionless time of 0.904 obtained from the calculation in which $N = 5$ and $\log \mu_p = 24.25$.

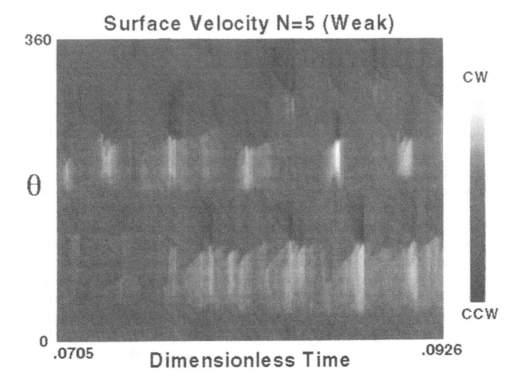

Figure 6
The evolution of the surface velocity for the calculation in which $N = 5$ and $\log \mu_p = 22.97$. The peak dimensionless speed is 1505.

The first is simply geometrical. When the primary downwellings merge, the convection planform shifts to smaller wavenumbers. Simple analysis (see appendix) shows that the wavelengths of the planform can have a large influence on the magnitude of the hoop speed with smaller wavenumbers being more efficient at exciting flow in the hoop. The second reason is that with fewer downwelling centers, drifting instabilities tend to initiate further away from the primary downwellings. Thus when they eventually merge with the primary downwellings, they are more developed and contribute a greater amount of buoyancy, resulting in larger hoop speed increases.

An example of a surface velocity profile, corresponding to $t = 0.90476$, is shown in Figure 5. This profile is reproduced as a simple graph because it delineates more clearly certain aspects of the velocity field than can be discerned from Figure 2. This profile shows that several sections of the hoop are traveling at near uniform speeds. These sections are the analogs of tectonic plates and manifest the ability of the non-Newtonian rheology to concentrate deformation into narrow regions. These plate-like structures cannot be generated with Newtonian rheology. One should

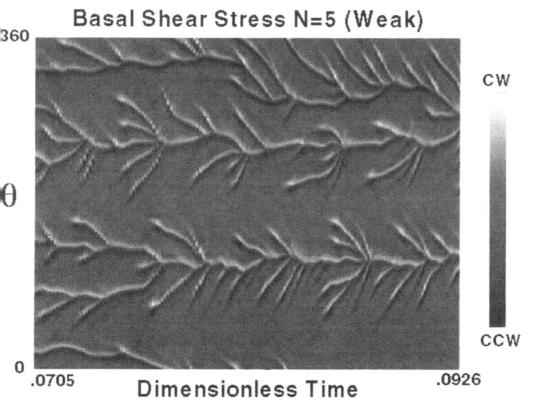

Figure 7
The evolution of the basal shear stress for the calculation in which $N = 5$ and $\log \mu_p = 22.97$.

note that the margins between the plate-like structures are sometimes large. In the case of the Earth, plate margins tend to be very small compared to the dimensions of the plate themselves. Smaller margins may be achieved (better plate-like behavior) with large power-law exponents (WEINSTEIN and OLSON, 1992).

Figures 6 and 7 illustrate the evolution of the velocity and shear stress fields obtained from a calculation in which the power-law index is also 5, but with a smaller rheological constant $\log \mu_p = 22.97$. Although the shear stress distribution initially is the same as that in Figure 3, it quickly diverges. Because the rheology is weaker, the velocity field (Figure 6) shows a little more structure. The effects of drifting instabilities are more visible in the hoop velocity field. They produce oblique lineations in the velocity field that track the motions of the instabilities themselves.

Large Power-law

The evolution of the velocity and shear stress fields for a power-law index of 17 and a hoop stiffness constant of log $\mu_p = 89.31$ are presented in Figures 8–9. As in the first calculation, the rheology is made strong enough so that the mean hoop speed is low. Through the first half of the calculation there were only a few brief events in which appreciable peak speeds occurred. These resulted from the merging of drifting instabilities with downwelling centers. The most prominent of these events occur about midway through the calculation near an azimuth of 45°. At this point in time, several drifting instabilities merge at a primary downwelling. Subsequently, two primary downwellings which straddle an azimuth of 180° merge. Thereafter, an era of high peak speeds begins and continues for the duration of the calculation.

Figures 10a,b show snapshots of the temperature field taken from before and after the era of large hoop speeds commences. The temperature field shown in Figure 10a, which was obtained during the era of small plate speeds, contains several downwelling instabilities which are roughly evenly distributed. The temper-

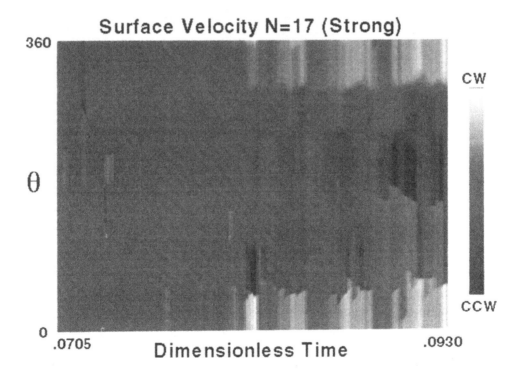

Figure 8
The evolution of the surface velocity for the calculation in which $N = 17$ and log $\mu_p = 89.31$. The peak dimensionless speed is 583.

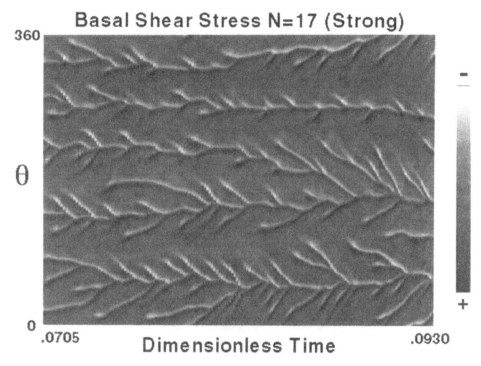

Figure 9

The evolution of the basal shear stress for the calculation in which $N = 17$ and $\log \mu_p = 89.31$.

ature field in Figure 10b illustrates the thermal structure in the Newtonian annulus when the hoop speed was at its maximum. In this case all but one of the instabilities are concentrated in the lower half of the annulus. The combination of the longer wavelength and the merging of drifting instabilities near an azimuth of 180° results in the high peak speed. The group of instabilities near an azimuth of 280° generates a clockwise motion in the hoop which produces a speed of about 1/2 the maximum speed.

The surface velocity profiles in Figures 10a,b are replotted in Figure 11. Now the difference in magnitude between the profiles can be fully appreciated. The mean speed of the profile from Figure 10b is 301, greater by a factor of 20 than the mean speed of the profile from Figure 10a. The increase in velocity is not simply due to an increase in the vigor of convection. In fact the r.m.s. shear stress obtained with the temperature field in Figure 10a (9326.0) is 7% less than the r.m.s. shear stress obtained with the temperature field in Figure 10b (8742.2). Clearly, a substantial difference in the mean hoop speed can be attributed to the difference in convection planform which is so evident in Figure 10.

For purposes of comparison, velocity profiles obtained from calculations in which the hoop is Newtonian and non-Newtonian with a power-law exponent of 5

Ra=2.5E6, R=40, N=17 (Strong)

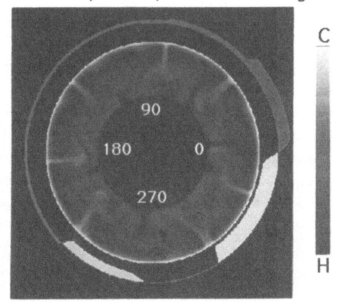

Ra=2.5E6, R=40, N=17 (Strong)

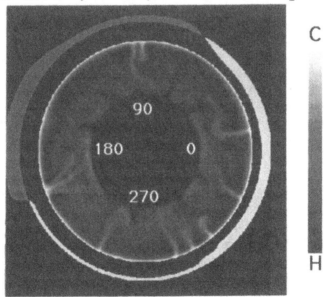

Figure 10

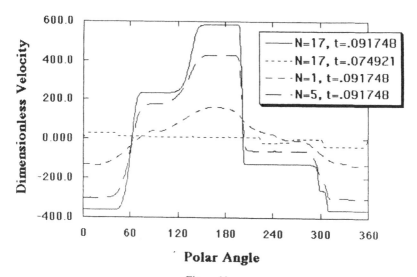

Figure 11

Plot of the surface velocity profiles obtained at times of 0.091748 and 0.074921. Two other curves are obtained using different rheologies ($N = 1,5$) for comparison.

are also plotted. Both of these profiles were calculated using the temperature field shown in Figure 10b. The Newtonian viscosity and the hoop stiffness constant for these cases were adjusted so that when the temperature field of Figure 10a was used, the mean speed of the hoop is the same as it is for the $N = 17$ case. The velocity profile obtained with the Newtonian hoop is devoid of plate-like structures and maintains a mean speed of only 1/4 of that obtained in the $N = 17$ case. The velocity profile for the $N = 5$ case is similar to the $N = 17$ profile, but the mean speed is 25% less and the plate-like structures are not as well defined. Aside from producing plate-like structures, non-Newtonian rheology can greatly enhance the velocity increases associated with changes in convection planform and vigor.

Figures 12 and 13 show the evolution of the velocity and shear stress fields for a calculation in which the power-law index is also 17, but the hoop stiffness constant is lowered to log $\mu_p = 81.98$. The integration was carried out over a shorter time span, as this calculation was more computationally expensive than the previous cases. The peak hoop speeds for this calculation exceed 1400. In this case,

Figure 10

Snapshots of the temperature field at $t = 0.074921$ when the mean hoop speed was small (a) and at $t = 0.091748$ when the mean hoop speed was large (b). The outer ring shows the velocity distribution. Light and dark areas indicate counter-clockwise and clockwise motions, respectively. The thickness indicates the speed scaled by the maximum speed obtained at that time. The small numbers in the interior indicate the polar angle.

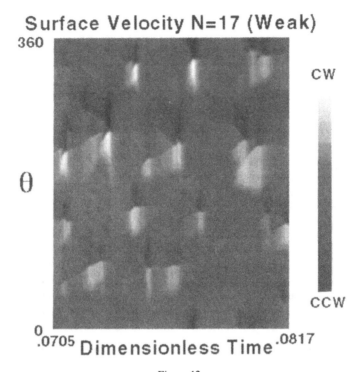

Figure 12
The evolution of the surface velocity for the calculation in which $N = 17$ and $\log \mu_p = 81.98$. The peak dimensionless speed is 1467.

the number of primary downwellings remain constant through time and the frequency of mergers is reduced. The shear stresses in this calculation are typically only 60% of those in the previous case. Such a reduction in shear may result in fewer boundary instabilities as the boundary becomes more stress-free. The velocity field is similar to Figure 6 in appearance and has a checker board like pattern. Sections of the hoop become highly mobile for time spans on the order of 0.001 dimensionless units, after which the velocity falls off sharply. This type of behavior appears to be fundamentally different than in the previous case.

Examples of individual velocity profiles are illustrated in Figure 14. These profiles were obtained from periods when the mean hoop speed was low and high. As is the case with non-Newtonian rheology, plate-like structures are readily visible. However, they tend not to be as well defined as the structures in the stiffer hoop. When the rheology is made softer, the plate-like structures become less distinct (WEINSTEIN and OLSON, 1992). The velocity profile from a calculation in which the hoop is Newtonian is also plotted in Figure 14. This calculation used the temperature field at $t = 0.079893$, which produced high hoop speeds in the non-Newtonian case. The Newtonian viscosity is 32, and for the temperature field at $t = 0.079893$,

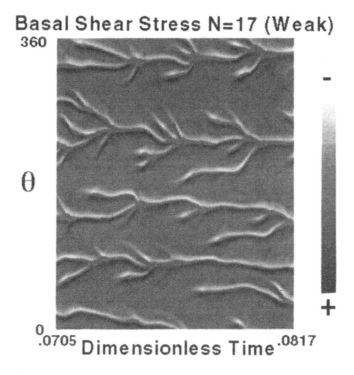

Basal Shear Stress N=17 (Weak)

Figure 13
The evolution of the basal shear stress for the calculation in which $N = 17$ and $\log \mu_p = 81.98$.

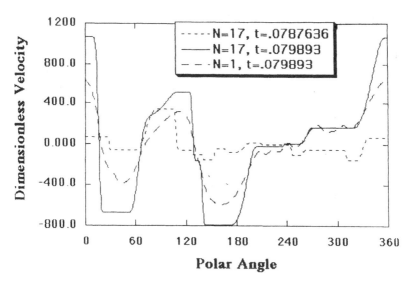

Figure 14
This plot shows the surface velocity profiles obtained at times of 0.0787636 and 0.07893. The curve, composed of long dashes, is obtained using a Newtonian rheology for the hoop.

the mean hoop speed is the same as found for the non-Newtonian hoop. Just as in the previous case, the Newtonian hoop is devoid of plate-like structures. However, the increase in hoop speed for the non-Newtonian case is only 70% larger than the increase found with Newtonian rheology. The smaller difference between the Newtonian and non-Newtonian rheology, with respect to the increase in hoop speed, is the result of the rheology being weaker. As the rheology is made weaker (μ_p is decreased), the potential for a substantial increase in velocity is less because the hoop is already behaving more like a stress-free boundary for either Newtonian or non-Newtonian rheology.

Conclusions

These calculations indicate the existence of at least two basic regimes of time-dependent behavior for the coupled non-Newtonian hoop—Newtonian cylindrical annulus system. These are the *stop and go* and the *transition* regimes.

The *stop and go* regime is found when the hoop rheology is relatively weak (smaller values of μ_p). In this regime, sections of the hoop become highly mobile during mergers of boundary-layer instabilities and then fall off rapidly after the mergers. This process occurs on relatively short time scales. When the rheology is stronger (larger values of μ_p) the *transition* regime is found. In the transition regime, changes in the planform of thermal convection in the Newtonian annulus may bring about periods inwhich the hoop speeds increase (or decrease). These calculations also suggest that non-Newtonian rheology can greatly amplify the changes in the mobility of the lithosphere that occur with changes in the vigor or planform of convection. Of the two regimes, the *stop and go* regime may be the most earth-like.

Studies of impact craters and other impact related features suggest a large mean age for the Venusian surface. The Venusian surface is perhaps 500 myr old, and a uniform age for the surface is not inconsistent with the available surface age data (PHILLIPS et al., 1992; STROM et al., 1994). Thus prior to 500 myr ago, the Venusian surface was apparently recycled on time scales of < 100 myr. The transition regime offers a possible mechanism for this type of tectonic history. A decrease in the vigor of Venusian convection or a change in the planform of convection (or both) may have led to a sharp decline, due to the effects of non-Newtonian rheology, in the mobility of the Venusian lithosphere.

Appendix: Effects of Wavelengths

Consider a shear stress distribution with the form:

$$\tau_{r\theta} = A \sin(k\theta) \tag{22}$$

where k is the wavenumber and A is the amplitude. Substitution of this expression in (16) yields:

$$\int_0^{2\pi} \frac{[\sigma_{\theta\theta}(0) + \frac{A}{kH}(1. - \cos(k\theta))]}{\left[|\sigma_{\theta\theta}(0) + \frac{A}{kH}(1. - \cos(k\theta))|\right]^{-(N-1)}} \, d\theta = 0. \tag{23}$$

After integrating the above expression one finds that for integer values of N such that $N \geq 1$, $\sigma_{\theta\theta}(0) = -(A/kH)^N$. Integration to find the velocity (17) yields another factor of $1/k$. For $N = 3$ the expression for the velocity is:

$$U(\theta) = -\left(\frac{A}{kH}\right)^3 \left[\frac{1}{k} \sin(k\theta) - \frac{1}{3k} \sin^3(k\theta)\right]. \tag{24}$$

The velocity falls off with increasing wavenumbers as $(1/k)^{(N+1)}$. Therefore the longer wavelength components of the shear stress are more efficient at exciting flow in the non-Newtonian hoop.

Acknowledgements

The author is grateful to Maria Zuber for encouragement and scientific conversations. The author also thanks NASA's Goddard Space Flight Center for the use of their Cray C-98 computer on which all of their calculations in this study were performed and the IUTAM organizing committee for inviting me to Beijing. The author also acknowledges support from NSF grant EAR–9102808 under which the code was developed during my tenure as an NSF postdoctoral fellow and a SOEST Young Investigator award from the University of Hawaii.

REFERENCES

BERCOVICI, D., SCHUBERT, G., and GLATZMEIR, G. A. (1992), *Three-dimensional Convection of an Infinite-Prandtl Number Compressible Fluid in a Basally Heated Spherical Shell*, J. Fluid Mech. *239*, 683–719.

BERCOVICI, D. (1993), *A Simple Model of Plate Generation from Mantle Flow*, Geophys. J. Int. *114*, 635–650.

BERCOVICI, D. (1995), *A Source-sink Model of the Generation of Plate Tectonics from non-Newtonian Mantle Flow*, J. Geophys. Res. *100*, 2013–2030.

CATHLES, L. M., *Viscosity of the Earth's Mantle* (Princeton University Press, Princeton 1975).

CHRISTENSEN, U. R. (1983), *Convection in a Variable Viscosity Fluid: Newtonian Versus Power-law*, Earth. Planet. Sci. Lett. *64*, 153–162.

CHRISTENSEN, U. R. (1984), *Convection with Pressure- and Temperature-dependent non-Newtonian Rheology*, Geophys. J. R. Astr. Soc. *77*, 343–384.

CHRISTENSEN, U. R., and HARDER, H. (1991), *3-D Convection with Variable Viscosity*, Geophys. J. Int. *104*, 213–226.

CSEREPES, L. (1982), *Numerical Simulations of non-Newtonian Mantle Convection*, Phys. Earth. Planet. Inter. *30*, 49–61.

DAVIES, G. F. (1988), *Role of the Lithosphere in Mantle Convection*, J. Geophys. Res. *93*, 10,451–10,466.

DEMETS, C., GORDON, R. G., ARGUS, D. F., and STEIN, S. (1990), *Current Plate Motions*, Geophys. J. Int. *101*, 425–478.

ELSASSER, W. M., *Convection and stress propagation in the upper mantle*. In *The Application of Modern Physics to the Earth and Planetary Interiors* (ed. Runcorn, S. K.) (Wiley Interscience, New York 1969) pp. 223–246.

HAGER, B. H., and O'CONNELL, R. J. (1981), *A Simple Model of Plate Dynamics and Mantle Convection*, J. Geophys. Res. *86*, 4843–4867.

JACOBY, W. R., and SCHMELING, H. (1982), *On the Effects of the Lithosphere on Mantle Convection and Evolution*, Phys. Earth Planet. Int. *29*, 305–319.

KING, S. D., and HAGER, B. H. (1990), *The Relationship between Plate Velocity and Trench Viscosity in Newtonian and Power-law Subduction Calculations*, Geophys. Res. Lett. *12*, 2409–2412.

MINSTER, J. B., and JORDAN, T. H. (1978), *Present-day Plate Motions*, J. Geophys. Res. *83*, 5331–5354.

PELTIER, W. R., *Mantle convection and viscosity*. In *Proceedings of the Enrico Fermi International School of Physics, Course LXXVIII* (eds. Dziewonski, A. M., and Boschi, E.) (North Holland, Amsterdam 1980) pp. 362–431.

PHILLIPS, R. J., RAUBERTAS, R. F., ARVIDSON, R. E., SARKAR, I. C., HERRICK, R. R., IZENBERG, N., and GRIMM, R. E. (1992), *Impact Craters and Venus Resurfacing History*, J. Geophys. Res. *97*, 15,923–15,948.

SOLOMON, S. C., SMREKAR, S. E., BINDSCHADLER, D. L., GRIMM, R. E., KAULA, W. M., McGILL, G. E., PHILLIPS, R. J., SAUNDERS, S., SCHUBERT, G., SQUYERS, S. W., and STOFAN, E. (1992), *Venus Tectonics: An Overview of Magellan Observations*, J. Geophys. Res. *97*, 13,199–13,255.

SOLOMON, S. C. (1993), *The Geophysics of Venus*, Phys. Today *47*, 48–55.

STROM, R. G., SCHABER, G. G., and DAWSON, D.D. (1994), *The Global Resurfacing of Venus*, J. Geophys. Res. *99*, 10,899–10,926.

SURKOV, Y. A., KIRNOZOV, F. F., GLAZOV, V. N., DUNCHENKO, A. G., TASTY, L. P., and SOBORNOV, O. P. (1987), *Uranium, Thorium and Potassium in the Venusian Rocks at the Landing Site of Vega 1 and 2*, Proc. Lunar Planet. Sci. Conf. *17th*, Part 2, J. Geophys. Res. *92* Suppl., E537–E540.

TURCOTTE, D. L. (1993), *An Episodic Hypothesis for Venusian Tectonics*, J. Geophys. Res. *98*, 17,061–17,068.

WEINSTEIN, S. A., and OLSON, P. L. (1992), *Thermal Convection with non-Newtonian Plates*, Geophys. J. Int. *109*, 481–487.

(Received September 7, 1994, revised April 6, 1995, accepted April 24, 1995)

PAGEOPH, Vol. 146, Nos. 3/4 (1996)

0033-4553/96/040573-15$1.50 + 0.20/0
© 1996 Birkhäuser Verlag, Basel

Mantle Flow with Existence of Plates and Generation of the Toroidal Field

ZHENG-REN YE,[1] XIN-WU ZHANG,[1] and CHUN-KAI TENG[1]

Abstract — The observed plate velocities contain two types of motions. The poloidal component is related to the formation of ridges and subduction zones and the toroidal field expresses the shearing of surface plates. One very important consideration in modeling flow in the earth's mantle is the existence and motion of the lithospheric plates. The motion of plates represents a large-scale circulation with strong viscous coupling to the mantle underneath. The mantle flow probably is neither a purely free convection driven by buoyancy forces due to nonadiabatic temperature gradients in the mantle nor a forced convection generated by boundary forces, but a mixed convection that combines the effects of boundary and buoyancy forces. We present, in this paper, the mixed convection model resulting in a surface velocity field that contains both the observed poloidal and toroidal components.

Key words: Free convection, forced convection, plate velocity field, mixed convection model.

1. Introduction

The observed plate velocities contain two types of motions and can be described by means of two scalar fields. The poloidal component is related to motions associated with ridges and subducted slabs, and the toroidal field represents shearing associated with transform faults. For the earth, the surface kinetic energy is known to be equally distributed between the two fields (HAGER and O'CONNELL, 1978). One of the aims of studying mantle dynamics is to explain the generation of these two fields. By introducing the lateral mantle density variations revealed by seismic tomography into dynamic mantle models, some authors have predicted the major features of the observed poloidal field. However, the generation of the toroidal field is still an open problem. FORTE and PELTIER (1987) suggested that the generation of the torroidal field might be accomplished by introducing lateral viscosity variations which then produced a nonlinear coupling of the poloidal and toroidal scalars in the momentum equation. Using a seismically inferred density heterogeneity in the mantle and slabs as input data and taking into account the existence of rigid and independent plates, RICARD and VIGNY (1989) developed a

[1] Institute of Geophysics, Chinese Academy of Sciences, Beijing 100101, China.

model to predict the surface motions and a satisfactory prediction was obtained for surface velocities and geoid.

One very important consideration in modeling flow in the earth's mantle is the existence and the motion of lithospheric plates. The motion of the plates implies a large-scale circulation due to mass flux from the moving lithospheric plates themselves and viscous coupling between the plates and underlying mantle. As a relatively independent region of the earth's dynamic system, the motion of the plates should have a substantial influence on material flow in the mantle (RICHTER and PARSONS, 1975; PARMENTIER and TURCOTTE, 1976; HAGER and O'CONNELL, 1981; YE and HONG, 1983). In other words, the boundary forces due to motions of the plates themselves will excite a forced convective motion in the mantle. The flow pattern in the mantle is expected to be strongly influenced by the motion of the lithospheric plate at the surface. It follows that the mantle flow is probably neither a purely free convection driven by buoyancy forces due to nonadiabatic temperature gradients in the mantle nor a forced convection generated by boundary forces, but a mixed convection that combines the effects of boundary and buoyancy forces.

In the following, we present the mixed convection model which can predict a surface velocity field containing not only the poloidal component but also the toroidal component, consistent with the observations.

2. Physical Model and Mathematical Formulation

The mantle is assumed to behave dynamically as an infinite Prandtl number Pr ($Pr = v/k$ where v is the kinematic viscosity and k is the thermal diffusivity) Boussinesq fluid shell overlaid by rigid plates and surrounding an inviscid core. It is also assumed that there is no radioactive heat source in the mantle. The nondimensional equations which describe the free convective motion are

$$\nabla \cdot \mathbf{v} = 0 \tag{1a}$$

$$-\nabla p + \nabla^2 \mathbf{v} + Ra\theta \mathbf{e_r} = 0 \tag{1b}$$

$$\frac{\partial \theta}{\partial t} + \mathbf{v} \cdot \nabla(\theta + \theta_c) = \nabla^2 \theta. \tag{1c}$$

In (1a)–(1c), the length scale, the velocity $\mathbf{v}$, temperature θ, time t and pressure p are nondimensionalized by the thickness of the shell d, k/d, $\Delta T = T_1 - T_2$ ($T_1 > T_2$. T_1, T_2 are temperatures at the inner and the outer boundaries, respectively), d^2/k and $d^2/\mu k$, respectively. θ_c ($\theta_c = r_2 r_1/(r_2 - r_1)r - r_1/(r_2 - r_1)$, where r_1, r_2 are nondimensional radii at the inner and the outer boundaries, respectively) is the temperature distribution in the purely conduction state which satisfies one-dimensional homogeneous Laplace equation and the boundary conditions of $\theta_c = 0$ at r_2 and $\theta_c = 1$ at r_1. Ra is the Rayleigh number.

Equation (1a) means velocity is a solenoidal field and thus can be written in terms of poloidal and toroidal functions which automatically satisfy the continuity equation (CHANDRASEKHAR, 1961),

$$\mathbf{v} = \nabla \times \wedge \Phi - \wedge W \tag{2}$$

where Φ is the poloidal field and W is the toroidal field. The vector operator $\wedge$ is defined as

$$\wedge = \mathbf{r} \times \nabla. \tag{3}$$

It can be shown that the velocity field is purely poloidal in the infinite Prandtl number limit, i.e., the toroidal field always vanishes (ZEBIB et al., 1980).

Substituting (2) into (1b) and taking into account that the toroidal component W vanishes, and applying $\wedge \cdot \nabla \times$ to the resulting equation, we obtained

$$\nabla^4 \wedge^2 \Phi = \frac{-Ra}{r} \wedge^2 \theta \tag{4}$$

where $\wedge^2 = \wedge \cdot \wedge$. Equations (4) and (1c) form the working equations.

Appropriate boundary conditions are: at the inner boundary (CMB) it is free to slip; at the outer boundary, a no-slip condition is imposed; and both are taken to be isothermal. The spherical harmonic expansion and the Galerkin method are used to solve the system for free convection (ZEBIB et al., 1980; YE et al., 1995).

Let us now turn to the aspect of the forced convection excited by plates themselves. In this case, the velocity field in the mantle is unaffected by the temperature field (TRITTON, 1977), which is quite different from the free convection, and can be described by a homogeneous Navier-Stokes equation which reads

$$-\nabla p + \nabla^2 \mathbf{v} = 0. \tag{5}$$

Substituting (2) into (5) and applying $\wedge \cdot \nabla \times$ and $\mathbf{r} \cdot \nabla \times$, respectively, to the resulting equation, we have

$$\nabla^4 \wedge^2 \Phi = 0, \tag{6}$$

$$\nabla^2 \wedge^2 W = 0. \tag{7}$$

Again, the spherical harmonic expansion method is used to solve the problem for the forced convection with the boundary conditions at the CMB shear stress-free and at the outer boundary defined by a set of angular velocities, which will be determined through torque balance.

Observation reveals that at present each plate is moving at a constant angular velocity, which means that plates are in a dynamically equilibrium state such that the net torque applied to each plate should vanish.

Both free convection and forced convection exert torques on the plates. If the direct interactions between the plates are negligible, that the net torque applied on each plate is zero implies the torque generated by the free convection must be balanced by the torque from the forced convection on each plate. In the mixed convection model, the flow in the mantle consists of the superposition of two dynamic processes. The first is free convection and the second is forced convection excited by moving plates which have known numbers and shapes but an unknown set of angular velocities. The magnitudes of the unknown rotation vectors are deduced from the balance of the torques imposed by the two dynamic processes. Here, we assume that there are no other forces besides those from the free convection and the forced convection. The function of the slab pull forces and the ridge push forces will be considered and discussed later.

The torque N_j, applied on the j-th plate can be expressed as

$$N_j = \iint (\mathbf{r} \times \tau_j) \, ds_j \tag{8}$$

where $\mathbf{r}$ is a radius vector and τ_j is the shear stress induced beneath the j-th plate by either the free convection or the forced convection. The double integration is carried out over the area of the j-th plate. A more detailed description of the calculation of torques is given in the Appendix.

3. Results

We considered 11 main plates for our investigation. They are the African, the American, the Arabian, the Caribbean, the Cocos, the Eurasian, the Indian, the Nazca, the Pacific and the Philippine plates. MINSTER and JORDAN (1978) have obtained an absolute motion model consisting of the present-day rotation rate vectors for each of the 11 surface plates. The Rayleigh number Ra is assumed to be 5×10^6 in the computation of the free convection which is compatible with the order of magnitude of the Rayleigh number estimated for the whole mantle convection.

Figure 1a shows the well-known observed plot of plate motions. Figure 1b shows the predicted plate motions calculated from the angular velocities for 11 plates, on the basis of the primary mixed convection model in which the other forces besides those from the free convection and the forced convection are not considered. Spherical harmonic coefficients up to degree 20 are summed. The

---

Figure 1
a) The observed earth's plate motions. b) The predicted earth's plate motions obtained from the model that does not consider slab pulls. c) The predicted earth's plate motions obtained from the model that considers the slab pull applied on the Pacific and the Nazca plates. The same scale for velocity vectors is used in the plots.

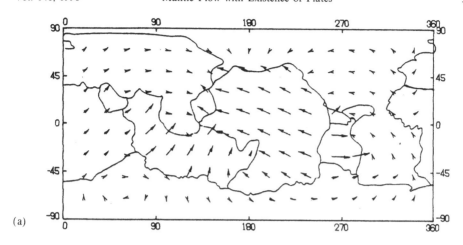

(a)

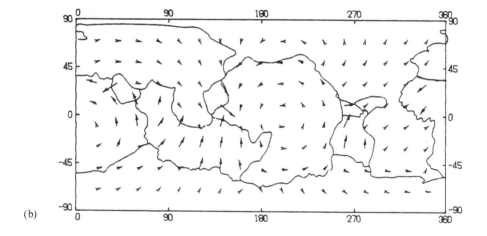

(b)

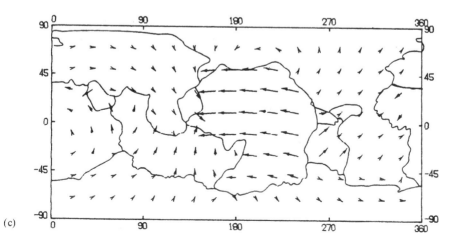

(c)

match between the observation and the prediction appears unsatisfactory, although the velocity vectors for some plates such as the Indian and the Eurasian plates display nearly the same direction and magnitude. The major discrepancies occur in the Pacific and the Nazca plates. The reason for the misfit might be that in our primary model we do not consider the other forces, e.g., the slab pull at subduction zones, which act on the plates. FORSYTH and UYEDA (1975) determined the relative strength of forces applied on the plates, given the observed motions and geometries of the lithospheric plates. The results indicated that the forces acting on the downgoing slabs (slab pull) were an order of magnitude stronger than any other force. The force coefficients of the slab pulls and effective lengths of trenches for the Pacific and the Nazca plates are used to calculate the torques applied on these two

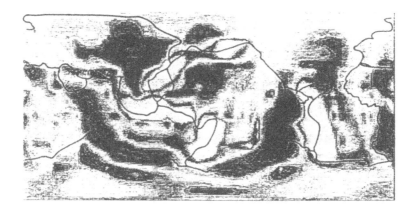

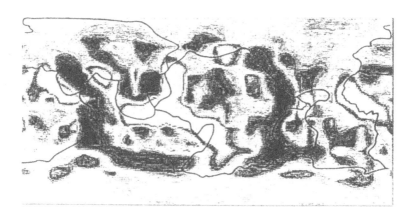

Figure 2

The observed (top) and the predicted (bottom) horizontal divergence of the surface plate velocities. Spherical harmonics up to degree and order 20 are included. Darker colors represent divergent motions whereas lighter colors represent convergent motions.

plates. When the resulting torques are added to the primary model the agreement manifests a significant improvement as shown in Figure 1c.

Figure 2 illustrates the comparison between the observed (top) and the predicted (bottom) horizontal divergence obtained from the model, including slab pulls acting on the Pacific and Nazca plates (the same below) which is related only to the poloidal component of the velocity field. It is clear that the main features in the observations can be matched in the prediction. For example, at the East Pacific Rise and the East Indian Ridge, both plots exhibit divergence motions. On the other hand, at the West Pacific Trenches, they display convergence. The observed (top) and the predicted (bottom) radial vorticities, which depend only on the toroidal field, are displayed in Figure 3. The patterns are quite similar. The left-lateral shear between the Indian-Australian and the Pacific plates, and right-lateral shear at the

Figure 3
As Figure 2, but for the radial vorticity. Darker color represents right-lateral shear and lighter color represents left-lateral shear.

San Andreas Fault are well predicted. However, there is the opposite sign predicted between the Nazca and the Antarctica plates compared to the observations.

In the previous figures we have shown the space characterizations of the poloidal and the toroidal fields. Another way to compare the two fields is by studying their power spectra. Included in Figures 4, 5 and 6 are the degree variances of the power spectra of the observed poloidal and toroidal fields as well as the computed ones plotted against the spherical harmonic degrees. Figure 4 illustrates the comparison between the predicted poloidal and the toroidal components. These two fields exhibit, at each degree except for degree two, an almost equipartition of kinetic energy, which is revealed in the power spectra of the spherical harmonic expansion of the observed plate velocities; this is more clearly presented in Figures 5 and 6.

Figure 7 displays, as an example, the flow pattern in the mantle obtained from the mixed convection model. It is a plot of the velocity vectors for the great-circle cross section through the Equator. The section passes through the African, the Indian, the Eurasian, the Pacific, the Nazca and the American plates successively, when moving counterclockwisely from the left side of the outer circle. It is remarkable that under the East Pacific Rise a strong upwelling flow from near the

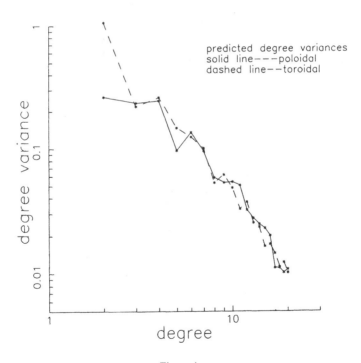

Figure 4

Degree variance of the predicted poloidal and toroidal components of the surface velocity field. Solid line—poloidal, dashed line—toroidal.

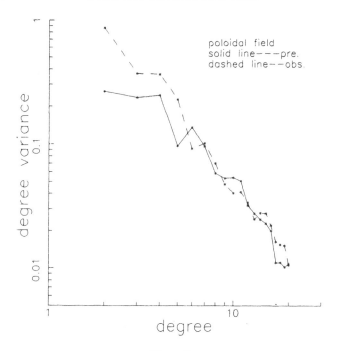

Figure 5
Degree variance of the observed and the predicted poloidal components of the surface velocity field.
Solid line — prediction, dashed line — observation.

core-mantle boundary arises and convective cells exist under the Nazca, the Pacific and the Indian plates.

Conclusions and Discussion

The purely free convection, driven by buoyancy due to the nonadiabatic temperature gradients in the mantle, or those flow models based on seismically inferred density heterogeneity, are unlikely to generate the toroidal component in the observed plate velocities, although they do well predict some of the major features of the poloidal field (PELTIER et al., 1989; YE et al., 1993). On the other hand, the existence and the motion of the lithospheric plates at the earth's surface can cause a forced convection in the mantle. Thus, a mixed convection model that combines the effects of buoyancy and boundary forces is developed. The geometry of the plates is given a priori, which means that we do not deal with the problem of their creation, and the unknown angular velocity for each plate is determined by the assumption that the plates are in a dynamic equilibrium state at the present day.

The agreement between the observed poloidal and toroidal fields and the fields predicted with the mixed convection model is encouraging. Both the observed spatial characteristics and the power spectra of the two fields are well predicted.

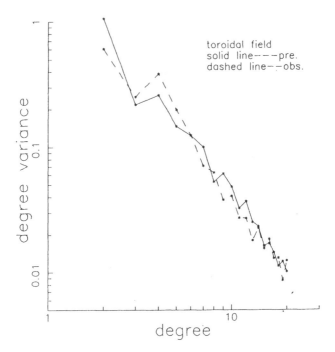

Figure 6
As Figure 5, but for the toroidal components.

The misfit shown in Figure 1b and the good fit in Figure 1c for surface velocities have demonstrated the function of the slab pull forces in modeling mantle flow and plate dynamics. Although the match between the observed and the predicted radial vorticity fields is generally good, the opposite sign is predicted for the border between the Nazca and the Antarctica plates. Also, there is a significant discrepancy of degree variance between observation and prediction at degree two. In the mixed convection presented here, we do not consider the ridge push forces although some investigators suggested that the force was at least comparable in magnitude to other forces and played an important role in driving the plates without the downgoing slabs (SOLOMON *et al.*, 1975). The difference between the predicted and the observed degree variance and the opposite sign could be the result of the lack of ridge push forces incorporated in the model computation. It is necessary in further study to examine the effect on the torque equilibrium of the ridge push forces. It can be found from Figure 7 that the effect of the free convection on the flow field in the mantle is small, where basically no characteristics of free convection are shown. This is consistent with the results from RICHTER (1973) and PARMENTIER and TURCOTTE (1976) that in dynamic models of convection with a moving boundary, the large-scale flow pattern is dominated by the motion of the boundary, even for low boundary velocities and a high Rayleigh number.

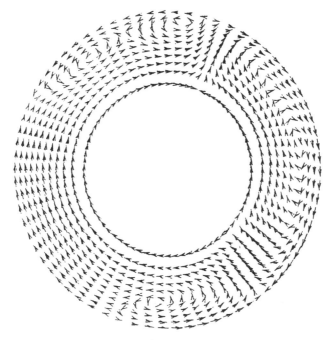

Figure 7

The normalized velocity vectors for the great-circle cross section through the Equator. The inner boundary is the core-mantle boundary and the outer boundary represents the earth's surface.

Thermal convection in the earth's mantle is the most likely cause of plate motions and the plate should actually be part of the mantle convection system. But detailed and realistic dynamic models which can predict or interpret the high complexity of the observed plate motions have so far been beyond both our understanding of the properties of the earth's interior and our computational ability. Thus, it is useful to make simpler models to investigate the effect and relative importance of complicated factors individually. The model developed by HAGER and O'CONNELL (1978, 1981) remained essentially kinematic since they neglected the effect of thermally generated buoyancy force in the earth's interior and included the observed surface plate velocities as the prior parameters. The mixed convection model presented here, on one hand, is similar to the model by RICARD and VIGNY (1989) in that both models have considered the function of the dynamic process in the mantle besides the function of plate motion, and superposed the two dynamic processes to obtain the complete solution. On the other hand, our model differs from Ricard's model in that the latter was based on seismically inferred density heterogeneity in the mantle and slabs to calculate the torque applied to plates. In the mixed convection model, the coupling between the free convection and the forced convection is through torque equilibrium, i.e., mechanically. There probably is another coupling that the forced convection causes the temperature perturbation which then affects the free convection. However, it is expected that this thermal coupling has a secondary order effect, for a very high Prandtl number and a relatively high Péclet number so that a concentration of temperature perturbation into the thermal boundary layer arises (TRITTON, 1977).

Acknowledgements

This research was supported by the National Science Foundation of China. The authors are grateful to W. Jacoby and an anonymous reviewer for their reviews on the manuscript, which are particularly helpful in making revision.

Appendix

The flow velocity and stress can be expressed in terms of vector spherical harmonics. The components of velocity and radial shear, in the southerly and easterly directions, read (cf. e.g., HAGER and O'CONNELL, 1978)

$$v_\theta = \sum_{l,m} (v_s^{lm} Y_l^\theta + v_t^{lm} Y_l^\phi) \tag{A1}$$

$$v_\phi = \sum_{l,m} (v_s^{lm} Y_l^\phi - v_t^{lm} Y_l^\theta) \tag{A2}$$

$$\tau_{r\theta} = \sum_{l,m} (\tau_s^{lm} Y_l^\theta + \tau_t^{lm} Y_l^\phi) \tag{A3}$$

$$\tau_{r\phi} = \sum_{l,m} (\tau_s^{lm} Y_l^\phi - \tau_t^{lm} Y_l^\theta) \tag{A4}$$

where

$$Y_l^\theta \equiv \frac{\partial Y_{lm}}{\partial \theta} \qquad Y_l^\phi \equiv \frac{1}{\sin \theta} \frac{\partial Y_{lm}}{\partial \phi}$$

and Y_{lm} are the surface spherical harmonics of degree l and order m normalized such that their root mean square is unity.

The coefficients in the vector spherical harmonic expansion of surface velocities, v_s^{lm} and v_t^{lm}, are given by appropriate integration over the unit sphere

$$v_s^{lm} = \frac{1}{4\pi L} \iint (v_\theta Y_l^\theta + v_\phi Y_l^\phi) \, ds \tag{A5}$$

$$v_t^{lm} = \frac{1}{4\pi L} \iint (v_\theta Y_l^\phi - v_\phi Y_l^\theta) \, ds \tag{A6}$$

where $L = l(l + 1)$.

v_s^{lm} and τ_s^{lm} are only associated with the poloidal field and v_t^{lm} and τ_t^{lm} are only associated with the toroidal field. Mathematically,

$$v_s^{lm} = -\frac{d\Phi^{lm}}{dr} - \frac{\Phi^{lm}}{r} \tag{A7}$$

$$\tau_s^{lm} = \mu\left[-\frac{d^2\Phi^{lm}}{dr^2} + \left(\frac{2-L}{r^2}\right)\Phi^{lm}\right] \tag{A8}$$

$$v_t^{lm} = W^{lm} \tag{A9}$$

$$\tau_t^{lm} = \mu\left(\frac{dW^{lm}}{dr} - \frac{W^{lm}}{r}\right) \tag{A10}$$

where Φ^{lm} and W^{lm} are the spherical harmonic coefficients of degree l and order m of the poloidal field Φ and the toroidal field W, respectively. μ is viscosity.

On the Cartesian coordinate system with basic vectors E_k in the directions of 0°N, 0°E; 0°N, 90°E and 90°N, 0°, the k-direction component of the torque applied on the j-th plate, N_{jk}, can be expressed as

$$N_{jk} = \iint (\mathbf{r} \times \tau_j) \cdot \mathbf{E}_k \, ds_j$$

$$= -R \iint \left(\tau_{r\theta} \sum_{n=1}^{3} B_{2n} A_{kn} + \tau_{r\phi} \sum_{n=1}^{3} B_{3n} A_{kn}\right) ds_j \tag{A11}$$

where R is the earth's mean radius and

$$A = \begin{pmatrix} 0 & \cos\theta & -\sin\theta\sin\phi \\ -\cos\theta & 0 & \sin\theta\cos\phi \\ \sin\theta\sin\phi & -\sin\theta\cos\phi & 0 \end{pmatrix}$$

$$B = \begin{pmatrix} \sin\theta\cos\phi & \sin\theta\sin\phi & \cos\theta \\ \cos\theta\cos\phi & \cos\theta\sin\phi & -\sin\theta \\ -\sin\phi & \cos\phi & 0 \end{pmatrix}.$$

Solving the system for the free convection mentioned in the text and carrying out the integration of (A11) over the area of the j-th plate, taking (A3), (A4) and (A8) into account, we obtain the k-direction component of the torque applied to the j-th plate by the free convection and express it as $N_{jk}^{(f)}, j = 1, 2, \ldots, M$; $k = 1, 2, 3$, where M represents the number of the plates.

The kinematics of M plates on a spherical earth imposes the same number of angular velocity vectors. Thus the surface velocity reads

$$v_\theta = R \sum_{j=1}^{M} (\mathbf{\Omega}_j \times \mathbf{e}_r) \cdot \mathbf{e}_\theta \delta(j; \theta, \phi)$$

$$= R \sum_{j=1}^{M} \sum_{k=1}^{3} B_{3k} \Omega_{jk} \delta(j; \theta, \phi) \tag{A12}$$

$$v_\phi = -R \sum_{j=1}^{M} \sum_{k=1}^{3} B_{2k} \Omega_{jk} \delta(j; \theta, \phi) \tag{A13}$$

where Ω_{jk} is the k-direction component of the angular velocity of the j-th plate. $\mathbf{e}_\theta$ and $\mathbf{e}_\phi$ represent the basic vectors in the southerly and easterly direction on the spherical coordinate, respectively. $\delta(j; \theta, \phi)$ is 1 or 0 depending on whether coordinates θ and ϕ point to the j-th plate or not.

The forced convection excited by the unknown angular velocities also generates a shear stress field acting on the plates which can be, by means of relations (A3)–(A7), (A9), (A12) and (A13), expressed as

$$\tau_{r\theta}^{(p)} = \frac{R}{4\pi} \sum_{j=1}^{M} \sum_{k=1}^{3} \sum_{l,m} \frac{1}{L} (a_l M_{jk}^s Y_l^\theta + b_l M_{jk}^t Y_l^\phi) \Omega_{jk} \tag{A14}$$

$$\tau_{r\phi}^{(p)} = \frac{R}{4\pi} \sum_{j=1}^{M} \sum_{k=1}^{3} \sum_{l,m} \frac{1}{L} (a_l M_{jk}^s Y_l^\phi - b_l M_{jk}^t Y_l^\theta) \Omega_{jk} \tag{A15}$$

where a_l, b_l are factors relating the harmonic coefficients of the surface velocities, v_s^{lm} and v_t^{lm}, to that of the shear stress, τ_s^{lm} and τ_t^{lm}, and are obtained by solving the forced convection system, and

$$M_{jk}^s = \iint (B_{3k} Y_l^\theta - B_{2k} Y_l^\phi) \delta(j; \theta, \phi) \, ds$$

$$M_{jk}^t = \iint (B_{3k} Y_l^\phi + B_{2k} Y_l^\theta) \delta(j; \theta, \phi) \, ds.$$

Substituting (A14) and (A15) into (A11), we obtain the k-direction component of the torque $N_{jk}^{(p)}$, applied to the j-th plate by the forced convection

$$N_{jk}^{(p)} = \sum_{p=1}^{M} \sum_{q=1}^{3} C_{jkpq} \Omega_{pq} \tag{A16}$$

where C_{jkpq} is a matrix.

$$C_{jkpq} = \frac{-R^2}{4\pi} \sum_{p,q} \sum_{l,m} \left\{ \frac{a_l M_{pq}^s}{L} \iint \left[\sum_n (B_{2n} Y_l^\theta + B_{3n} Y_l^\phi) A_{kn} \right] \delta(j; \theta, \phi) \, ds \right.$$
$$\left. + \frac{b_l M_{pq}^t}{L} \iint \left[\sum_n (B_{2n} Y_l^\phi - B_{3n} Y_l^\theta) A_{kn} \right] \delta(j; \theta, \phi) \, ds \right\}.$$

If the torques generated by any other forces such as slab pull, ridge push are negligible, the condition that the net torque applied to each plate vanish means $N_{jk}^{(p)} = -N_{jk}^{(f)}$, $(j = 1, \ldots, M; k = 1, 2, 3)$. Consequently the angular velocities Ω_{jk} can be deduced from (A16). When the slab pull is considered, the summation of three torques, i.e., the torque due to the free convection, the forced convection and the slab pull, acting on the plate is zero.

REFERENCES

CHANDRASEKHAR, S., *Hydrodynamic and Hydromagnetic Stability* (Oxford Clarendon Press, 1961).

FORSYTH, D., and UYEDA, A. (1975), *On the Relative Importance of the Driving Forces of Plate Motion*, Geophys. J. R. Astr. Soc. *43*, 163–200.

FORTE, A. M., and PELTIER, W. R. (1987), *Plate Tectonics and Aspherical Earth Structure: The Importance of Poloidal-toroidal Coupling*, J. Geophys. Res. *92* (B5), 3645–3679.

HAGER, B. H., and O'CONNELL, R. J. (1978), *Subduction Zone Dip Angles and Flow Driven by Plate Motion*. Tectonophysics *50*, 111–133.

HAGER, B. H., and O'CONNELL, R. J. (1981), *A Simple Global Model of Plate Dynamics and Mantle Convection*, J. Geophys. Res. *86*, 4843–4867.

MINSTER, J. B., and JORDAN, T. H. (1978), *Present-day Plate Motions*, J. Geophys. Res. *83*, 5331–5354.

PARMENTIER, E. M., and TURCOTTE, D. L. (1976), *Studies of Finite Amplitude non-Newtonian Thermal Convection with Application to Convection in the Earth's Mantle*, J. Geophys. Res. *81*, 1839–1846.

PELTIER, W. R., JARVIS, G. T., FORTE, A. M., and SOLHEIM, L. P., *The radial structure of mantle general circulation*. In *Mantle Convection* (ed. Peltier, W. R.) (Gordon and Breach, New York, 1989) pp. 765–816.

RICARD, Y., and VIGNY, C. (1989), *Mantle Dynamics with Induced Plate Tectonics*, J. Geophys. Res. *94* (B12), 17543–17559.

RICHTER, F. M. (1973), *Dynamical Models for Sea Floor Spreading*, Rev. Geophys. Space Phys. *11*, 223–287.

RICHTER, F. M., and PARSON, B. (1975), *On the Interactions of Two Scales of Convection in the Mantle*, J. Geophys. Res. *80*, 2529–2541.

TRITTON, D. J., *Physical Fluid Dynamics* (Van Nostrand Reinhold Company, 1977).

YE, Z. R., and HONG, M. D. (1983), *The Action of the Mantle Asthenosphere on the Plates: Driving or Dragging*, Acta Geophys. Sinica. *26*, Suppl. 651–660 (in Chinese, with English abstract).

YE, Z. R., TENG, C. K., and BAI, W. M. (1993), *A Mantle Convection Model to Fit in with the Surface Observations*, Phys. Earth Planet. Inter. *76*, 35–41.

YE, Z. R., TENG, C. K., and ZHANG, X. W. (1995), *Coupling between Mantle Convection and Lithospheric Plates (I) the Free Convection in a Spherical Shell*, Acta Geophys. Sinica. *38*, 174–181 (in Chinese, with English abstract).

ZEBIB, A., SCHUBERT, G., and STRAUS, J. M. (1980), *Infinite Prandtl Number Thermal Convection in a Spherical Shell*, J. Fluid Mech. *97*, 257–277.

(Received October 14, 1994, revised March 28, 1995, accepted May 26, 1995)

PAGEOPH, Vol. 146, Nos. 3/4 (1996)

0033-4553/96/040589-32$1.50 + 0.20/0

Constraints on Melt Production Rate Beneath the Mid-ocean Ridges Based on Passive Flow Models

YONGSHUN JOHN CHEN[1]

Abstract —We present a model for computing the total melt production rate from the decompression partial melting region beneath a mid-ocean ridge, and the maximum oceanic crustal thickness created at the ridge axis assuming an ideal melt migration mechanism. The calculations are based on a self-consistent numerical model for the thermal structure and steady-state mantle flow field at a mid-ocean ridge. The model includes the effect of decreasing the melt production rate within the partial melting region by melt extraction as the residual mantle matrix becomes increasingly difficult to melt. Thus the melt fraction depends not only on temperature and pressure determined by the location beneath the ridge axis (the Eulerian description) but also on the accumulated melt extraction since the upwelling mantle matrix enters the partial melting region determined by the location along the flow-line path (the Langrangian description). This effect has been neglected by previous models. The model can predict the size of the melting region and the locations of the boundaries between mantle, residual mantle, and the partial melting region for a given spreading rate, also the distribution of the melt depletion and the mean melting depth. Given the observed average thickness of oceanic crust (~ 6 km), which is relatively independent of spreading rate, the model results also provide a constraint on the overall efficiency of melt migration to the ridge axis; the efficiency must decrease from 100% at 10 mm/yr to about 60% at fast spreading rates (> 50 mm/yr). Although this reduction may be partially due to the increasing size of the melting region with increasing spreading rate, it still requires less efficient melt migration near the ridge axis at fast spreading rate. We found that the calculated crustal thickness is very sensitive to the mantle temperature. For a normal mantle temperature of 1350°C, the model can generate the observed 6 km oceanic crust over the global range of spreading rates, while the anomalous thicker crusts of the Iceland hotspot and the Reykjanes Ridge are related to higher mantle temperatures associated with the hotspot. Finally, by comparing our model results with previous ones we found that neglecting variations of the melting relations of the residual mantle matrix with melt removal will overestimate the crustal thickness by at least a factor of 1.7.

Key words: Melt production rate, fractional melting, melt depletion, crustal thickness.

Introduction

The most voluminous volcanism on the earth occurs at mid-ocean ridges (MOR). Seafloor spreading generates about 20 km^3 per year of mid-ocean ridge

[1] College of Oceanic and Atmospheric Sciences, Oregon State University, Corvallis, OR 97331, U.S.A.

basalt (MORB), which is the dominant volcanic rock type on the earth. It has long been suggested that decompression partial melting beneath the mid-ocean ridges occurs in response to the upwelling of the mantle beneath the ridge. Decompression partial melting is due to a faster release of pressure than the decrease in temperature in comparison with the Clapeyron slope. For simplicity, we will use the work "melting" throughout the text although we are referring to partial melting.

It is believed that most of the melt generated in the decompression melting region separates from the residual mantle and migrates to the ridge axis to form the oceanic crust. In general, the thermal and flow structures beneath a mid-ocean ridge as well as the thermodynamic properties of the mantle should control the extent of partial melting of the upwelling mantle as it ascends beneath the ridge axis. The extent of melting and the melt migration mechanism should, in turn, govern both the thickness of the oceanic crust and the chemistry of MORB. Thus, variations in crustal thickness and MORB chemistry should correlate with each other and with physical parameters including the mantle temperature, spreading rate, and melt separation depth.

To account for the observed thickness, composition, and chemistry of the oceanic crust and the resulting ridge axis morphology, three processes of the crustal genesis at the mid-ocean ridges have to be addressed. First, we need to calculate the melt production rate to be extracted from the upwelling mantle within the partial melting region. Second, melt migration mechanisms (e.g., PHIPPS MORGAN, 1987; SPIEGELMAN and McKENZIE, 1987; SPARKS and PARMENTIER, 1991) are required to transport the melt from the partial melting region to the ridge axis to form the oceanic crust. Third, crustal flow models such as those of PHIPPS MORGAN and CHEN (1993a,b) and HENSTOCK et al. (1993) are needed to investigate the effects of magma injection and crustal accretion within the crust.

This study attempts to explore the first process, the rate of melt production from the upwelling mantle using current knowledge of the mid-ocean ridge dynamics. Assuming that all the melt extracted from the mantle can reach the ridge axis to form the new oceanic crust, the model results such as the crustal thickness are compared with the observations. The differences between the model predictions and the observations will provide useful constraints on melt migration mechanisms.

It is currently thought (AHERN and TURCOTTE, 1979) that practically all melt migrates upward much faster than the mantle upwelling. To estimate the total amount of melt (crustal thickness) separated from the upwelling mantle, we assume an ideal fractional melting taking place within the partial melting region. That is, once melt has been generated, it is assumed to separate rapidly from the matrix and move towards the ridge axis to form the new crust. Hence calculated crustal thickness puts an upper bound to the observed crustal thickness and it provides a constraint on thermal and mantle flow models, and the overall efficiency of melt migration mechanisms.

Following a brief review of previous models, I first derive the equations for estimating the melt production rate from the upwelling mantle. I then examine two steady-state thermal and mantle flow models. The first is an analytic model, which includes both the half-space cooling thermal model and a simple mantle flow model, a case of two triangular rigid plates moving away from the ridge axis. The second is a numerical model which divides the oceanic lithosphere into the brittle and ductile layers. Both models predict an increase in computed crustal thickness with spreading rate because of the increase in the size of the partial melting region and it should provide information on the overall efficiency of melt migration models. The variations in melt fraction, melt-separation depth, and separation conditions (P and T at the time of melt separation) with spreading rate are calculated from the model and they should offer useful information on the global trends of composition and chemistry of mid-ocean ridge basalts.

Previous Models

For the past three decades, the extent of melting resulting from adiabatic decompression of mantle material has been frequently calculated, using constraints from experimental results of thermodynamic properties (solidus and liquidus temperatures), and mantle temperature estimates (e.g., CAWTHORN, 1975; AHERN and TURCOTTE, 1979; STOLPER et al., 1981; McKENZIE, 1984, 1985; KLEIN and LANGMUIR, 1987, 1989; SCOTT and STEVENSON, 1989; WHITE and McKENZIE, 1989a). In addition, several authors have estimated the thickness of the oceanic crust based on the amount of melt produced during mantle upwelling beneath the mid-ocean ridges, although this depends on the mantle flow regime (e.g., REID and JACKSON, 1981; SLEEP and WINDLEY, 1982; McKENZIE and BICKLE, 1988; KLEIN and LANGMUIR, 1987; PHIPPS MORGAN and FORSYTH, 1988). Due to the complexity of melting and melt migration processes, there is no definitive method for the calculation of melt production rate from the upwelling mantle beneath the mid-ocean ridges. The main difficulty is that the thermodynamic properties (solidus and liquidus temperatures) of the upwelling mantle vary with its composition as melt is extracted. The melting experiments in the laboratory do not extract the melt from the rock samples, therefore, the estimates of the solidus and liquidus temperatures at different pressures do not account for the composition changes. Thus, the melt fraction calculated based on these experimental data will lead to overestimation of melt production as the residual mantle becomes increasingly difficult to melt.

WHITE and McKENZIE (1989b) estimated that "mantle at 1340°C should start to melt when it rises to a depth of about 50 km; by the time it reaches the surface an average of 25% of the rock should have melted. The result will be the oceanic crust of the observed thickness." Apparently, the result of this model is derived as $H_c = 50$ km $\cdot 0.25 - 0.0/2 = 6.25$ km if a linear variation of the melt fraction with depth is assumed. While the model both explains the basic observation and is simple, it has two main problems. First, it assumes an unphysical flow model which results in a single "melting column" beneath a ridge axis. Second, it overestimates the amount of melt extracted from this column since it does not account for the effects of the composition changes as melt is extracted.

Another example of the previous models is that developed by PHIPPS MORGAN and FORSYTH (1988). Assuming a pressure governed solidus, $T_m = 1100°C + y \cdot 3.25°C/km$, they used a relation for the total extent of melting $\phi = (T - T_m)/600°C$, where 600°C is the temperature interval from incipient to total melting of mantle material. Thus they derived the melt production rate $\dot{\phi} = V \cdot \nabla \phi$ for a steady state which relates the melt production rate at any location within the partial melting region to the dot product of the upwelling velocity and the gradient of the total extent of melting at that point. While the above formulation of the melt production rate is correct they assume that the melting relation does not change as melt is removed which can be seen by the pressure-only governed solidus and the relation of the melt production rate to the solidus.

In summary, previous models have not accounted for the effects of the composition changes of the upwelling mantle on the melt production rate. Including this effect is the new feature of the model in this study. We will compare our model predictions with previous model results to see how much our model can improve calculations of melt production rate beneath the ridge axis, hence of the crustal thickness.

Model Formulation for Calculating Melt Production Rate

Here, we present an explicit method for calculating the total melt production rate from the partial melting region beneath a mid-ocean ridge. It includes the effects of the variations in solidus and liquidus temperatures of the residual mantle rocks with the melt extraction. The procedure of the modeling includes two steps.

First, we need to calculate the melt production rate for the case of batch melting, that is, all the melt is retained within the solid mantle matrix keeping a constant composition of the upwelling mantle. MCKENZIE and BICKLE (1988) have obtained the analytic expressions for the solidus T_s and liquidus T_l temperatures as a function of pressure for the peridotite mantle which agreed well with the experimental observations. The expressions they obtained for T_s and T_l are

$$P = (T_s - 1100)/136 + 4.968 \times 10^{-4} \exp[1.2 \times 10^{-2}(T_s - 1100)] \qquad (1)$$

$$T_l = 1736.2 + 4.343 P + 180 \tan^{-1}(P/2.169) \qquad (2)$$

where P is the pressure in GPa and T_s and T_l in °C. They fit the solidus temperature and liquidus temperature with a mean error of 6°C and 7°C, respectively (see

MCKENZIE and BICKLE, 1988, Figure 5). By defining a dimensionless temperature T'

$$T' = \frac{T - (T_s + T_l)/2}{T_l - T_s} \tag{3}$$

they obtained an expression for the fraction of melt F of the peridotite at a temperature T and pressure P;

$$F = 0.5 + T + (T'^2 - 0.25)(0.4256 + 2.988T'). \tag{4}$$

This expression fits the experimental data well with a mean error of 3% (see Figure 6 of MCKENZIE and BICKLE, 1988). Note that these expressions were derived from the experimental data for the peridotite mantle without any melt removal. They can be used to compute the fraction of melt for a batch melting case. In order to compute the amount of melt separated from the residual mantle with changing

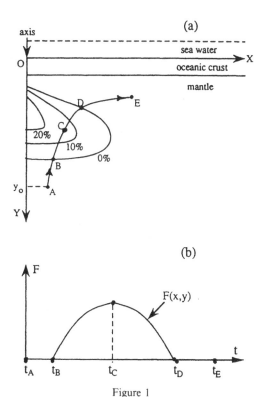

Figure 1
(a) A sketch of a decompression melting process beneath a mid-ocean ridge axis. Fractions of melt for the mantle rock are shown as contours. The curve from A to E represents a flow line of the upwelling mantle flow. (b) The fraction of melt $F(S)$ for a no-melt-separation case, that is, all the melt is retained in the system shown as a function of S, the distance along the state path from point A in (a).

composition (fractional melting), additional considerations have to be made in the next step.

Second, we derive the equations for the melt production rate from the upwelling mantle including the effect of composition changes of the residual mantle assuming an ideal fractional melting taking place in the partial melting region beneath the ridge axis.

Figure 1a shows a schematic cross section of a mid-ocean ridge where a decompression partial melting region exists beneath the ridge as the consequence of the rapid mantle upwelling. For a given temperature distribution, the fraction of melt $F(x, y)$ within the upwelling mantle can be calculated through (1)–(4) for the case of batch melting and is illustrated as contours in Figure 1a.

We will focus on a small mass per unit time $\dot{W} = \rho_m \cdot u_y \, dx \, dz$ passing through point $A(x, y_0)$ and determine the amount of melt $d\dot{W}_m(x) = D(x) \, d\dot{W}$ that can be separated from it after it goes through the partial melting region (past point D). Here ρ_m is the mantle density, u_y the upwelling velocity at point A, and y_0 is so chosen that the point A is always below the melting region. The melt depletion $D(x)$ is defined as the percentage of the melt being removed from the initial mass. Here dz is along the third dimension perpendicular to the $x - y$ plane and is chosen to be a unit length. Thus it will be dropped from the following equations. Assume that the upwelling mantle flow velocity is known and so are the flow streamlines. Thus the fraction of melt $F(x, y)$ along the flowline from A to E (Figure 1a) is known as shown in Figure 1b. According to equation (A11) in the appendix the amount of melt removed from this small mass $d\dot{W}_m(x)$ after it passes through the partial melting region can be calculated in terms of the maximum fraction of melt (F_{max}) along its entire flowline from A to E, where

$$F_{max} = \max F(x, y) \quad A \to E. \tag{5}$$

Once $d\dot{W}_m(x)$ is known, for a steady state, the total melt production rate from the upwelling mantle within the entire melting region is simply

$$\dot{W}_m = \int_0^\infty d\dot{W}_m(x) = \int_0^\infty \rho_m u_y(x, y_0) D(x) \, dx \tag{6}$$

in which W_m is in kg/yr per unit length of ridge, and $D(x)$ can be determined by substituting (5) into (A11). That is,

$$D(x) = 1 - e^{-F_{max}}. \tag{7}$$

Assuming that all the melt have reached the ridge axis to form the oceanic crust (ideal melt migration), the thickness of the oceanic crust can be calculated by equating the accretion rate of new crust per unit length of ridge at the ridge axis $(\rho_c H_c u)$ to equation (6). This gives

$$H_c = \frac{\rho_m}{\rho_c} \int_0^\infty \frac{u_y(x, y_0)}{u} D(x) \, dx \tag{8}$$

where ρ_c and H_c are the density and thickness of the crust, respectively, and u is the half-spreading rate in mm/yr. The density of crust can be calculated if the melting fraction and melting depth are known (see KLEIN and LANGMUIR, 1987). For purposes of calculations the values in Table 1 are assumed in the following analyses unless specified otherwise.

Thermal and Mantle Flow Models

Equations (6) and (8) relate the total melt production rate and the crustal thickness to $D(x)$ which depends on ridge axis thermal structure, and $u_y(x, y_0)$, the material supply rate from the mantle to the chosen depth y_0. Next, we examine two thermal and mantle flow models; the first is analytical and the second is numerical.

Half-space Cooling Model with an Analytic Mantle Flow Field

We first apply the above analysis to a half-space cooling thermal model and a simple analytic mantle flow model. Assuming a uniform horizontal velocity, the temperature distribution of a half-space cooling model is

$$T = T_m \operatorname{erf}\left(\frac{y}{\sqrt{2\kappa t}}\right) \qquad (9)$$

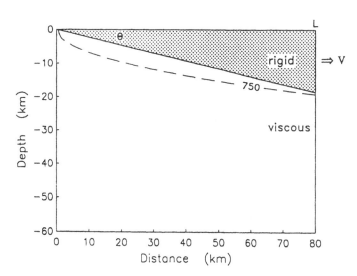

Figure 2

Schematic cross section of a mantle flow model with a uniform viscous fluid under two triangular rigid plates (only one plate is shown), which are moving horizontally away from the ridge axis with a constant velocity V as shown. The 750°C isotherm of a half-space cooling model is also shown as a dashed curve for reference.

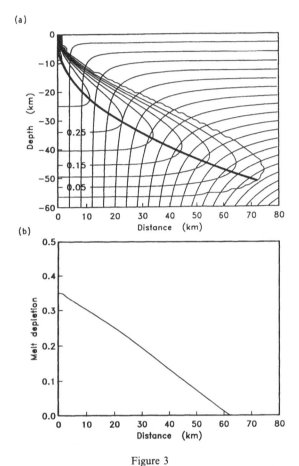

Figure 3
(a) The flow lines of the analytic solutions of the mantle flow problem as shown in Figure 4 for a half-spreading rate of 10 mm/yr. Also shown are the contours of $F(x, y)$ computed from the half-space cooling model similar to Figure 3a. The thicker solid line is the location (Y_{max}) of the maximum of $F(x, y)$ for each flow line passing through the melting region. (b) Calculated D versus the distance from the ridge axis.

where T_m is the mantle temperature, κ the thermal diffusivity, and t is the age of the plate defined by the distance from the ridge and the spreading rate, $t = x/u$. For ages less than 80 Ma, there is no difference between this model and a plate cooling model (e.g., MCKENZIE, 1967; MORGAN, 1975; PARSONS and SCLATER, 1977). At the ridge axis, the temperature is constant ($T = T_m$ at $x = 0$) right up to the surface. Although the half-space cooling thermal model is not consistent with the analytic flow model given below, they are used here mainly to demonstrate qualitatively the major results and the procedure of calculating melt production beneath ridge axis if realistic thermal and flow models are given.

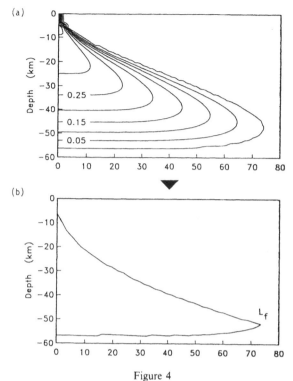

Figure 4

(a) The same contours of $F(x, y)$ as shown in Figure 7a. The zero value contour (close to zero) describes the size of the melting region should all the melt be retained within the upwelling mantle. (b) The size of the melting region should melt be allowed to be removed from the upwelling mantle. It is bounded on top by the $Y_{max}(x)$ and at the bottom by the zero value contour of $F(x, y)$.

Consider a flow problem with a uniform viscous fluid under two triangular rigid plates which are moving horizontally away from the ridge axis with a constant velocity V as shown in Figure 2. The steady-state solution of the viscous flow field in Cartesian coordinates x, y as shown in Figure 2 takes the form (CHEN and MORGAN, 1990a)

$$u_x = U\left(\tan^{-1}\left(\frac{x}{y}\right) - \frac{xy}{x^2 + y^2}\right) \tag{10a}$$

$$u_y = U\left(\sin^2 \theta - \frac{y^2}{x^2 + y^2}\right) \tag{10b}$$

where

$$U = V \bigg/ \left(\frac{\pi}{2} - \theta - \sin \theta \cos \theta\right) \tag{10c}$$

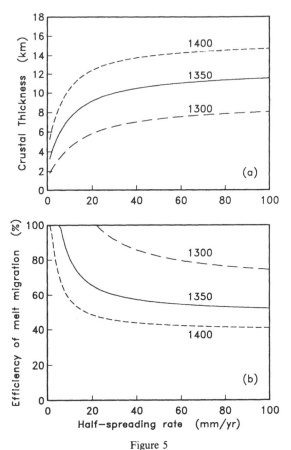

Figure 5

(a) Computed crustal thickness versus half-spreading rates for three different deep mantle temperatures, 1300°C (long dashed), 1350°C (solid), and 1400°C (short dashed). (b) The overall efficiency factor β of a melt migration mechanism versus spreading rates for the three deep mantle temperatures. The β is required to reduce the model crustal thicknesses shown in (a) to the observed 6 km oceanic crust thickness.

and θ is the angle between the lower boundary of the rigid plate and the x axis. To discuss the melting problem in terms of the analytic solutions we use the temperature solution (9) of the half-space cooling model, although it is derived from a uniform horizontal velocity distribution. The angle θ in (10) is determined by the depth of the 750°C isotherm at a given distance L from the axis as shown in Figure 2. Thus θ decreases with increasing spreading rate V for a fixed distance L. Although the triangular rigid plate fits the "thermal" rigid plate (the area above the 750°C isotherm) poorly near the axis, it provides an analytic upwelling velocity field everywhere.

The fraction of melt $F(x, y)$ determined by the temperature field (9) using (1)–(4) is shown as contours in Figure 3a. Also shown are the mantle flow lines computed from the velocity field (10) for $V = 10$ mm/yr. They are the streamlines which started evenly at the depth of 60 km. As discussed in the previous section, the total melt production rate or the crustal thickness can be calculated by (6) or (8) where $D(x)$ is calculated by equation (7) in terms of the maximum of $F(x, y)$ that a small patch at point $A(x, y_0)$ has encountered after flowing through the partial melting region (Figure 1a). That is,

$$D(x) = 1 - e^{-F(x, Y_{max}(x))}. \qquad (11)$$

As shown in Figure 3b, the calculated $D(x)$ decreases from 0.35 at the ridge axis to zero at a distance about 62 km from the axis. The location of the maximum of $F(x, y)$ for each streamline passing through the partial melting region (Y_{max}) is shown as a thicker line in Figure 3a. Once the fractional melting takes place, the region that contributes to the melt production is the area between (Y_{max}) and the lower boundary of the melting region (the contour of $F = 0$), based on the previous analysis.

To clarify the picture we summarize the above discussions as follows: First, the melting region is the area where $F(x, y) > 0$ (Figure 4a) should all the melt be retained within the upwelling mantle. Second, the region that contributes to melt production is reduced to the area between Y_{max} and the lower boundary of the melting region (Figure 4b) should melt be extracted from the upwelling mantle matrix. This is important for melt migration models since it provides the size of the melt source region and the distribution of the melt production rate within the source region.

The computed crustal thickness H_c from this model is shown as a solid curve in Figure 5a for the global range of half-spreading rates. Seismic studies have revealed a rather uniform thickness of oceanic crust (6 km) independent of spreading rate (CHEN, 1992). Here the computed crustal thickness increases with spreading rate. For the commonly accepted average mantle temperature of 1350°C, the model predicts a crustal thickness of 7.4 km at 10 mm/yr and 11 km at 60 mm/yr. Therefore, this model with a mantle temperature of 1350°C satisfies one of the criterion, that it can generate more than 6 km of oceanic crust in the range of the global spreading rates. To create the observed uniform thickness (6 km) of oceanic crust, the model requires an overall efficiency factor β of a melt migration model as a function of the spreading rate to be

$$\beta(u) = 6/H_c(u) \qquad (12)$$

in which H_c is in km and $\beta(u)$ is shown as a solid curve in Figure 5b. The β describes the percentage of the total melt to be delivered to the ridge axis or the overall efficiency of melt transport to the ridge axis, forming new crust. In general, β decreases with increasing spreading rate. It decreases rapidly from 100% at 10 mm/yr to 60% at 30 mm/yr, then decreases gradually to 55% at 60 mm/yr.

The computed results, of course, depend on the mantle temperature T_m, which controls $F(x, y)$. The calculated crustal thicknesses for two other choices of the mantle temperature are also shown in Figure 5a as a long dashed curve ($T_m =$ 1300°C) and a short dashed curve ($T_m = 1400$°C), respectively. The predicted crustal thickness increases/decreases for hotter/cooler mantle temperature which can also be used to explain the anomalous crustal thicknesses observed at the ridges associated with a hotspot (the Iceland and the Reykjanes Ridge) or a "coldspot" (the Australian-Antarctic Discordance Zone). We believe that the results derived from this model have the correct trend within the range of the global spreading rates since the model represents roughly the "correct" thermal structure and the mantle upwelling flow field near the ridge axis. More accurate results require more accurate thermal and mantle flow models which have to involve numerical calculations as described in the next section.

Numerical Model of Thermal and Mantle Flow Field

In the second model, the mantle flow field beneath mid-ocean ridges is determined by the plate spreading rate and the rheological behavior of the oceanic lithosphere which depends strongly on its thermal structure. On the other hand, the thermal structure of mid-ocean ridges is influenced by the mantle flow field in terms of the heat transfer by advection. Therefore, a numerical scheme is needed to solve the coupled problem between the thermal regime and the mantle flow field. The model of CHEN and MORGAN (1990b) includes the effect of lithospheric thickening on the upwelling mantle flow, the heat of magmatic crustal accretion at the ridge axis, and the hydrothermal cooling due to seawater circulation through the crust. In addition, it models the oceanic lithosphere as a brittle plate, with maximum shear stress limited by BYERLEE's (1978) strength, which overlies a viscous mantle fluid with a realistic power-law rheology.

For the purpose of this study, the numerical model employed here will be the same as the model of mantle half-space with an overlying crust in CHEN and MORGAN (1990b) except that a simple uniform viscosity model is adopted for the oceanic lithosphere. The power-law rheology is important in their study since it generates a weak region in the lower crust beneath the ridge axis for fast spreading rates. This is an important factor causing the transition in ridge axis topography and gravity with spreading rate (CHEN and MORGAN, 1990a,b). Our interest in this study is the temperature structure in the decompression partial melting region deeper than 10 km beneath the ridge axis. Therefore, employing a uniform viscosity would not introduce much error for the temperature field within the partial melting region since the temperature structure near the ridge axis is mostly influenced by magmatic heat input and hydrothermal cooling (PHIPPS MORGAN et al., 1987; CHEN and MORGAN, 1990b; PHIPPS MORGAN and CHEN, 1993a,b).

We use a finite-element program, similar to the RIDGE described in CHEN (1989) to simultaneously solve for the temperature and the mantle flow fields. The oceanic lithosphere is modeled as a viscous fluid with a uniform viscosity

(a)

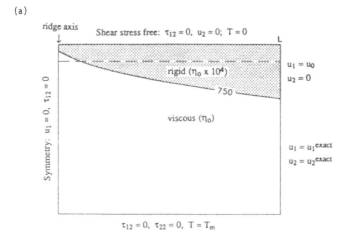

(b)

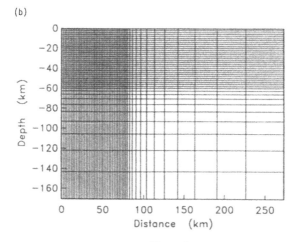

Figure 6

Finite-element mesh (b) and schematic cross section of the numerical model (a), with the x, y coordinate axes centered on the crest of the ridge as shown. The oceanic lithosphere is modeled as a viscous fluid with a uniform viscosity (η_0) overlying a rigid plate (with a higher viscosity) and they are separated by the 750°C isotherm. The analytic solutions (u_1^{exact}, u_2^{exact}) for a triangular plate problem given by (13) define the boundary conditions on the right-hand side. The other boundary conditions are also shown here.

($\eta_0 = 10^{19}$ Pa-s) underlying a brittle plate with a higher viscosity (4 orders of magnitude larger) as shown in Figure 6a. The brittle plate is assigned to have a rigid horizontal motion and is separated from the mantle flow by an isotherm (750°C). We use 750°C for consistency with previous rheological models (CHEN, 1988; CHEN and MORGAN, 1990a,b) and with seismic observations of the oceanic lithosphere (WIENS and STEIN, 1984; BERGMAN and SOLOMON, 1984). The temperature of

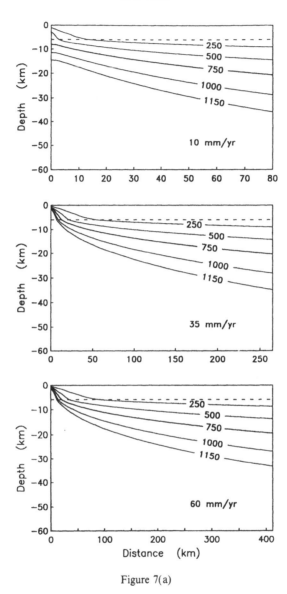

Figure 7(a)

magmatically emplaced material in the crust at the ridge axis is adjusted to be 320°C higher than the intrusion temperature (1200°C) to represent the heat of fusion of molten basalt. The extra heat transfer due to hydrothermal circulation is represented as an increased thermal conductivity within the crust. An uneven grid mesh (shown in Figure 6b) is used to achieve both the maximum accuracy within the melting region (where 40 by 30 grid mesh is adopted) and the minimum error introduced by the boundary conditions on the right and bottom sides of the

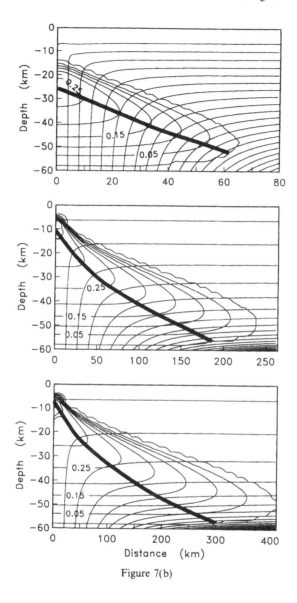

Figure 7(b)

computed region. Details about the finite-element model can be found in CHEN and MORGAN (1990b).

The thermal models shown in Figure 7a, from top to bottom for three different half-spreading rates (10, 35, and 60 mm/yr), were calculated for the case where the thermal conductivity of the crust was enchanced by a factor of three due to hydrothermal cooling, which was suggested in CHEN and MORGAN's (1990b) study for generating the correct transition in ridge axis topography and gravity. Dashed

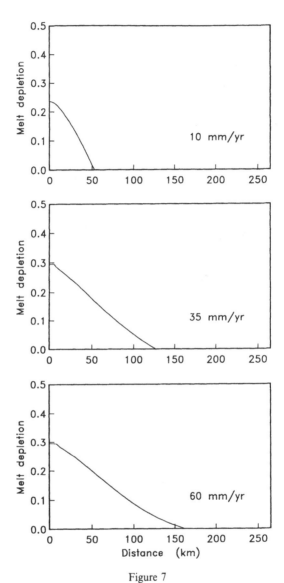

Figure 7
The thermal models are shown in (a), the contours of fraction of melt ($F(x, y)$) and the mantle flow lines in (b), and the melt depletion $D(x)$ in (c). They are computed from the numerical model for half-rates of 10, 35, and 60 mm/yr. An oceanic crust (6 km) is included in this model and is shown as dashed lines in (a). The location $Y_{max}(x)$ of the maximum of $F(x, y)$ for each flow line passing through the melting region is shown as thicker solid lines in (b).

lines represent the Moho (at 6 km) and the temperature contours at 250°C interval except the 1150°C of the bottom isotherm. At the slow spreading rate (10 mm/yr), the 750°C isotherm reaches a depth of 9 km at the ridge axis which is consistent with the observed focal depths of microseismicity near 23°N on the mid-Atlantic

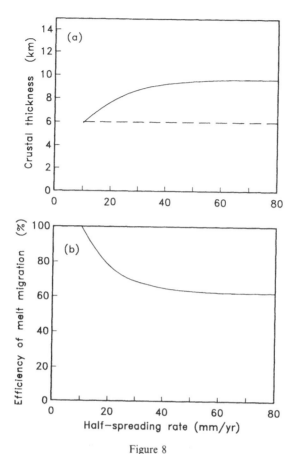

Figure 8
(a) Computed crustal thickness versus half-spreading rates. The observed uniform thickness (6 km) of crust with spreading rates is shown as a dashed line. (b) The overall efficiency factor β of a melt migration mechanism versus the half-spreading rates.

ridge (TOOMEY et al., 1985). At fast spreading rates, the 1150°C isotherm reaches the upper crust which is consistent with the depth to the top of an axial magma chamber detected by previous seismic studies (DETRICK et al., 1987).

The fraction of melt $F(x, y)$ can be calculated by substituting the temperature field (shown in Fig. 7a) into equation (4) and the results are shown as contours in Figure 7b for the three spreading rates. Also shown in Figure 7b are the mantle flow lines which started evenly at the depth of 60 km. Following the procedure described in previous section, the total melt production rate or the crustal thickness created from the melt can be calculated by (6) or (8). Again, $D(x)$ is calculated by (11) in terms of the maximum of $F(x, y)$ that a small patch at (x, y_0) has encountered after flowing through the melting region. The calculated $D(x)$ is shown in Figure 7c for the three spreading rates where we choose $y_0 = 60$ km. The location $Y_{max}(x)$ of the maximum of $F(x, y)$ for each flow line passing through the melting

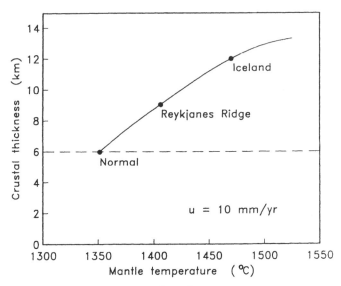

Figure 9
Computed crustal thickness versus mantle temperature for a typical slow half-spreading rate of 10 mm/yr. We choose the model parameters such that the accepted normal mantle temperature of 1350'C can produce the observed 6-km thickness of the oceanic crust. A thicker crust of 9 km or 12 km observed at the Reykjanes Ridge or the Iceland Ridge requires a higher mantle temperature of 1405°C or 1470°C.

region is shown as thicker solid lines in Figure 7b. Again, for the case of a fractional melting the region that contributes the melt production rate becomes the area between $Y_{max}(x)$ and the lower boundary of the melting region (the contour of $F = 0$) as shown in Figure 7b.

The computed crustal thicknesses from this model are shown in Figure 8a for the global range of plate half-spreading rates. Similar to the first analytic model, the calculated H_c increases with spreading rate. For the parameters we choose here, the model predicts a crustal thickness of 6 km at 10 mm/yr and 9.6 km at 60 mm/yr. Therefore, this model satisfies the criterion that it must generate more than 6 km oceanic crust for the range of global spreading rates. An overall efficiency factor β of a melt migration model is required to create the observed uniform thickness (6 km) of oceanic crust. It is calculated by equation (12) and shown in Figure 8b as a function of the half-spreading rate. Again, β decreases with increasing spreading rate. It decreases rapidly from 100% at 10 mm/yr to 70% at 30 mm/yr, then decreases gradually to 63% at 60 mm/yr. Noticing the increases in the size of the melting region with spreading rate (Figure 7b) one should expect that the focusing effect of a melt migration mechanism becomes weaker as the spreading rate increases since it becomes more difficult to transport melts to the ridge axis from larger horizontal distances.

The computed results, of course, depend on the mantle temperature which strongly influences the $F(x, y)$. Similar to the second model, the predicted crustal thickness will increase (or decrease) for hotter (or cooler) mantle temperature which can be used to explain the anomalous thicker (thinner) crustal thicknesses observed at ridges associated with a hotspot (the Iceland and the Reykjanes Ridge) or a "coldspot" (the Australian-Antarctic Discordance Zone).

For example, the Reykjanes Ridge is located on the north Mid-Atlantic Ridge south of the Iceland hotspot. It spreads at a half-rate of 10 mm/yr and has a thicker crust (~ 9 km) (BUNCH and KENNETT, 1980) than the 6 km average (CHEN, 1992). Its thicker crust and "hot" thermal structure due to the Iceland hotspot have been used successfully to explain its peculiar ridge axial topography; an axial block (a typical fast-ridge-axis topography) instead of a rift valley (a typical slow-ridge-axis topography) (CHEN and MORGAN, 1990a,b; PHIPPS MORGAN and CHEN, 1993b). Figure 9 shows the calculated crustal thickness versus mantle temperature for the half-rate of the Reykjanes Ridge with all the other model parameters unchanged. The normal crustal thickness of 6 km is also shown as a short dashed line for reference. The crustal thickness is very sensitive to the mantle temperature: H_c increases from 6 km at 1350°C to 13 km at 1500°C. The crust doubles in thickness when mantle temperature increases 120°C from the normal mantle temperature. To account for the observed crustal thickness 9 km for the Reykjanes Ridge or over 12 km for the Iceland region, we only need to increase the mantle temperature by about 55°C or over 120°C which can be achieved by the Iceland hotspot. This result is consistent with the decreasing mantle temperature away from the Iceland hotspot suggested from analyes of mid-ocean ridge basalts (KLEIN and LANGMUIR, 1987). The sensitivity of the crustal thickness to the mantle temperature can provide useful information about deep mantle temperatures in terms of the observed crustal thickness although we need to develop melt migration models to precisely relate the calculated crustal thickness to observations.

Discussion

Mean Melt Depletion and Mean Melting Depth

The major elements and the isotopic and trace elements of MORB are mainly governed by melting conditions in the partial melting region: melting depletion and melting depth for given thermal and mantle flow models. For simplicity, we assume that all the melt have been removed along the curve of $Y_{max}(x)$ as shown in Figure 7b. Here, $Y_{max}(x)$ is the location of the maximum $F(x, y)$ for each streamline passing through the melting region and is shown as a thicker curve in Figure 7b for three different spreading rates. We define the mean melt depletion (D_{mean}), that is,

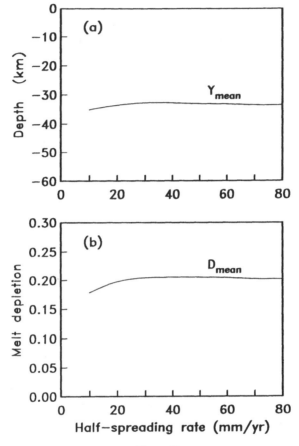

Figure 10
The mean melting depth (a) and the mean melt depletion (b) versus half-spreading rates.

the average degree of melting and the mean melting depth as the weighted average
by choosing the $u_y D(x)$ as the weighting function

$$D_{mean} = \frac{\int_0^{L_f} D(x)u_y D(x) \, dx}{\int_0^{L_f} u_y D(x) \, dx} \tag{13}$$

$$Y_{mean} = \frac{\int_0^{L_f} Y_{max}(x)u_y D(x) \, dx}{\int_0^{L_f} u_y D(x) \, dx} \tag{14}$$

in which L_f is the length of $Y_{max}(x)$ measured from the ridge axis as shown in Figure 4b. The D_{mean} and Y_{mean} versus the spreading rate are shown in Figure 10 and both are not sensitive to the spreading rate which is consistent with the conclusion of FORSYTH (1993), who has shown that the mean melting depth is at two-thirds of the initial melting depth Z_0; here Z_0 is about at 55 km as shown in Figure 7b. The mean melting depth shown in Figure 10a decreases slightly with increasing spreading rate. It decreases from 35.1 km at 10 mm/yr to 33.6 km at 70 mm/yr (only 1.5 km decrease). This is because of variations with spreading rate

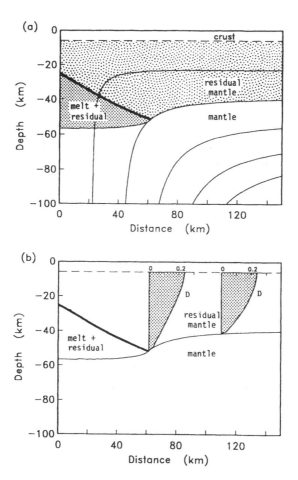

Figure 11

Models of the uppermost mantle at the mid-ocean ridges. The partial melting region is shown as a dark-shaded area and the residual mantle as a light-shaded area in (a). The lower boundary of the residual mantle is defined by the flow line passing the tip of the partial melting region. The boundary ($Y_{max}(x)$) that separates the partial melting region from the residual mantle is shown as a thicker line. Two vertical profiles of melt depletion are shown in (b) at two locations within the residual mantle region.

in the size of the melting region and the locations of $Y_{max}(x)$. Note that $Y_{max}(x)$ is shallower near the ridge axis at fast spreading rates than at slow rates.

Uppermost Mantle Structure

Figure 11 shows our models of the uppermost mantle within the oceanic lithosphere. Although the scales of the axes are for a mid-ocean ridge with a half-spreading rate of 10 mm/yr, the general picture can be applied to all mid-ocean ridges. The partial melting region (melt + residual) is shown as the dark-shaded area and the residual mantle as the light-shaded area in Figure 11a, with the curve of $Y_{max}(x)$ between them. The upwelling mantle flow lines are also shown in Figure 11a. Two typical vertical profiles of the melt depletion are shown in Figure 11b at two locations within the residual mantle region.

By combining this model with a compaction mechanism, we can determine the vertical distribution of the composition and the density of the residual mantle. Qualitative models of the layered structure in the uppermost mantle (depleted mantle over undepleted mantle) and the partial melting region beneath mid-ocean ridges have existed for decades (e.g., OXBURGH and PARMENTIER, 1977; OXBURGH, 1980). However, the quantitative model as shown in Figure 11 is new. Here, the model calculates the exact locations of the boundaries between the mantle, residual mantle, and the partial melting region and the vertical profiles of the melt depletion within the residual mantle for given thermal and mantle flow models.

Efficiency of Melt Migration

In order to generate the uniform thickness (6 km) of oceanic crust, we need a melt migration model with the ability to transport melts from the melting region to the ridge axis as shown in Figure 8b. The efficiency drops from 100% at 10 mm/yr to about 63% at fast spreading rate. We suggested earlier that this is due to the increase in the size of the melting region with spreading rate (Figure 7b) and the difficulty in transporting melt from larger horizontal distances to the ridge axis.

To investigate this effect, let us assume that the melt generated beyond a distance (L_{cut}) can no longer be migrated to the ridge axis; instead it is retained with the mantle matrix. Figure 12 shows the calculated crustal thickness (short dashed line) versus spreading rate with the cutoff limit ($L_{cut} = 60$ km). The result in Figure 8a (corresponding to $L_{cut} = \infty$) is also shown as a solid line, and the uniform crustal thickness (6 km) is shown as a long dashed line for reference. Note that 60 km is the size of the melting region for the 10 mm/yr-half-spreading rate (Figure 7b), that is, we have fixed the size of the melt production region for all different spreading rates by choosing $L_{cut} = 60$ km. Although the crustal thickness at fast spreading rates is reduced significantly they are still thicker than 6 km. This

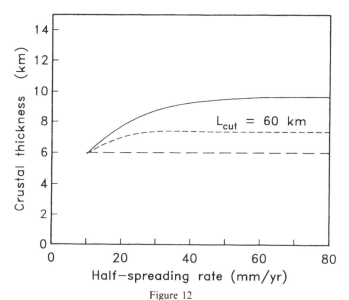

Figure 12

The calculated crustal thickness versus half-spreading rates with the cutoff limit $L_{cut} = 60$ km (dash line) and $L_{cut} = \infty$ (solid line).

indicates that the melt generated near the ridge axis (<60 km) should not totally be extracted at fast spreading rates. From the results shown in Figure 12, one must conclude that the ability to extract the melt within the melting region must decrease with increasing spreading rate even within the area near the ridge axis (<60 km) to create the observed uniform crustal thickness. This is an important constraint to all melt migration models if we accept the thermal and mantle flow models of mid-ocean ridges developed to date.

Comparison with Previous Models

In order to investigate the effect of neglecting variations of melting relations with melt removal in previous models, we computed the total melt production rate beneath the ridge axis (hence the crustal thickness) using both our and previous models. As one example, here we like to compare our model results with the PHIPPS MORGAN and FORSYTH's (1988) model since the formulation of their model is correct and explicit except neglecting variations of the melting relations with melt removal.

According to their model, the melt production rate $\dot{\phi}$ at any place (x, y) beneath the sea floor is

$$\dot{\phi}(x, y) = \mathbf{u}(x, y) \cdot \nabla F(x, y) \tag{15}$$

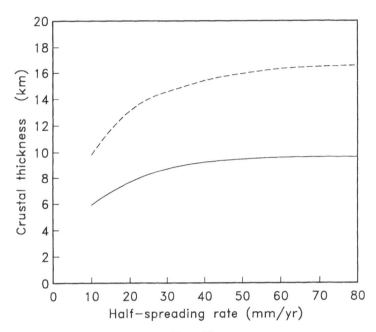

Figure 13
Computed crustal thickness versus half-spreading rates from the model presented in this study (solid line) and the model of PHIPPS MORGAN and FORSYTH (1988) (dashed line). It is shown that the PHIPPS MORGAN and FORSYTH's model (neglecting the effect of changing melting relations with composition changes of the mantle residual because of melt removal) overestimates the crustal thickness by a factor of 1.7 for the global range of spreading rates comparing with the model in this study (including this effect).

where $\mathbf{u}(x, y)$ is the upwelling velocity vector of the mantle matrix. We assume the fraction of melt $F(x, y)$ to be that of the mantle for the P-T conditions at (x, y) so that it does not vary when the composition of the residual mantle matrix changes. Thus $F(x, y)$ can be calculated by equation (4) and it depends on the temperature $T(x, y)$ and pressure $P(y)$ at that location. Assuming that the ideal melt migration takes place, the total melt production rate per unit length of ridge can be calculated by integrating equation (15) over the region S where $\dot{\phi}$ is positive;

$$\dot{W}_m = \int\int_S \rho_m \mathbf{u}(x, y) \cdot \nabla F(x, y) \, dx \, dy. \tag{16}$$

Similar to equation (8), the crustal thickness is

$$H_c = \frac{\rho_m}{\rho_c} \int\int_S \frac{\mathbf{u}(x, y) \cdot \nabla F(x, y)}{u} \, dx \, dy \tag{17}$$

where u is the sea-floor half-spreading rate. Note that $\dot{W}_m$ has to be zero if S includes the entire region of $F \geq 0$. This is because by doing so, we simply do not extract any melt from the upwelling mantle. Following each flow line melt is first generated in the region $\dot{\phi} \geq 0$, then it is recrystallized with the same amount of mass to the mantle matrix in the region $\dot{\phi} \leq 0$ (after passing the maximum of $F(x, y)$). Therefore, the net melt production rate should be zero after the upwelling mantle passes through the region of $F \geq 0$. We use this argument to check the accuracy of the numerical integration of equations (16) or (17). The grid mesh for the numerical integration is refined until that the integral over the region of $\dot{\phi} \geq 0$ equals the integral over the region of $\dot{\phi} \leq 0$.

To compare this model with our presented model in this study, the crustal thickness for different spreading rates calculated using equation (17) is shown as a dashed curve in Figure 13 where our model result (Figure 8a) is replotted as a solid curve. As we expected the previous model predicts a thicker crust than our model since it neglects the effect of the composition changes of the residual mantle. Once lower melting temperature ingredients are removed from the upwelling mantle the residual mantle matrix which consists of higher melting temperature constituents will melt less than the mantle at the same P-T conditions. Thus neglecting the effect of the composition changes with melt removal will always overestimate the total melt production rate. This is clearly seen from Figure 13 and the equation (A10) in the Appendix.

One interesting result we found from Figure 13 is that the predicted crustal thickness (dashed) of the PHIPPS MORGAN and FORSYTH's (1988) model is about 1.7 thicker than our model crustal thickness (solid) for the global range of spreading rates. As we stated earlier our model predicts the maximum crustal thickness because we assume an ideal melt migration mechanism. Including a realistic melt migration model and its interactions with the melt production model will decrease the crustal thickness predicted by our model here (see below). Therefore, we can conclude that neglecting variations of the melting relations with melt extraction will lead to overestimation of the total melt production rate and thus the crustal thickness by at least a factor of 1.7.

Limitations of the Model

Finally, we discuss the limitations of our model due to the complexities of the magmatism at mid-ocean ridges. First of all, given the temperature and mantle flow fields, our model calculates the total melt production rate by integrating the melt removed from a small mass along its flow line over all the flow lines passing the partial melting region. The model includes the effects of the variations in the solidus and liquidus temperatures of the residual mantle with the changes in its composition. However, the model does not consider the interactions between the flow lines. That is, in a real situation, the melt generated from a flow line at a further distance from the axis must migrate through the partial melting region to the ridge axis and this will affect the melt generation along the flow lines it passes. An intuitive conclusion is that the total melt production rate will be reduced by this effect. Here we avoid this complexity of the interactions between melt production and melt migration mechanisms by assuming that melt extracted from flow lines will add to the new oceanic crust at the ridge axis.

Second, melting of the upwelling mantle rocks will affect the temperature field (via release of latent heat) which has not been included in the thermal models used in this study. Third, the removal of melt from the mantle matrix and the resulting mantle matrix compaction will create an effective mass sink within the mantle. The mantle flow to fill in this mass deficit in the melting region has not been included in the mantle flow models used in this study. The normal stress-supported axial topography produced by melt compaction has been studied by PHIPPS MORGAN *et al.* (1987) for a case of a line sink beneath a spreading center. We believe further study of combining their solutions with the results of this study is needed to investigate the effects of both the melt production and the melt compaction on the mantle flow field and the axial topography.

Summary

We present a model for computing the total melt production rate from the decompression partial melting region beneath a mid-ocean ridge, and the maximum oceanic crustal thickness created at the ridge axis assuming an ideal melt migration mechanism. The calculations are based on a self-consistent numerical model for the thermal structure and steady-state mantle flow field at a mid-ocean ridge. The model includes the effect of decreasing the melt production rate within the partial melting region by melt extraction as the residual mantle matrix becomes increasingly difficult to melt. Thus the melt fraction depends not only on temperature and pressure determined by the location beneath the ridge axis (the Eulerian description), but also on the accumulated melt extraction since the upwelling mantle matrix enters the partial melting region determined by the location along the flow-line path (the Lagrangian description). The last effect has been neglected by previous models. The model can predict the size of the melting region and the locations of the boundaries between mantle, residual mantle, and the partial melting region for a given spreading rate, also the distribution of the melt depletion and the mean melting depth.

Given the observed thickness of oceanic crust (~ 6 km), which is relatively independent of spreading rate, the model results provide a constraint on the overall efficiency of melt migration to the ridge axis; the efficiency must decrease from 100% at 10 mm/yr to about 60% at fast spreading rates (>50 mm/yr). Although this reduction may be partially due to the increasing size of the melting region with increasing spreading rate, it still requires less efficient melt migration near the ridge axis at fast spreading rate.

The calculated crustal thickness is sensitive to the mantle temperature. For a normal mantle temperature of 1350°C, the model can generate the observed 6–km oceanic crust over the global range of spreading rates, while the anomalous thicker crusts of the Iceland hotspot and the Reykjanes Ridge must be related to higher mantle temperatures associated with the hotspot.

Finally, by comparing our model results with previous ones we found that neglecting variations of the melting relations with melt extraction will lead to overestimation of the total melt production rate and thus the crustal thickness by at least a factor of 1.7.

Future studies should include modeling the melt migrations beneath the ridge axis, analyzing the chemistry of mid-ocean basalts to compare the available data, and calculating variations in composition and density of the residual mantle with depth. The investigation should cooperate with a compaction model which can also influence the ridge axis topography. Further, the analysis described in this paper for the magmatism at mid-ocean ridges can be applied to the magmatism at continental rift zones due to lithospheric stretching or hotspots.

The same two-dimensional numerical test can be performed for the rift zones to see how much stretching (or horizontal velocity) is needed to induce partial melting. Note that the thermal structure is much different between a conductive-cooling-only lithosphere and a conductive-cooling lithosphere with upwelling mantle flow. Once the thermal structure and the mantle flow field at the rift zones are known, we can calculate the melt production rate and compare it with the observed crustal thickness at the rift zones or to the volume of the flood basalts; which may provide quantitative information about the evolution of the rift zones.

Appendix

In this Appendix, we present the derivation of Equations (5)–(7) in the text. We will consider a rock partial melting process shown in Figure A1. The curve in Figure A1a is a state path in the P-T plane which represents a rock partial melting process from state A to state E. Assume the fraction of melt F (the ratio of the total amount of melt W_m retained in the system to the initial mass of the rock sample W_0) follows the curve in Figure A1b. That is, the rock reaches the solidus temperature at state A and starts to melt; the maximum amount of melt retained in the system is reached at state C, and then it decreases toward state D when the solidus is passed again and all the melt is completely recrystallized. From state D to state E the condition is well below the solidus and no melt is present in the system. Note that none of the melt is removed from the rock sample in the process and in practice, the melting curve in Figure A1b can usually be determined in the laboratory. Here the mass of melt present at any state is

$$W_m = FW_0. \tag{A1}$$

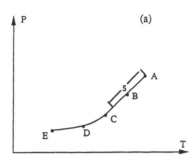

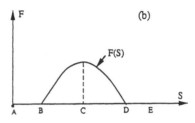

Figure A1

(a) A state path A → E in (a) represents a general melting process. (b) The fraction of melt $F(S)$ for the case of retaining all the melt is shown as a function of S, the distance along the state path from point A in (a).

When melt is allowed to be removed from the system, however, this simple relationship breaks down, or the relationship can still hold as

$$W_m = DW_0. \qquad (A2)$$

But now D (defined as melt depletion) is different from F because of the continuous change of the residual rock composition from the initial composition. Thus, D at a state depends on the composition of the residual rock at that state and is usually not known.

The question is can the amount of melt removed from the system be estimated given F? Next we try to answer this question for an ideal fractional melting process, that is, we assume that any melt generated during the process will be removed from the system instantaneously.

Let us define a variable s which represents the path along the curve in the P-T plane as shown in Figure A1. Then for the case of retaining all the melt as done in the laboratory, we rewrite equation (A1) as

$$\frac{dW_m(s)}{ds} = \frac{dF(s)}{ds} W_0. \qquad (A3)$$

Let us assume a similar equation holds for the case of an ideal fraction melting as

$$\frac{dW_m(s)}{ds} = \frac{dF_f(s)}{ds} W_s \tag{A4}$$

where W_s is the mass of the residual rock which is the total mass remaining in the system due to melt removal. The unknown function $F_f(s)$ is different from $F(s)$ because the composition of the residual rock at state s differs from the initial composition which controls $F(s)$. On the other hand, we have $W_s' = -W_m'$, here the prime represents the derivative respect to the variable s. Therefore, (A4) can be written as

$$W_s'(s) = -F_f'(s)W_s(s). \tag{A5}$$

Integration of (A5) over the state path gives

$$W_s(s) = W_0 \, e^{-\int_{s_B}^{s} F_f'(s)\, ds} = W_0 \, e^{-F_f(s)} \tag{A6}$$

where the following relations are applied: $W_s'(s_B) = W_0$ and $F_f(s_B) = 0$. The amount of melt removed from the system at state s is

$$W_m(s) = W_0 - W_s(s) = W_0[1 - e^{-F_f(s)}]. \tag{A7}$$

The melt depletion D is defined as the ratio of the mass of melt removed to the initial mass of the rock

$$D(s) = \frac{W_m(s)}{W_0} = 1 - e^{-F_f(s)}. \tag{A8}$$

The derivation above is exact except that $F_f(s)$ is not known. Next, let us introduce a major assumption

$$F_f'(s) = -F'(s) \tag{A9}$$

during the melting period from state B to state C. Then from (A8) we have

$$D(s) = 1 - e^{-F(s)} \tag{A8}$$

$$= F(s) - \frac{F(s)^2}{2!} + \frac{F(s)^3}{3!} - \cdots \tag{A9}$$

where Equation (A9) is obtained by expanding (A8) using Taylor series expansion. One conclusion we can draw from (A9) immediately is that

$$D(s) < F(s) \tag{A10}$$

since $F(s) < 1$, that is, the amount of melt removed from the system is always less than the amount of melt retained in the system along the same state path in the P-T plane. For example, for a reasonable value of $F = 0.35$, that is a 35% of melt will exist within the system, $D = 1 - e^{-0.35} = 0.29$. It says that only 29% of the rock sample will melt and be removed from the system. It is about 83% ($0.29/0.35 = 83\%$) of the melt retained with the rock sample that can be removed for the case of an ideal fractional melting.

For the melting process with curve of $F(s)$ shown in Figure A1, the total melt removed from the rock with an initial mass W_0 is

$$W_m = W_0 D = W_0[1 - e^{-F_{max}}] \tag{A11}$$

where $F_{max} = F(s_c)$. Thus, for the purpose of calculating the total melt removed from the system only the value of F_{max} is required as long as $F(s)$ increases monotonously from state B to state E and the exact form of $F(s)$ is not required to be known.

To understand the physical picture behind (A10), let us compare (A3) with (A4). Even though we introduced the assumption of (A9) later, the fundamental difference between (A3) and (A4) is that the total mass within the system in (A4) equals W_s (instead of W_0) which decreases continuously with the removal of melt. Thus the increment of melt produced during the small path ds given by (A4) is less than that given by (A3). On the other hand, it is believed that in general, $F_f'(s) < F'(s)$, therefore (A8) gives an upper bound to the melt removal from the rock sample during a partial melting process.

The above analysis can be directly applied to the mid-ocean ridge systems to estimate the total melt production rate from the depression partial melting region beneath the ridge axis.

Acknowledgments

The author was supported by the Ida and Cecil Green Fellowship during the time this work was done and the first draft was written. Special thanks go to J. Bernard Minster and John Orcutt, who helped make possible the stay of the author at the Institution of Geophysics and Planetary Physics, Scripps Institution of Oceanography. I have benefited from discussions with Andrew Woods, who was my officemate in the Green Scholar office, David Sandwell, and Jason Phipps Morgan. David Sandwell improved the manuscript. I thank Dr. Aki and Prof. Wang for organizing the IUTAM Symposium (September 1994, Beijing) and editing this special issue. The completion of the final version was partly supported by NSF OCE-9300512.

REFERENCES

AHERN, J. L., and TURCOTTE, D. L. (1979), *Magma Migration Beneath an Ocean Ridge*, Earth Planet. Sci. Lett. *45*, 115–122.

BERGMAN, E. A., and SOLOMON, S. C. (1984), *Source Mechanisms of Earthquakes near Mid-ocean Ridges from Body Waveform Inversion: Implications for the Early Evolution of Oceanic Lithosphere*, J. Geophys. Res. *89*, 11415–11441.

BUNCH, A. W. H., and KENNETT, B. L. N. (1980), *The Crustal Structure of the Reykjanes Ridge at 59°30'N*, Geophys. J. Roy. Astr. Soc. *61*, 141–166.

BYERLEE, J. D. (1978), *Friction of Rocks*, Pure and Appl. Geophys. *116*, 615–626.

CAWTHORN, R. G. (1975), *Degrees of Melting in Mantle Diapirs and the Origin of Ultrabasic Liquids*, Earth Planet. Sci. Lett. *27*, 113–120.

CHEN, Y. (1988), *Thermal Model of Oceanic Transform Faults*, J. Geophys. Res. *93*, 8839–8851.

CHEN, Y. (1989), *Dynamics of Mid-ocean Ridge Systems*, Ph.D. Thesis, Princeton University, Princeton, N.J., 221 pp.

CHEN, Y. (1992), *Oceanic Crustal Thickness versus Spreading Rate*, Geophys. Res. Lett. *19*, 753–756.

CHEN, Y., and MORGAN, W. J. (1990a), *Rift Valley/No Rift Valley Transition at Mid-ocean Ridges*, J. Geophys. Res. *95*, 17,751–17,581.

CHEN, Y., and MORGAN, W. J. (1990b), *A Nonlinear-rheology Model for Mid-ocean Ridge Axis Topography*, J. Geophys. Res. *95*, 17,583–17,604.

DETRICK, R. S., BUHL, P., VERA, E., MUTTER, J., ORCUTT, J., MADSEN, J., and BROCHER, T. (1987), *Multi-channel Seismic Imaging of a Crustal Magma Chamber along the East Pacific Rise*, Nature *326*, 35–41.

FORSYTH, D. W. (1993), *Crustal Thickness and the Average Depth and Degree of Melting in Fractional Melting Models of Passive Flow Beneath Mid-ocean Ridges*, J. Geophys. Res. *98*, 16,073–16,079.

HENSTOCK, T. J., WOODS, A. W., and WHITE, R. S. (1993), *The Accretion of Oceanic Crust by Episodic Sill Intrusion*, J. Geophys. Res. *98*, 4143–4154.

KLEIN, E. M., and LANGMUIR, C. H. (1987), *Global Correlations of Ocean Ridge Basalt Chemistry with Axial Depth and Crustal Thickness*, J. Geophys. Res. *92*, 8089–8115.

KLEIN, E. M., and LANGMUIR, C. H. (1989), *Local Versus Global Variations in Ocean Ridge Basalt Composition: A Reply*, J. Geophys. Res. *94*, 4241–4252.

McKENZIE, D. P. (1967), *Some Remarks on Heat Flow and Gravity Anomalies*, J. Geophys. Res. *72*, 6261–6273.

McKENZIE, D. P. (1984), *The Generation and Compaction of Partially Molten Rock*, J. Petrology *25*, 713–765.

McKENZIE, D. P. (1985), *The Extraction of Magma from the Crust and Mantle*, Earth Planet. Sci. Lett. *74*, 81–91.

McKENZIE, D. P., and BICKLE, M. J. (1988), *The Volume and Composition of Melt Generated by Extension of the Lithosphere*, J. Petrology *29*, 625–679.

MORGAN, W. J., *Heat flow and vertical movements of the crust*. In *Petroleum and Global Tectonics* (Fischer, A. G., and Judson, S., eds.) (Princeton University Press, Princeton, N. J. 1975).

OXBURGH, E. R., and PARAMENTIER, E. M. (1977), *Compositional and Density Stratification in Oceanic Lithosphere — Causes and Consequences*, J. Geol. Soc. Lond. *133*, 343–355.

OXBURGH, E. R., *Heat flow and magma genesis*. In *Physics of Magmatic Processes* (Hargraves, R. B., ed.) (Princeton University Press, Princeton, N. J. 1980).

PARSONS, B., and SCLATER, J. G. (1977), *An Analysis of the Variation of Ocean Floor Bathymetry and Heat Flow with Age*, J. Geophys. Res. *82*, 803–827.

PHIPPS MORGAN, J. (1987), *Melt Migration Beneath Mid-ocean Spreading Centers*, Geophys. Res. Lett. *14*, 1238–1241.

PHIPPS MORGAN, J., and CHEN, Y. J. (1993a), *The Genesis of Oceanic Crust: Magma Injection, Hydrothermal Circulation, and Crustal Flow*, J. Geophys. Res. *98*, 6283–6297.

PHIPPS MORGAN, J., and CHEN, Y. J. (1993b), *The Dependence of Ridge-axis Morphology and Geochemistry on Spreading Rate and Crustal Thickness*, Nature *364*, 706–708.

PHIPPS MORGAN, J., PARMENTIER, E. M., and LIN, J. (1987), *Mechanisms for the Origin of Mid-ocean Ridge Axial Topography: Implications for the Thermal and Mechanical Structure of Accreting Plate Boundaries*, J. Geophys. Res. *92*, 12823–12836.

PHIPPS MORGAN, J., and FORSYTH, D. W. (1988), *Three-dimensional Flow and Temperature Perturbations due to a Transform Offset: Effects on Oceanic Crustal and Upper Mantle Structure*, J. Geophys. Res. *93*, 2955–2966.

REID, I. D., and JACKSON, H. R. (1981), *Oceanic Spreading Rate and Crustal Thickness*, Mar. Geophys. Res. *5*, 165–172.

SCOTT, D. R., and STEVENSON, D. J. (1989), *A Self-consistent Model of Melting, Magma Migration and Buoyancy-driven Circulation Beneath Mid-ocean Ridges*, J. Geophys. Res. *94*, 2973–2988.

SLEEP, N. H., and WINDLEY, B. F. (1982), *Archean Plate Tectonics: Constraints and Inferences*, J. Geol. *90*, 363–379.

SPARKS, D. W., and PARMENTIER, E. M. (1991), *Melt Extraction from the Mantle Beneath Spreading Centers*, Earth Planet. Sci. Lett. *105*, 368–377.

SPIEGELMAN, M., and MCKENZIE, D. (1987), *Simple 2-D Models for Melt Extraction at Mid-ocean Ridges and Island Arcs*, Earth Planet. Sci. Lett. *83*, 137–152.

STOLPER, E., WALKER, D., HAGER, B. H., and HAYS, J. F. (1981), *Melt Segregation from Partially Molten Source Regions the Importance of Melt Density and Source Region Size*, J. Geophys. Res. *86*, 6261–6271.

TOOMEY, D. R., SOLOMON, S. C., PURDY, G. M., and MURRAY, M. H. (1985), *Microearthquakes Beneath the Median Valley of the Mid-Atlantic Ridge near 23°N: Hypocentres and Focal Mechanisms*, J. Geophys. Res. *90*, 5443–5458.

WHITE, R. S., and MCKENZIE, D. P. (1989a), *Magmatism at Rift Zones: The Generation of Volcanic Continental Margins and Flood Basalts*, J. Geophys. Res. *94*, 7685–7729.

WHITE, R. S., and MCKENZIE, D. P. (1989b), *Volcanism at Rifts*, Scientific American, 62–71, Jul.

WIENS, D. A., and STEIN, S. (1984), *Intraplate Seismicity and Stresses in Young Oceanic Lithosphere*, J. Geophys. Res. *89*, 11442–11464.

(Received December 5, 1994, revised/accepted June 11, 1995)

PAGEOPH, Vol. 146, Nos. 3/4 (1996)

0033-4553/96/040621-28$1.50 + 0.20/0

Dynamics of the Mid-ocean Ridge Plate Boundary: Recent Observations and Theory

YONGSHUN JOHN CHEN[1]

Abstract—The global mid-ocean ridge system is one of the most active plate boundaries on the earth and understanding the dynamic processes at this plate boundary is one of the most important problems in geodynamics. In this paper I present recent results of several aspects of mid-ocean ridge studies concerning the dynamics of oceanic lithosphere at these diverging plate boundaries. I show that the observed rift valley to no-rift valley transition (globally due to the increase of spreading rate or locally due to the crustal thickness variations and/or thermal anomalies) can be explained by the strong temperature dependence of the power law rheology of the oceanic lithosphere, and most importantly, by the difference in the rheological behavior of the oceanic crust from the underlying mantle. The effect of this weaker lower crust on ridge dynamics is mainly influenced by spreading rate and crustal thickness variations. The accumulated strain pattern from a recently developed lens model, based on recent seismic observations, was proposed as an appealing mechanism for the observed gabbro layering sequence in the Oman Ophiolite. It is now known that the mid-ocean ridges at all spreading rates are offset into individual spreading segments by both transform and nontransform discontinuities. The tectonics of ridge segmentation are also spreading-rate dependent: the slow-spreading Mid-Atlantic Ridge is characterized by distinct "bulls-eye" shaped gravity lows, suggesting large along-axis variations in melt production and crustal thickness, whereas the fast-spreading East-Pacific Rise is associated with much smaller along-axis variations. These spreading-rate dependent changes have been attributed to a fundamental differences in ridge segmentation mechanisms and mantle upwelling at mid-ocean ridges: the mantle upwelling may be intrinsically plume-like (3-D) beneath a slow-spreading ridge but more sheet-like (2-D) beneath a fast-spreading ridge.

Key words: Oceanic spreading center, mantle upwelling, rift valley to no-rift valley transition, ridge segmentation.

1. Introduction

The 65,000-km-long mid-ocean ridge system was identified as a diverging plate boundary in the global plate tectonics (e.g., MORGAN, 1968; MCKENZIE, 1967; LEPICHON *et al.*, 1973) where new ocean floor is created. It is one of the most active plate boundaries on the earth and its processes affect profoundly the internal structure of the oceanic lithosphere that covers two-thirds of the earth's surface.

The early exploration of the ocean basins showed that the global ridge system is segmented (over hundreds of kilometers) by transform faults which left remark-

---

[1] College of Oceanic and Atmospheric Sciences, Oregon State University, Corvallis, OR 97331, U.S.A.

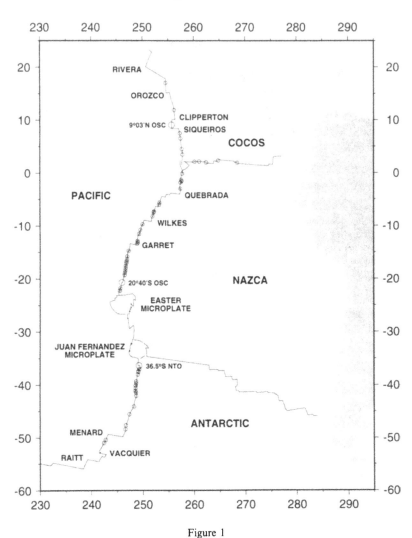

Figure 1
Location of the major plate boundaries of the Southeast Pacific. The occurrence of Overlapping Spreading Centers (OSCs) and nontransform offsets (NTOs) is indicated by the small circles along the East Pacific Rise based primarily on data from MACDONALD *et al.* (1986) for north of the Easter microplate and from LONSDALE (1994) for south of the Juan Fernandez microplate. The three offsets of 9°03′N OSC, 20°40′S OSC, and 36.5°S NTO discussed in the text are shown as slightly larger circles.

able scarps in the ocean floor, known as the fracture zones. With the availability of the rapidly-improving multibeam capability during the last two decades, this early view of the global mid-ocean ridge system has been rapidly changing. It is now recognized that these oceanic spreading centers are offset not only by large transforms but also by nontransform offsets with smaller offset length (e.g., MACDONALD and FOX, 1983; LONSDALE, 1983; SEMPERE *et al.*, 1990). As an example, Figure 1 shows the oceanic spreading centers in the East Pacific. Some of these nontransform offsets are constantly evolving and are propagating along the ridge axis, leaving oblique trails in the ridge flanks (e.g., LONSDALE, 1985; MACDONALD et al., 1986).

Although understanding the dynamics of the mid-ocean ridge system is a difficult and challenging problem, significant progress has been made in both observations and theory. For the last two decades, mid-ocean ridges have been the subject of intense international studies; most recent examples include the decade-long InterRidge initiative that incorporates several national programs such as RIDGE (U.S.), BRIDGE (U.K.), French RIDGE, InterRidge Japan, and others. In this paper as well as at the IUTAM Symposium (September 1994, Beijing), I present results of several aspects of mid-ocean ridge studies concerning the dynamics of oceanic lithosphere at these diverging plate boundaries. The discussions will include the dynamics of axial rifting and tectonics of ridge segmentation and associated problems such as 3-D versus 2-D mantle upwelling and passive versus active upwelling at these oceanic spreading centers.

2. Dynamics of Rifting and Thermal Structure of Mid-ocean Ridges

Transition in Ridge Axis Topography and Gravity with Spreading Rate

It has been known since the early expeditions in the 1960s that the topographic expression of the oceanic spreading centers varies at global scale and that such variations are dominantly spreading-rate dependent. The contrast between the observed axial topography in the Atlantic and the Pacific has generated an interesting debate about the representative topographic expression of these oceanic spreading centers (e.g., HEEZEN et al., 1959; HILL, 1960; MENARD, 1960). The Mid-Atlantic Ridge (MAR), spreading at slow half-rates of 10–20 mm/yr, has a median rift valley, 1–2 km deep and 15–30 km wide, whereas the East Pacific Rise (EPR), spreading at fast half-rates of 40–80 mm/yr, has a contrasting axial topographic high, 100–200 m high and 1–2 km wide (Fig. 2a). This well-observed transition in the axial topography as a function of spreading rate shown in Figure 2b (MENARD, 1967; MACDONALD, 1982, 1986) is consistent with the observations of satellite gravity across the mid-ocean ridge axes (Fig. 2c). The peak to trough vertical deflection of the geoid across the ridge axis was found to have a contrasting character between slow and fast spreading rates (SMALL and SANDWELL, 1989). The gravity signatures usually have high amplitude gravity troughs and are highly variable for half-rates less than 30 mm/yr, indicating a dynamic origin of the rugged topography including the rift valleys at these slow-spreading ridges. Fast-spreading ridges are characterized by uniform, low amplitude gravity highs indicating a state of nearly local isostatic equilibrium (SMALL and SANDWELL, 1989).

This first-order dependence of the ridge axis topography on spreading rate has been the focus of intensive theoretical studies over the past two decades and

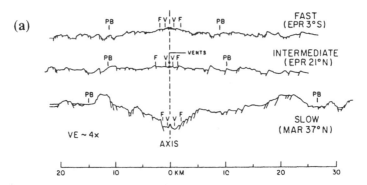

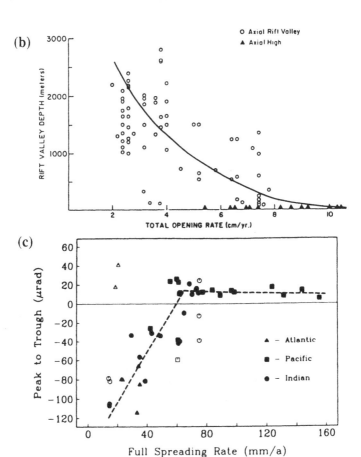

Figure 2

Spreading-rate dependence of ridge axis topography and gravity. (a) Typical profiles across fast, intermediate, and slow-spreading ridges show the transition from a rift valley at slow-spreading rates to an axial high at fast spreading rates (MACDONALD, 1982). (b) Depth of rift valley versus spreading rate (MACDONALD, 1986). The calculated axial relief (solid line) from our model matches the observations very well (CHEN and MORGAN, 1990b). (c) Peak to trough vertical deflection of the geoid across the ridge axis versus full spreading rate from SMALL and SANDWELL (1989).

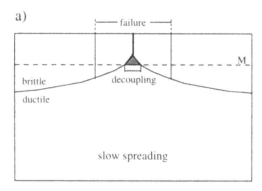

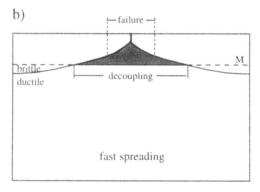

Figure 3
A conceptual model of a passive spreading center for two different spreading rates (CHEN and MORGAN, 1990a). (a) At a slow-spreading ridge, the coupling between the viscous flow field and the strong lithospheric plate will cause a failure zone, an axial rift valley would be formed due to stretching/necking of the cold lithosphere. (b) At a fast spreading ridge, however, the hot lower crust beneath the ridge axis (dark shaded area) is rheologically weak and can effectively decouple the thin lithosphere of the upper crust (only 1–2 km thick) from underlying mantle flows. Thus the size of the hot lower oceanic crust with low ductile strength determines the transition from a well-developed rift valley at slow ridges to no-rift valley at fast ridges.

numerous models have been motivated by this observation (e.g., SLEEP, 1969; LACHENBRUCH, 1973, 1976; TAPPONNIER and FRANCHETEAU, 1978; MADSEN et al., 1984; PHIPPS MORGAN et al., 1987; LIN and PARMENTIER, 1989). While these models address the ridge dynamics at either slow-spreading ridges or fast-spreading ridges, none of them were focused on the transition in the contrasting characters of the Mid-Atlantic Ridge and the East Pacific Rise nor did they offer a mechanism to explain this transition with spreading rate. For example, PHIPPS MORGAN et al. (1987) suggested that the stretching of a strong oceanic lithosphere may be the origin of median valley topography as first proposed by TAPPONNIER and FRANCHETEAU (1978). On the other hand, MADSEN et al. (1984) proposed an isostatic model of a thin elastic plate with a low-density magma region beneath the plate to account for the axial topographic high and the rise crest gravity anomaly of fast-spreading ridges.

We have proposed a dynamic model of rift formation which is especially focused on the transition in observed ridge axis morphology and gravity with spreading rate (CHEN and MORGAN, 1990a,b). The important new feature in this model is the inclusion of an oceanic crust in the model and the emphasis on its weaker rheological strength than the underlying mantle at moderate and high temperatures. We have attributed the contrasting ridge-axis topography to both the spreading-rate dependence of the axial thermal structure and the associated rheology of the oceanic lithosphere at mid-ocean ridge plate boundary, and most importantly, the difference in the rheological behavior of the oceanic crust from the underlying mantle. For example, the viscosity of an oceanic crustal rock such as a gabbro at 800°C is at least a few orders of magnitude less than that of the mantle at the same temperature.

The basic mechanism is shown in Figure 3. At slow-spreading ridges (Fig. 3a), the cold crust is coupled to the underlying strong mantle to create a relatively thick lithospheric plate (>6 km) at the ridge axis; an axial rift-valley is then formed as the result of failure due to stretching and necking of the cold lithosphere as suggested by the earlier models (TAPPONNIER and FRANCHETEAU, 1978; PHIPPS MORGAN et al., 1987). At fast-spreading ridges (Fig. 3b), however, the hot lower crust beneath the ridge axis (dark shaded area) is rheologically weak and can effectively decouple the thin lithosphere of the upper crust (only 1–2 km thick) from underlying mantle flows. As a consequence, the dominant form of plate failure at fast ridges is dike intrusion within a narrow neovolcanic zone at ridge-axis rather than the formation of a rift valley over a broad zone (~30 km) of deformation. Thus the size of the hot lower oceanic crust with low ductile strength determines the transition from a well-developed rift valley at slow ridges to no-rift valley at fast ridges. While the size of this hot lower crustal region (decoupling zone) is mainly controlled by the spreading rate, it could also be influenced by local crustal thickness variation (magma supply) and thermal anomaly such as due to the proximity to a hotspot (e.g., the Reykjanes Ridge). Therefore, the crustal thickness and mantle temperature are also important parameters in this new model to explain local variations in ridge axis topography along a ridge segment. A thicker crust or a hotter thermal structure will increase the size of this decoupling zone and a transition from the Atlantic type to the Pacific type of ridge axis topography could occur at a slow-spreading ridge such as at the Reykjanes Ridge and at the unusually shallow ridge segment of 33°S, Mid-Atlantic Ridge (KUO and FORSYTH, 1988).

This mechanism has been implemented using a finite element program of "RIDGE2D" which solves the coupled problems between the ridge thermal regime and the upwelling mantle flow field for the oceanic lithosphere with a power-law rheology for both the crust and the mantle (CHEN, 1989; CHEN and MORGAN, 1990b). It includes the effects of lithospheric thickening on the upwelling mantle

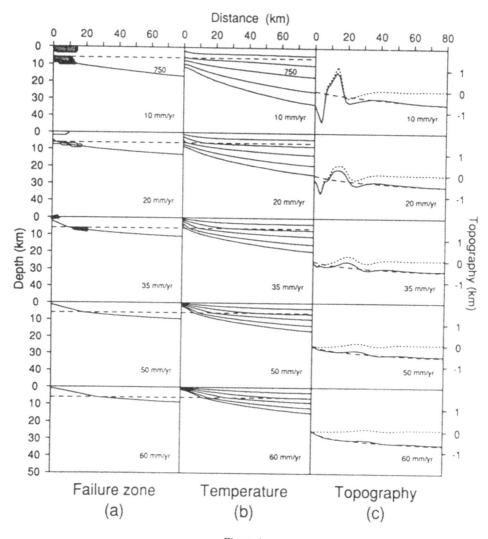

Figure 4

Results from the mantle-plus-crust model for half-rates of 10, 20, 35, 50, and 60 mm/yr. Shown on the left are the computed failure zones (shaded area) and the 750°C isotherm for reference, in the middle are the isotherms contoured at a 250°C interval, and on the right are the computed topography profiles: the total topography (solid) is the sum of the dynamic topography (dotted) and the thermal isostatic topography (dashed). The dynamic topography is induced by the stresses of the viscous flow and the thermal isostatic topography is due to the square-root-of-age cooling effect (after CHEN and MORGAN, 1990b).

flow, the heat of magmatic crustal accretion at the ridge axis, and the hydrothermal cooling due to seawater circulation through the crust. These nonlinear finite element calculations (Fig. 4) have successfully confirmed our conceptual model (Fig. 3): the pronounced rift valley at slow-spreading ridges is absent at fast-spreading ridges

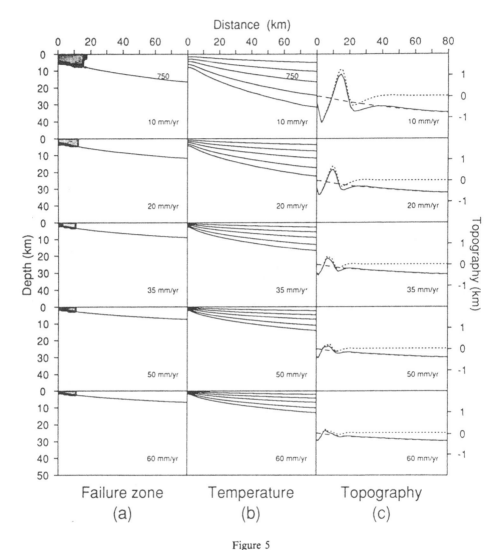

Figure 5
Results from the mantle-only model for half-rates of 10, 20, 35, 50, and 60 mm/yr. The rest are the same
as in Figure 4 (after CHEN and MORGAN, 1990b).

because of the decoupling of the lower, weaker crust at faster spreading rates. The predicted relief of the rift valley (solid line) decreases with increasing spreading rate and agrees well with the bathymetry observations (Fig. 2b).

Without including the power-law rheology of the oceanic crust as shown in Figure 5, however, the mantle-only model calculations show that while the rift valley becomes less pronounced with increasing spreading rate (right column), it does not disappear at fast-spreading rates as displayed in Figure 4 for the case with an oceanic crust. The diminishing form of the rift valley correlates with the

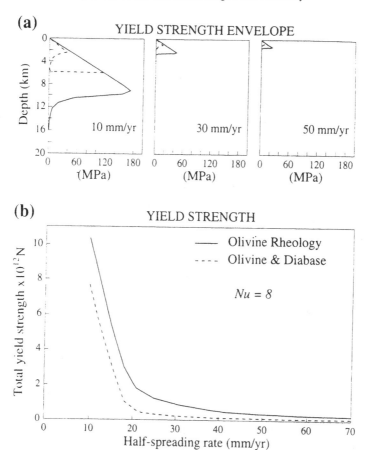

(a) YIELD STRENGTH ENVELOPE

(b) YIELD STRENGTH

Figure 6

(a) Model yield strength envelopes of the oceanic lithosphere at the ridge axis with an olivine rheology (solid lines) and a diabase-plus-olivine rheology (dashed) for 10, 30, and 50 mm/yr (half-rates) (CHEN and MORGAN, 1990b). (b) Depth-integrated total axial yield strength as a function of spreading rate for the mantle-only model (solid) and the mantle-plus-crust model (dashed). There is a strong decrease in the total axial yield strength for both models at spreading half-rates below 20 mm/yr because of the strong temperature dependence of the power-law rheology. This indicates that while the stretching of a strong lithosphere may be responsible for the formation of a rift valley at slow-spreading ridges, the absence of a median valley at fast-spreading ridges cannot be explained solely by the sharp decrease of the total yield strength at fast-spreading rates (Figs. 4 and 5). It is the presence of a hot lower crust with weaker rheology than the mantle that decouples the thin upper lithosphere from the mantle flow below and prevents the formation of the rift valley.

decreasing size of the failure zone and a thinner lithosphere shown in the left column of Figure 5. Also notice the absence of the failure zone of the lithosphere at fast-spreading rates when the crust is included (Fig. 4). Comparison of Figures 4 and 5 reveals the importance of the oceanic crust in causing the rift valley to no-rift valley transition against spreading rate as proposed in our simple conceptual model (Fig. 3). On the other hand, Figure 6 illustrates that there is an abrupt increase in the integrated axial lithosphere strength with decreasing spreading rate below 20 mm/yr for both the mantle-only model and mantle-plus-crust model. The yield strength envelope of the oceanic lithosphere is shown on the top panel of Figure 6 for three different, half-spreading rates: 10, 30, and 60 mm/yr, respectively. This sharp increase is due to the strong temperature dependence of the power-law rheology of both the crust and the mantle. Yet without including the crust, the mantle-only model fails to predict the no-valley topography at fast-spreading ridges due to the presence of a failure zone within the relatively thin brittle lithosphere (Fig. 5). We concluded that such a rift valley to no-rift valley transition (globally due to an increase of spreading rate or locally due to the crustal thickness variations and/or thermal anomalies) is caused not only by the strong temperature dependence of the power-law rheology of the oceanic lithosphere, but also, most importantly by the different rheological behavior of the oceanic crust from the underlying mantle (CHEN and MORGAN, 1990a,b). Thus the success of this model in explaining the transition from rift-valley to no-rift-valley ridge axis topography confirmed the important role of the oceanic crust in shaping the dynamic flow beneath a mid-ocean ridge and ridge axial topography, indicating the crustal thickness variation is also another important factor in ridge dynamics besides the spreading rate differences.

Finally, the observed sharp transition in axial gravity signature (Fig. 2c) can be understood as the combination of effects of decreasing topographic amplitude and increasing isostatic compensation with increasing spreading rate (SMALL and SANDWELL, 1989; SMALL et al., 1989). We will refer to the model discussed above (CHEN and MORGAN, 1990a,b) as a dike model since uniform crustal accretion along a vertical, crustal-size dike is assumed to distinguish it from the lens model which will be discussed next.

Figure 7

(a) Schematic model of a mid-ocean ridge spreading center derived from Oman ophiolite studies (SMEWING, 1981). As shown on the right side, isotropic gabbros just beneath the sheeted dike complex grade into gabbros with a (weakly developed) near vertical dip which becomes more developed and more shallowly dipping deeper into the gabbro section. This dip structure was used by SMEWING (1981) to infer that gabbro layering reflects cumulate deposition on the floor of the large magma chamber sketched here. (b) Seismic model of magma chamber structure for the East Pacific Rise. Molten magma is concentrated in a lens that is approximately 1 km wide which resides at the base of the sheeted dike complex. Beneath the magma lens and extending to mid-crustal depths is a broader region of rock at elevated temperatures which contains a few percent partial melt. (c) Schematic theoretical model depicting the flowlines, accumulated strain, and layering predicted by lower crustal flow away from a shallow injection lens at the sheeted dike/gabbro interface (PHIPPS MORGAN and CHEN, 1993a). The left side of the figure shows predicted layering "isochrons" generated from crustal flow away from a magma lens at the base of the sheeted dike complex. The right-hand side shows several typical crustal flow lines and the accumulated strain along each flowline. Strain is most intense in the lowermost part of the gabbro section (after PHIPPS MORGAN et al., 1994).

(a)   Gabbro Layering & Genisis -- Oman Ophiolite

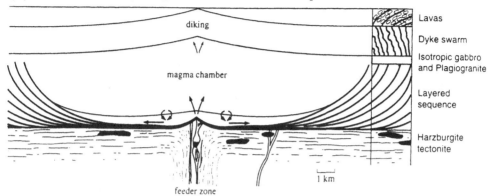

(b)   Seismic Image of Axial Crustal Structure

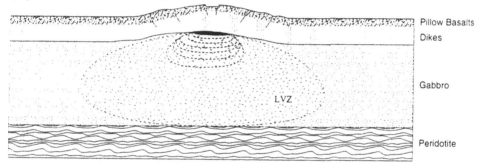

(c)   Magma Lens Model for Crustal Flow and Layering

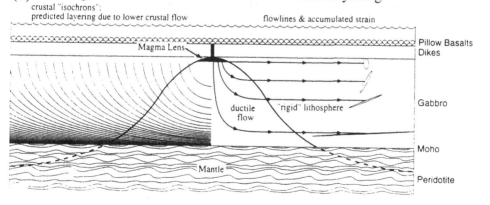

Transition in Magma Lens Depth against Spreading Rate and Oceanic Crustal Genesis

Ophiolite studies have provided us with detailed information of the oceanic crustal structure and reasonable interpretation of the seismically determined layering structure of the oceanic crust. In particular, the well-studied Oman ophiolite, believed to be created at a fast-spreading mid-ocean ridge (NICOLAS, 1989), has a structural sequence through the gabbro section from a less developed vertical dip beneath the sheeted dike/gabbro contact to well-developed, moho parallel layers above the gabbro-peridotite contact as shown on the right column of Figure 7a. This changing layer dips through the gabbro section has been used to suggest the shape of a crustal-size magma chamber beneath a fast-spreading ridge (Fig. 7a) in which these layers are results of cumulate deposits (SMEWING, 1981).

However, these early scenarios of a crustal-size magma chamber have not been confirmed by recent seismic observations at mid-ocean ridges. On the contrary, seismic reflection studies regarding the fast-spreading East Pacific Rise have found a rather small magma body about 1–2 km beneath the axis which is narrow (~ 1 km wide) and thin (few hundreds of meters thick) in cross-axis section, and it seems to be continuous along the axis (DETRICK *et al.*, 1987; MUTTER *et al.*, 1988; HARDING *et al.*, 1989, 1993; VERA *et al.*, 1990; KENT *et al.*, 1990). This thin melt lens is underlain by a broader (~ 6 km wide and 2–4 km thick) low velocity region of hot rock that may include ~ 3–5% of melt (Fig. 7b). The depth of the magma lens was reported to be deeper at intermediate spreading Juan de Fuca Ridge (~ 3 km) (MORTON *et al.*, 1987; ROHR *et al.*, 1988) than the 1–2 km depths along the fast-spreading East Pacific Rise (PURDY *et al.*, 1992). While these strong seismic reflectors of the magma lens can be traced over tens of kilometers along the EPR, such seismic signature has not been detected at slow-spreading Mid-Atlantic Ridges (DETRICK *et al.*, 1990).

Motivated by these recent seismic studies of fast- and slow-spreading ridges and the field observations of gabbro layering sequence at the Oman ophiolite site, we have extended our dike model by including a crustal-accretion lens in the upper crust which is similar to the earlier "magma chamber" and crustal flow model of SLEEP (1975). We have shown (PHIPPS MORGAN and CHEN, 1993a,b, 1994) that magma injection into a small magma lens at the base of the sheeted-dike complex has the potential to accommodate both the gabbro layering patterns seen in the ophiolite record and the small magma lens observed in seismic studies at fast-spreading ridges (Fig. 7c). The accumulated strain of this crustal flow field is small and almost vertical at the depth of the lens; it becomes progressively more intense and closer to horizontal as the Moho is approached. From the depth of the lens down to the moho the amplitude of the strain increases a few orders of magnitude and the strain axis rotates from vertical to within a few degrees from horizontal. We proposed that the gabbro fabric seen in Oman ophiolite reflects the accumulated strain pattern resulting from a lower crustal flow away from a quasi-steady-state, shallow-level intrusion zone (PHIPPS MORGAN and CHEN, 1993a), rather than being due to cumulates settling from a large axial magma chamber (SMEWING, 1981). The model also successfully predicts the observed pattern in the variation of the depth to the lens with spreading rate (see Figure 7 of PHIPPS MORGAN and CHEN, 1993a), that is, the deepening of the magma lens with decreasing spreading rate (PURDY *et al.*, 1992) and the sharp transition from a mid-crustal level lens to no lens for half-spreading rates less than ~ 20 mm/yr (DETRICK *et al.*, 1990).

In a recent JGR comment, NICOLAS (1994) argued that the observed dips within the Oman gabbro layer should dip away from the paleoaxis instead of toward the paleoaxis as shown in SMEWING (1981) based on his team's extensive field observations in Oman. While the issue of the exact location of the paleoaxis at the Oman ophiolite is important and should be resolved among the ophiolite community, a recent seismic reflection study (EITTERIM *et al.*, 1994) has found ridgeward (eastward) dipping lower crustal reflectors in the central Pacific crust that was created at the fast-spreading paleo-East Pacific Rise. The sense of eastward dip of both the seismic reflectors in the lowermost (~ 2 km) crust and the general pervasive ridgeward dipping seismic fabric in the data of the ~ 6100 km long crust (parallel to the crustal flow line) are consistent with our lens model. We believe that when the issue raised in the Oman ophiolite studies is resolved, together with the seismic reflection data they will offer the potential to distinguish between a passive ridge upwelling from an active, strongly diapiric process at a mid-ocean ridge (NICOLAS, 1994; PHIPPS MORGAN and CHEN, 1994).

Local Variability: Dependence on Crustal Thickness

Besides the fundamental spreading rate dependence, exceptions in ridge axis topography have long been noted. The often cited example is the slow-spreading (10 mm/yr half-rate) Reykjanes Ridge which has a typical Pacific type topography rather than an axial rift valley. Its thicker crust (8–10 km) (BUNCH, 1980; BUNCH and KENNETH, 1980; SMALLWOOD *et al.*, 1995) is believed to result from the excess magma supply from the adjacent Iceland hotspot. As illustrated in Figure 10 of CHEN and MORGAN (1990b), while a ridge spreading at this typical slow rate has a pronounced rift valley for the average crustal thickness of 6 km, a ridge with crust twice as thick would not develop a rift valley. Their model calculation for the Reykjanes Ridge with a thicker crust (9 km) and an anomalously hotter mantle temperature (1500°C) predicts a Pacific type topography. The anomalously thicker crust and hotter thermal structure, due to proximity to the Iceland hotspot, provide a large decoupling chamber in the lower crust which enables the Reykjanes Ridge to behave like a fast-spreading ridge: no rift valley and very little dynamic topography. Similar to the fast-spreading ridges, local isostasy takes place in the lower crust and buoyant forces bend the plate upward to produce the observed axial high.

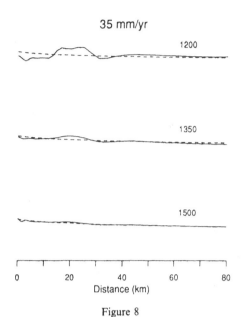

Figure 8
Computed ridge axis topography with a 6 km crust for an anomalous low mantle temperature of 1200°C
(top), a normal mantle temperature of 1350°C (middle), and an anomalous high mantle temperature of
1500°C (bottom). A cold mantle results in a rift valley at this intermediate spreading rate which would
explain the rift valley to no-rift valley transition in the observed ridge axis topography at the Australian
Antarctic Discordance (AAD) (CHEN and MORGAN, 1990b). A thinner crust would reduce the negative
mantle temperature anomaly to create the rift valley (after CHEN and MORGAN, 1990b).

Another example is the intermediate spreading (30–35 mm/yr half-rate) Australian Antarctic Discordance (AAD) along the Southeast Indian Ridge, which has
a median valley in contrast to neighboring, axial-high ridge-segments to the east
and west (HAYES and CONNOLLY, 1972; WEISSEL and HAYES, 1972, 1974; SEM-
PÉRÉ et al., 1991). The depth of the seafloor younger than 10 m.y. is more than
500 m deeper than expected from the normal square-root-of age relationship and
this depression is believed to be associated with the unusually cold thermal structure
in the underlying mantle (WEISSEL and HAYES, 1974), a scenario which is consistent with the chemistry of dredged basalts in this region (KLEIN and LANGMUIR,
1987) and seismic surface wave study (FORSYTH et al., 1987). Figure 8 shows three
model calculations for three different mantle temperatures of 1200°C, 1350°C, and
1500°C (CHEN and MORGAN, 1990b). With a lower mantle temperature of 1200°C,
the model predicts a rift valley at this intermediate spreading rate for a 6-km thick
crust. A thinner crust would enhance the relief of the rift valley and would also
require a smaller low mantle temperature anomaly to produce this transition.

Our model (CHEN and MORGAN, 1990a,b) has been tested at the well surveyed
area of 31°–36°S, along the South Mid-Atlantic Ridge, and has been able to
successfully predict the along-axis variation in axial topography resulting from the

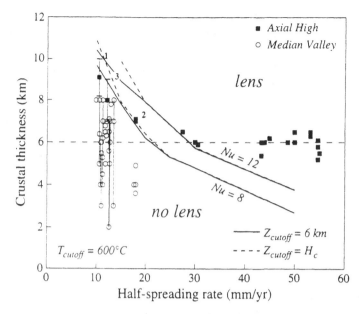

Figure 9

Minimum crustal thickness for the presence of a steady-state magma lens within the crust versus the half-spreading rate for two Nusselt numbers of 8 and 12 (solid curves). Crustal thickness data with an axial high morphology are shown by filled squares and those with a rift valley morphology are designated by open circles. Circles connected by solid lines illustrate the range of crustal thickness variation seen along several fracture zone profiles. All data are a near-ridge subsample of the data set reported by CHEN (1992) except for those points annotated by superscripts: 1, Reykjanes Ridge; 2, Mid-Atlantic Ridge at 31°S; and 3, Mid-Atlantic Ridge at 39°N (after PHIPPS MORGAN and CHEN, 1993b).

local along-axis variations in the magma supply (crustal thickness) (NEUMANN and FORSYTH, 1993).

To further address this issue, we have used our lens model to systematically investigate the effect of the along-axis variations in magma supply (crustal thickness) on the axial thermal structure and ridge axis topography at a given spreading rate (PHIPPS MORGAN and CHEN, 1993b). Our model calculations shown in Figure 9 indicate that a 'threshold' crustal thickness exists at a given spreading rate about which small changes in crustal thickness (magma supply) can produce a dramatic change in axial thermal structure and the axial morphology. It is worthy to note that variations in axial thermal structure (lens versus no lens) are most sensitive at intermediate spreading rates to small fluctuations in magma supply around the normal rate which leads to an average crustal thickness of ~ 6 km (CHEN, 1992). We found that for a normal mantle temperature of 1350°C, a reduction in crustal thickness from 6 km to 4.5 km within the AAD is sufficient to change the ridge axis crust from an axial magma lens structure at ~ 3.5 km depth to a 'slow-spreading' thermal structure with no quasi-steady-state magma lens present. A lower mantle temperature will require a smaller crustal thickness reduction for the transition to occur. Preliminary report from a seismic refraction experiment conducted in 1994 at the segments of the AAD and the adjacent Southeast Indian Ridge (SEIR) suggested that the crust is thinner at the AAD and a magma lens is detected underlying the SEIR axis but is absent at the AAD (TOLSTOY et al., 1995). On the other hand, the Reykjanes Ridge appears to need a greater crustal thickness perturbation from 'normal' 6 km crustal thicknesses to sustain a quasi-steady-state magma lens; our results suggest a minimum 9–10.5 km crustal thickness to sustain a magma lens within the crust while a higher mantle temperature may reduce this threshold crustal thickness.

3. Ridge Segmentation/3-D versus 2-D Mantle Upwelling

Ridge Segmentation

Marine geophysical surveys since the 1980s, using multi-beam swath-mapping and other state-of-the-art capabilities, have revealed increasingly three-dimensional nature of seafloor spreading processes at the mid-ocean ridge plate boundaries. It was found that mid-ocean ridges at all spreading rates are offset into individual spreading segments (10s to 100s of km long) by both transform faults and nontransform discontinuities (e.g., MACDONALD et al., 1991). The magmatic and tectonic segmentation is particularly apparent along slow-spreading ridges, such as the Mid-Atlantic Ridge (MAR), where new evidence for segmentation has been discovered recently by detailed geophysical mapping of bathymetry and gravity (KUO and FORSYTH, 1988; KONG et al., 1988; LIN et al., 1990; SEMPÉRÉ et al., 1990; BLACKMAN and FORSYTH, 1991; FOX et al., 1991; GRINDLAY et al., 1991; SHAW, 1992), seismic experiments (TOOMEY et al., 1985, 1988; KONG et al., 1992), and near-bottom structural geology (e.g., KARSON and DICK, 1983; KARSON et al., 1987). In the Pacific, it is ascertained that these long segments separated by major transforms along the fast-spreading East Pacific Rise are also divided into numerous small segments (Fig. 1) by overlapping spreading centers (OSCs) (MACDONALD and FOX, 1983; MACDON-ALD et al., 1984, 1986, 1991) and nontransform offsets (NTOs) (LONSDALE, 1983, 1985, 1986, 1989). Some of these offsets propagate along the ridge crest and leave oblique trails on the rise flanks such as the 20°S NTO (REA, 1978) and the 36.5°S NTO (LONSDALE, 1994), similar to the propagating rifts at the intermediate spreading Galapagos Ridge described by HEY et al. (1980), and the huge, dueling propagator system (~120 km offset) at 29°S (JOHNSON et al., 1994) with complex tectonic deformation similar to a microplate structure (HEY et al., 1994).

Along-axis Variations

There are systematic along-axis variations in zero-age seafloor depth associated with the ridge segmentation that are believed to reflect along-axis variations in the magmatic supply (e.g., MACDONALD et al., 1991; SEMPÉRÉ et al., 1990). The ridge axis generally deepens progressively toward transform and large nontransform offsets (e.g., FOX and GALLO, 1984; MACDONALD et al., 1984, 1986) which indicates that magma supply should decrease toward large discontinuities, resulting in a thinner crust at these offsets. Significant crustal thinning has been observed seismically at major fracture zones along the slow-spreading Mid-Atlantic Ridges. For example, an extremely thin crust, 2–3 km thick, was found at the Kane fracture zone (DETRICK and PURDY, 1980; CORMIER et al., 1984). DETRICK et al. (1993) presented an excellent review of the seismic crustal structure at the major fracture zones in the Atlantic and they have concluded that large-offset transforms along the slow-spreading Mid-Atlantic Ridge are associated with anomalous crustal structures, often a few kilometers thinner than the average 6 km crust. Anomalously thin crust has also been inferred from both seismic and gravity studies at smaller ridge axis discontinuities along the Mid-Atlantic Ridge (KUO and FORSYTH, 1988; LIN et al., 1990; TOLSTOY et al., 1993; DETRICK et al., 1995).

At fast spreading ridges, it has been reported that the deepening of the rise crest toward large offsets correlates with a narrowing of its cross section and, sometimes, with the disappearance of the axial magma chamber (AMC) reflectors (MACDONALD and FOX, 1988; SCHEIRER and MACDONALD, 1993). For example, an increase of the residual gravity anomaly toward the 20°40'S overlapping spreading center (20 km offset), correlated with the deepening of the rise crest depth by about 400 m toward the OSC, was interpreted by CORMIER et al. (1994) to be consistent with a 500-m thinner crust at the OSC. They concluded that the 20°40'S OSC is associated with a reduced magmatic budget for at least the past 1.5 Ma. As an exception, seismic evidence does not support the hypothesis of reduced melt supply near the 9°03'N OSC; the crustal thickness inferred from the seismic data increases toward the offset (HARDING et al., 1993; BARTH and MUTTER, 1991), indicating an enhanced melt supply near the offset. Detailed analysis and modeling of the multichannel seismic reflection data at and near the 9°03'N OSC have revealed an asymmetric distribution of melt beneath this relatively small (9 km) offset (KENT et al., 1993). The new model proposed by KENT et al. (1993), based on seismic observations at the 9°03'N OSC, argued that the offset is underlain by a large low velocity zone with few percent partial melt similar to the previous LONSDALE's model (1983).

Spreading Rate Dependence

As in the case of ridge-axis rift structure, the tectonics of ridge segmentation are also spreading-rate dependent: the slow-spreading Mid-Atlantic Ridge is characterized by distinct "Bulls-eye" shaped gravity lows associated with large zero-age-seafloor-depth variations, suggesting large along-axis variations in melt production (supply) and crustal thickness, whereas the fast-spreading East Pacific Rise is

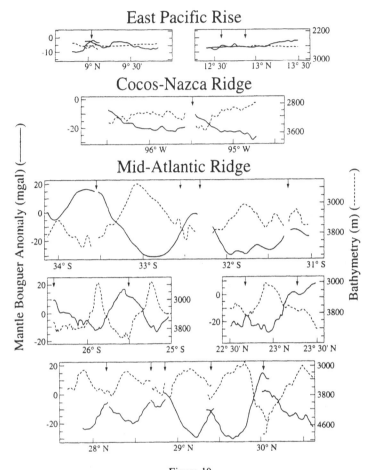

Figure 10
Along-axis profiles of seafloor depth (dashed lines) and mantle Bouguer anomaly (solid lines) at fast (EPR), intermediate (Cocos-Nazca), and slow (MAR) spreading ridges. Arrows mark the location of transforms and nontransform offsets. C-Cox; M-Moore; RG-Rio Grande; K-Kane; A-Atlantis (after LIN and PHIPPS MORGAN, 1992).

associated with much smaller along-axis variations and smooth topography as shown in Figure 10 (LIN and PHIPPS MORGAN, 1992).

At several locations along the slow-spreading Mid-Atlantic Ridge, pronounced large "Bulls-eye" mantle Bouguer anomaly (MBA) lows are observed at the center of the segments bounded by either transform or nontransform offsets such as the survey areas of 31°–34°S (KUO and FORSYTH, 1988), 25°–27.5°S (BLACKMAN and FORSYTH, 1991), the MARK area at 23°N (MORRIS and DETRICK, 1991), 27°–31°N (LIN et al., 1990), and 33°–40°N (DETRICK et al., 1995). In particular, a large circular-shaped MBA low is centered on the unusually shallow ridge segment at both 26°S and 33°S; the median valley virtually disappears at the center of both segments. The origin of gravity bull's eyes and the spatial distribution and relative importance of magmatism and tectonism, with respect to ridge segmentation at slow-spreading ridges, is one of the three major problems identified and recommended for RIDGE segment-scale studies in the next 3–5 years during the 1994 RISES Workshop at Wakefield, Massachusetts.

In contrast, fast-spreading East Pacific Rise has been found to be associated with subdued along-axis mantle Bouguer anomaly variations (MADSEN et al., 1990; LIN and PHIPPS MORGAN, 1992; CORMIER et al., 1994). For example, an amplitude of 10 mGal in along-axis mantle Bouguer anomaly variation is observed for the 160-km-long 7°12′–8°38′S segment (WANG and COCHRAN, 1993) and ENRIQUEZ and CHEN (1995) observed less than 10 mGal along-axis MBA variation for the segment north of the Menard transform at 49.5°S.

Compared to the continental crust, the oceanic crust is known to be simple: not only because of its simple layered structure but also due to its small variations in the total crustal thickness. A compilation of recent reliable seismic observations (CHEN, 1992) has not shown a systematic increase of the average crustal thickness with spreading rate (Fig. 9). Instead, the data show large variations in crustal thickness at slow-spreading rates and small variations at fast-spreading rates. This spreading-rate dependence of the crustal thickness variations is consistent with the global gravity studies and supports the speculation of a transition from a 3-D structure of crustal accretion at slow-spreading ridges to a 2-D accretion pattern at fast ridges (Fig. 11).

3-D versus 2-D Mantle Upwelling

These spreading-rate dependent changes in both zero-age seafloor depths and MBA variations have been attributed to a fundamental difference in ridge segmentation mechanisms and mantle upwelling at mid-ocean ridges (Fig. 11): the mantle upwelling may be intrinsically plume-like (3-D) beneath a slow-spreading ridge but more sheet-like (2-D) beneath a fast-spreading ridge (LIN and PHIPPS MORGAN, 1992). Such a transition in mantle upwelling may occur if the relative importance of passive upwelling over buoyant upwelling increases with increasing spreading rate (PARMENTIER and PHIPPS MORGAN, 1990). Alternatively, small amplitude, three-dimensional upwelling may occur at a fast-spreading ridge, but its effect on crustal thickness variations may be significantly reduced by along-axis melt flows along a persistent low-viscosity crustal magma lens (PHIPPS MORGAN, 1991; LIN and PHIPPS MORGAN, 1992).

There has been a debate recently whether this transition of 3-D versus 2-D mantle upwelling as a function of spreading rate takes place. Theoretical studies showed that mantle upwelling, driven by density variations from melt extraction, has a spreading-rate-dependent transition between 3-D and 2-D upwelling structure (PARMENTIER and PHIPPS MORGAN, 1990), suggesting that the ridge segmentation

FAST SPREADING RIDGE   SLOW SPREADING RIDGE

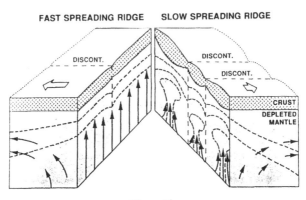

Figure 11
Sketch of a spreading-rate dependent mantle upwelling model (LIN and PHIPPS MORGAN, 1992). Solid arrows show mantle flow directions and open arrows are the plate spreading velocities. Dashed lines in the mantle show the isotherms. Global gravity studies indicate that a fundamental difference in the mantle upwelling and the crustal accretion pattern at the ridge axes: the mantle upwelling may be intrinsically plume-like (3-D) beneath a slow-spreading ridge (right) but more sheet-like (2-D) beneath a fast-spreading ridge (left) (after LIN and PHIPPS MORGAN, 1992).

may have a different origin at slow- and fast-spreading rates. On the other hand, a recent 3-D gravity study of the fast-spreading East Pacific Rise between 9°17′ and 9°27′N found 35% larger residual gravity anomaly near the axial depth minimum at 9°50′N, suggesting a 3-D pattern of mantle upwelling focused at the depth minimum of the ridge segment (WANG et al., 1996).

4. Passive Versus Active Spreading Centers

Our view of the mantle upwelling beneath the mid-ocean ridges has changed from the early notion of a hot rising jet which is part of the mantle convection (e.g., OXBURGH, 1980) and the location of this diverging plate boundary is controlled by the convecting mantle flow below. As pointed out by MCKENZIE and BICKLE (1988), this scenario of an active oceanic spreading center contains major problems. For example, if Africa and Antarctic are surrounded by active spreading ridges, then where does the upwelling material go and where is the downwelling limb of the mantle convection cell? Does the convecting system migrate with the propagating ridges? When a ridge jumps over a distance of hundreds of kilometers, does the hot jet move also, and if so how? These serious problems disappear with the current view that the mid-ocean ridges are passive spreading centers where the new oceanic crust is created, resulting from passive mantle upwelling as two plates are pulling apart at the plate boundary. The location of the plate boundary is mainly controlled by a preexisting weakness of the plate which is pulling apart by remote plate driving forces.

This concept of passive spreading centers has been the focus of a recent debate based on differences between recent global seismic tomography studies (SU *et al.*, 1992; ZHANG and TANIMOTO, 1992). The 3-D global *S*-wave velocity model of ZHANG and TANIMOTO (1992) shows that, while both mid-ocean ridges and hotspots are underlain by low-velocity anomalies in the upper mantle, the underlying structures are distinctly different. The anomalies associated with the mid-ocean ridges are limited to the upper 100 km below the surface, whereas the anomalies underlying the hotspots are at depths between 100 km and 200 km. The limited depth extent of the low velocity anomalies beneath the mid-ocean ridges in ZHANG and TANIMOTO's (1992) model is consistent with a passive mantle upwelling (induced flow) in the upper mantle at these oceanic spreading centers. However, SU *et al.* (1992) argued that such a shallow origin for the mid-ocean ridge anomalies is inconsistent with the observed travel-time residuals for the seismic phase SS. Their independent, 3-D, global, *S*-wave velocity model shows that the velocity anomalies associated with the mid-ocean ridges extend to a depth of 300 km, and some of these anomalies may even extend into the lower mantle (SU *et al.*, 1992, 1994). Their model and others (WOODWARD and MASTERS, 1992) have suggested a deeper origin for the low velocity anomalies beneath the mid-ocean ridges, implying an active mantle upwelling as part of the global mantle convection. While the debate could be resolved by higher resolution, 3-D, global, seismic velocity models, the outcome will certainly put strong constraints on one of the most important questions in global plate tectonics—whether mid-ocean ridges are passive or active spreading centers.

Locally, even in a passive system, there may be a dynamic component of mantle upwelling due to various buoyancy forces such as density anomaly caused by variable degrees of melt extraction and a small percent of melt retention within the mantle. The mechanism of melt delivery to the ridge axis has been debated for the past few years and various mechanisms have been proposed to focus the melt from a wider (>40 km) partial melting region at depths of 30–60 km to the moho (6 km depth) beneath the narrow neovolcanic zone (~2-km wide) (RABINOWICZ *et al.*, 1987; PHIPPS MORGAN, 1987; BUCK and SU, 1989; PHIPPS MORGAN and FORSYTH, 1988; SCOTT and STEVENSON, 1989; SOTIN and PARMENTIER, 1989; CORDERY and PHIPPS MORGAN, 1992; SPARKS and PARMENTIER, 1991). Theoretical studies indicate that both the lateral extent of the partial melting region beneath the ridge axis and the total melt production rate increase with spreading rate (e.g., FORSYTH, 1993; CHEN, this issue).

Finally, there is still a great deal to learn about the dynamics of the mid-ocean ridge plate boundaries. For instance, while the link of the subduction zone to whole mantle convection is clear as indicated by deep earthquakes down to 680 km depth, seismic velocity anomalies extending into the lower mantle, and the relationship between velocity anomalies and the geoid (e.g., HAGER, 1984; HAGER and CLAYTON, 1989; WOODWARD and MASTERS, 1992; GRAND, 1994; SU *et al.*, 1994), the role of oceanic spreading centers in the mantle convection is still debatable.

Another challenging problem is to understand the uniform thickness of oceanic crust, a well-known observation for the past three decades, in terms of the melt production and melt migration beneath the ridge axes. Seismic observations to date (CHEN, 1992; WHITE et al., 1992) have documented a rather uniform average crustal thickness for most of the spreading rate range (> 10 mm/yr half-rate), although the decrease of the crustal thickness at very slow-spreading rates (< 10 mm/yr half-rate) is debatable. An excellent example of uniformity is reported by EITTREIM et al. (1994), who reported that seismic reflection data along a 6100-km-long flow line of crustal generation in the central Pacific manifest a nearly constant Moho reflection time of 2 s, independent of the spreading rate and age of the seafloor. The few observations of a very thin crust at super-low spreading ridges might either indicate a significant thinning of the crust with spreading rate (REID and JACKSON, 1981; WHITE et al., 1992) or reflect an enhanced crustal thickness variation related to the ridge segmentation (CHEN, 1992), since the ridge axis at these extremely slow rates is heavily segmented as inferred by recent high-quality global satellite gravity data (MCADOO and MARKS, 1992; SANDWELL and SMITH, 1994). To advance our understanding of this important plate boundary, more data need to be collected, and importantly, realistic theoretical ridge models need to be developed. For example, in a recent three-dimensional gravity study conducted at a relatively large (45 km offset) nontransform offset at 36.5°S of the Pacific-Antarctic EPR (CHEN et al., 1995) it was discovered that the mantle Bouguer anomalies do not follow the well-observed ridge crest into the overlapping zone, and a lag seems to exist between the sub-seafloor structure and the surface structure related to the propagation of the offset. The authors called for a realistic three-dimensional, ridge-NTO thermal model to better understand and interpret the mantle Bouguer anomalies at these nontransform offsets.

Acknowledgments

I thank Dr. Aki and Prof. Wang for organizing the IUTAM Symposium (September 1994, Beijing) and editing this special issue. Special thanks are extended to Prof. Wang for his invitation and encouragement of my composition of this manuscript. This paper is an extension of the abstract reported at the IUTAM Symposium. A large portion of the research results reported in this manuscript is the result of the work I collaborated with either Jason Morgan or Jason Phipps Morgan. I also thank Jian Lin for his contribution to the IUTAM abstract. Don Forsyth and Min Liu provided helpful reviews which improved the manuscript. This work was partly supported by NSF OCE-9300512 and the College of Oceanic and Atmospheric Sciences at Oregon State University.

REFERENCES

BARTH, G. A., and MUTTER, J. C. (1991), *Patterns of Crustal Thickness Produced along the East Pacific Rise, 8°50′N to 9°50′N* (abstract), EOS Trans. AGU *72*, 490.

BLACKMAN, D. K., and FORSYTH, D. W. (1991), *Isostatic Compensation of Tectonic Features of the Mid-Atlantic Ridge: 25–27°30′S*, J. Geophys. Res. *96*, 11,741–11,758.

BUCK, W. R., and SU, W. (1989), *Focused Mantle Upwelling below Mid-ocean Ridges due to Feedback between Viscosity and Melting*, Geophys. Res. Lett. *16*, 641–644.

BUNCH, A. W. H. (1980), *Crustal Development of the Reykjanes Ridge from Seismic Refraction,* J. Geophys. *47*, 261–264.

BUNCH, A. W. H., and KENNETH, B. L. N. (1980), *The Crustal Structure of the Reykjanes Ridge at 59°30′N*, Geophys. J. R. Astron. Soc. *61*, 141–166.

CHEN, Y., *Dynamics of Mid-ocean Ridge Systems*, Ph.D. Thesis (Princeton University, Princeton, N.J. 1989) pp. 221.

CHEN, Y. (1992), *Oceanic Crustal Thickness versus Spreading Rate*, Geophys. Res. Lett. *19*, 753–756.

CHEN, Y. J. (1995), *Constraints on the Melt Production Rate beneath the Mid-ocean Ridges Based on Passive Flow Models*, Pure and Appl. Geophys., Pure Appl. Geophys. *146*, 589–620.

CHEN, Y., and MORGAN, W. J. (1990a), *Rift Valley/No Rift Valley Transition at Mid-ocean Ridges*, J. Geophys. Res. *95*, 17,571–17,581.

CHEN, Y., and MORGAN, W. J. (1990b), *A Nonlinear-rheology Model for Mid-ocean Ridge Axis Topography*, J. Geophys. Res. *95*, 17,583–17,604.

CHEN, Y. J., ENRIQUEZ, K. D., and LONSDALE, P. (1995), *A Gravity Study of the Large Nontransform Offset at 36.5°S of the East Pacific Rise*, J. Geophys. Res., submitted.

CORDERY, M. J., and PHIPPS MORGAN, J. (1992), *Melting and Mantle Flow beneath a Mid-ocean Spreading Center*, Earth Planet. Sci. Lett.

CORMIER, M. H., DETRICK, R. S., and PURDY, G. M. (1984), *Anomalously Thin Crust in Oceanic Fracture Zones: New Seismic Constraints from the Kane Fracture Zone*, J. Geophys. Res. *89*, 10,249–10,266.

CORMIER, M. H., MACDONALD, K. C., and WILSON, D. S. (1994), *A Three-dimensional Gravity Analysis of the East Pacific Rise from 18° to 21°30′S*, J. Geophys. Res., submitted.

DETRICK, R. S., and PURDY, G. M. (1980), *The Crustal Structure of the Kane Fracture Zone from Seismic Refraction Studies*, J. Geophys. Res. *85*, 3759–3779.

DETRICK, R. S., MADSEN, J. P., BUHL, P. E., VERA, J., MUTTER, J., ORCUTT, J., and BROCKER, T. (1987), *Multichannel Seismic Imaging of an Axial Magma Chamber along the East Pacific Rise between 4°N and 13°N*, Nature *326*, 35–41.

DETRICK, R. S., MUTTER, J., BUHL, P. E., and KIM, I. I. (1990), *No Evidence from Multichannel Reflection Data for a Crustal Magma Chamber in the MARK Area on the Mid-Atlantic Ridge*, Nature *347*, 61–64.

DETRICK, R. S., WHITE, R. S., and PURDY, G. M. (1993), *Crustal Structure of North Atlantic Fracture Zones*, Rev. Geophys. *31*, 439–458.

DETRICK, R. S., NEEDHAM, H. D., and RENARD, V. (1995), *Gravity Anomalies and Crustal Thickness Variations along the Mid-Atlantic Ridge between 33°N and 40°N*, J. Geophys. Res. *100*, 3767–3787.

EITTREIM, S. L., GNIBIDENKO, H., HESLEY, C. E., SLITER, R., MANN, D., and RAGOZIN, N. (1994), *Oceanic Crustal Thickness and Seismic Character along a Central Pacific Transect*, J. Geophys. Res. *99*, 3139–3145.

ENRIQUEZ, K. D., and CHEN, Y. J. (1995), *A Three-dimensional Gravity Study of the Pacific-Antarctic East Pacific Rise/Menard Transform Intersection*, J. Geophys. Res. *100*.

FORSYTH, D. W. (1993), *Crustal Thickness and the Average Depth and Degree of Melting in Fractional Melting Models of Passive Flow beneath Mid-ocean Ridges*, J. Geophys. Res. *98*, 16,073–16,079.

FORSYTH, D. W., EHRENBARD, R. L., and CHAPIN, S. (1987), *Anomalous Upper Mantle beneath the Australian-Antarctic Discordance*, Earth Plant. Sci. Lett. *84*, 471–478.

FOX, P. J., and GALLO, D. G. (1984), *A Tectonic Model for Ridge-transform-ridge Plate Boundaries: Implications for the Structure of Oceanic Lithosphere*, Tectonophysics *104*, 205–242.

Fox, P. J., Grindlay, N. R., and Macdonald, K. C. (1991), *The Mid-Atlantic Ridge (31°S – 34°30'S): Temporal and Spatial Variations of Accretionary Processes*, Mar. Geophys. Res. *13*, 1–20.

Grand, S. (1994), *Mantle Shear Structure beneath the Americas and Surrounding Oceans*, J. Geophys. Res. *99*, 11,591–11,621.

Grindlay, N. R., Fox, P. J., and Macdonald, K. C. (1991), *Second-order Ridge Axis Discontinuities in the South Atlantic: Morphology, Structure, and Evolution*, Mar. Geophys. Res. *13*, 21–49.

Hager, B. H. (1984), *Subducted Slabs and the Geoid: Constraints on Mantle Rheology and Flow*, J. Geophys. Res. *89*, 6003–6015.

Hager, B. H., and Clayton, R. W., *Constraints on the structure of mantle convection*, In *Mantle Convection: Plate Tectonics and Global Dynamics* (ed. Peltier, W. R.) (Gordon and Breach, New York 1989).

Harding, A. J., Orcutt, J. A., Kappus, M. E., Vera, E. E., Mutter, J. C., Buhl, P., Detrick, R. S., and Brocher, T. M. (1989), *Structure of Young Oceanic Crust at 13°N on the East Pacific Rise from Expanding Spread Profiles*, J. Geophys. Res. *89*, 12,163–12,196.

Harding, A. J., Kent, G. M., and Orcutt, J. A. (1993), *A Multichannel Seismic Investigation of Upper Crustal Structure at 9°N on the East Pacific Rise: Implications for Crustal Accretion*, J. Geophys. Res. *98*, 13,925–13,944.

Hayes, D. E., and Connolly, J. R., *Morphology of the southeast Indian Ocean*. In *Antarctic Oceanography II: The Australian-New Zealand Sector, Antarctic Res. Ser. 19* (ed. Hayes, D. E.) (AGU, Washington 1972) pp. 125–145.

Heezen, B., Tharp, M., and Ewing, M. (1959), *The Floors of the Oceans. 1. The North Atlantic*, Geol. Soc. Am. Spec. Paper 65.

Hey, R. N., Duennebier, F. K., and Morgan, W. J. (1980), *Propagating Rifts on Mid-ocean Ridges*, J. Geophys. Res. *85*, 3647–3558.

Hey, R. N., Johnson, P. D., and Martinez, F. (1994), *Protoplate Tectonics along the Fastest Seafloor Spreading Center (abstract)*, EOS Trans. AGU *75*, 657.

Hill, M. N. (1960), *A Median Valley of the Mid-Atlantic Ridge*, Deep-Sea Res. *6*, 193–205.

Johnson, P. D., Hey, R. N., and Martinez, F. (1994), *Recent Development of the Large-scale Overlapping Ridge System, EPR 29°S, Revealed by a New GLORI-B Processing Technique (abstract)*, EOS Trans. AGU *75*, 590.

Karson, J. A., and Dick, H. J. B. (1983), *Tectonics of Ridge-transform Intersections at the Kane Fracture Zone*, Mar. Geophys. Res. *6*, 51–98.

Karson, J. A., Thompson, G., Humphris, S. E., Edmond, J. M., Byran, W. B., Brown, J. R., Winters, A. T., Pockalny, R. A., Casey, J. F., Campbell, A. C., Klinkhammer, G., Palmer, M. R., Kinzler, R. J., and Sulanowska, M. M. (1987), *Along-axis Variations in Seafloor Spreading in the MARK area*, Nature *328*, 681–685.

Kent, G. M., Harding, A. J., and Orcutt, J. A. (1990), *Evidence for a Smaller Magma Chamber beneath the East Pacific Rise at 9°30'N*, Nature *344*, 650–653.

Kent, G. M., Harding, A. J., and Orcutt, J. A. (1993), *Distribution of Magma Beneath the East Pacific Rise near the 9°03'N Overlapping Spreading Center from Forward Modeling of Common Depth Point Data*, J. Geophys. Res. *98*, 13,971–13,995.

Klein, E. M., and Langmuir, G. H. (1987), *Global Correlations of Ocean Ridge Basalt Chemistry with Axial Depth and Crustal Thickness*, J. Geophys. Res. *92*, 8089–8115.

Kong, L. S. L., Detrick, R. S., Fox, P. J., Mayer, L. A., and Ryan, W. B. F. (1988), *The Morphology and Tectonics of the MARK Area from SEA BEAM and SEA MARC II Observations (Mid-Atlantic Ridge 23°N)*, Mar. Geophys. Res. *10*, 59–90.

Kong, L. S., Solomon, S. C., and Purdy, G. M. (1992), *Microearthquake Characteristics of a Mid-ocean Ridge Along-axis High*, J. Geophys. Res. *97*, 1659–1685.

Kuo, B.-Y., and Forsyth, D. W. (1988), *Gravity Anomalies of the Ridge-transform System on the South Atlantic between 31 and 34.5°S: Upwelling Centers and Variations in Crustal Thickness*, Mar. Geophys. Res. *10*, 205–232.

Lachenbruch, A. H. (1973), *A Simple Mechanical Model for Oceanic Spreading Center*, J. Geophys. Res. *78*, 3395–3413.

LACHENBRUCH, A. H. (1976), *Dynamics of a Passive Spreading Center*, J. Geophys. Res. *81*, 1883–1902.

LEPICHON, X., FRANCHETEAU, J., and BONNIN, J., *Plate Tectonics* (Elsevier, Amsterdam 1973).

LIN, J., and PARMENTIER, E. M. (1989), *Mechanisms of Lithospheric Extension at Mid-ocean Ridges*, Geophys. J. R. Astron. Soc. *96*, 1–22.

LIN, J., and PHIPPS MORGAN, J. (1992), *The Spreading Rate Dependence of Three-dimensional Mid-ocean Ridge Gravity Structure*, Geophys. Res. Lett. *19*, 13–16.

LIN, J., PURDY, G. M., SCHOUTEN, H., SEMPÉRÉ, J.-C., ZERVAS, C. (1990), *Evidence from Gravity Data for Focused Magmatic Accretion along the Mid-Atlantic Ridge*, Nature *344*, 627–632.

LONSDALE, P. (1983), *Overlapping Rift Zones at the 5.5°S Offset of the East Pacific Rise*, J. Geophys. Res. *88*, 9393–9406.

LONSDALE, P. (1985), *Nontransform Offsets of the Pacific-Cocos Plate Boundary and their Traces on the Rise Flank*, Bull. Geol. Soc. Am. *96*, 313–327.

LONSDALE, P. (1986), *Comments on "East Pacific Rise from Siqueiros to Orozco Fracture Zones: Along-strike Continuity of Axial Neovolcanic Zone and Structure and Evolution of Overlapping Spreading Centers" By K. Macdonald, J. Sempéré and P. J. Fox*, J. Geophys. Res. *91*, 10,493–10,499.

LONSDALE, P. (1988), *Geography and History of the Louisville Hotspot Chain in the Southwest Pacific*, J. Geophys. Res. *93*, 3078–3104.

LONSDALE, P. (1989), *Segmentation of the Pacific-Nazca Spreading Center 1°N–20°S*, J. Geophys. Res. *94*, 12,197–12,225.

LONSDALE, P. (1994), *Geomorphology and Structural Segmentation of the Crest of the Southern (Pacific-Antarctic) East Pacific Rise*, J. Geophys. Res. *99*, 4683–4702.

MACDONALD, K. C. (1982), *Mid-ocean Ridges: Fine Scale Tectonic, Volcanic, and Hydrothermal Processes within the Plate Boundary Zone*, Ann. Rev. Earth Planet. Sci. *10*, 155–190.

MACDONALD, K. C., *The crest of the Mid-Atlantic Ridge: Models for crustal generation processes and tectonics*. In *The Geology of North America: The Western North Atlantic Region, v. M* (eds. Vogt, P., and Tucholke, B.) (Geological Society of America, Boulder 1986) pp. 51–68.

MACDONALD, K. C., and FOX, P. J. (1983), *Overlapping Spreading Centers on the East Pacific Rise: Discussion and Reply*, Nature *303*, 549–550.

MACDONALD, K. C., and FOX, P. J. (1988), *The Axial Summit Graben and Cross-sectional Shape of the East Pacific Rise as Indicators of Axial Magma Chambers and Recent Volcanic Eruptions*, Earth Planet. Sci. Lett. *88*, 119–131.

MACDONALD, K. C., SEMPÉRÉ, J. C., and FOX, P. J. (1984), *East Pacific Rise from Siqueiros to Orozco Fracture Zones: Along-strike Continuity of Axial Neovolcanic Zone and Structure and Evolution of Overlapping Spreading Centers*, J. Geophys. Res. *89*, 6049–6069.

MCDONALD, K. C., SEMPÉRÉ, J. C., and FOX, P. J. (1986), *Reply: The Debate Concerning Overlapping Spreading Centers and Mid-ocean Ridge Processes*, J. Geophys. Res. *91*, 10,501–10,511.

MCDONALD, K. C., SCHEIRER, D. S., and CARBOTTE, S. M. (1991), *Mid-ocean Ridges: Discontinuities, Segments and Giant Cracks*, Science *253*, 986–994.

MADSEN, J. A., FORSYTH, D. W., and DETRICK, R. S. (1984), *A New Isostatic Model for the East Pacific Crest*, J. Geophys. Res. *89*, 9997–10,015.

MADSEN, J. A., DETRICK, R. S., MUTTER, J. C., BUHL, P., and ORCUTT, J. C. (1990), *A Two- and Three-dimensional Analysis of Gravity Anomalies Associated with the East Pacific Rise at 9°N and 13°N*, J. Geophys. Res. *95*, 4967–4987.

MCADOO, D. C., and MARKS, K. M. (1992), *Gravity Fields of the Southern Ocean from Geodat Data*, J. Geophys. Res. *97*, 3247–3260.

MCKENZIE, D. P. (1967), *Some Remarks on Heat Flow and Gravity Anomalies*, J. Geophys. Res. *72*, 6261–6273.

MCKENZIE, D. P., and BICKLE, M. J. (1988), *The Volume and Composition of Melt Generated by Extension of the Lithosphere*, J. Petrol. *29*, 625–679.

MENARD, H. W. (1960), *The East Pacific Rise*, Science *132*, 1737–1746.

MENARD, H. W. (1967), *Seafloor Spreading, Topography, and the Second Layer*, Science *157*, 923–924.

MORGAN, W. J. (1968), *Rises, Trenches, Great Faults, and Crustal Blocks*, J. Geophys. Res. *73*, 1959–1982.

MORRIS, E., and DETRICK, R. S. (1991), *Three-dimensional Analysis of Gravity Anomalies in the MARK Area, Mid-Atlantic Ridge 23°N*, J. Geophys. Res. *96*, 4355–4366.

MORTON, J. L., SLEEP, N. H., NORMARK, W. R., and TOMKINS, D. H. (1987), *Structure of the Southern Juan de Fuca Ridge from Seismic Reflection Records*, J. Geophys. Res. *92*, 11,315–11,326.

MUTTER, J. C., BARTH, G. A., BUHL, P., DETRICK, R. S., ORCUTT, J., and HARDING, A. (1988), *Magma Distribution across Ridge-axis Discontinuities on the East Pacific Rise from Multichannel Seismic Images*, Nature *336*, 156–158.

NEUMANN, G. A., and FORSYTH, D. W. (1993), *The Paradox of the Axial Profile: Isostatic Compensation along the Axis of the Mid-Atlantic Ridge?* J. Geophys. Res. *98*, 17,891–17,910.

NICOLAS, A., *Structures of Ophilites and Dynamics of Oceanic Lithosphere* (Kluwer Academic Publishers 1989).

NICOLAS, A. (1994), *Comment on "The Genesis of Oceanic Crust: Magma Injection, Hydrothermal Circulation, and Crustal Flow" by Jason Phipps Morgan and Y. John Chen*, J. Geophys. Res. *99*, 12,029–12,030.

NICOLAS, A., REUBER, I., and BENN, K. (1988), *A New Magma Chamber Model Based on Structural Studies in the Oman Ophiolite*, Tectonophysics *151*, 87–105.

OXBURGH, E. R., *Heat flow and magma genesis*, In *Physics of Magmatic Processes* (ed. Hargraves, R. B.) (Princeton University Press, Princeton, N.J. 1980) pp. 161–199.

PARMENTIER, E. M., and PHIPPS MORGAN, J. (1990), *Spreading Rate Dependence of Three-dimensional Structure in Oceanic Spreading Centers*, Nature *348*, 325–328.

PHIPPS MORGAN, J. (1987), *Melt Migration beneath Mid-ocean Spreading Centers*, Geophys. Res. Lett. *14*, 1238–1241.

PHIPPS MORGAN, J. (1991), *Mid-ocean Ridge Dynamics: Observations and Theory*, U. S. Nat. Rep. Int. Union Geod. Geophys. Rev. Geophys. *29*, Suppl., 807–822.

PHIPPS MORGAN, J., and CHEN, Y. J. (1993a), *The Genesis of Oceanic Crust: Magma Injection, Hydrothermal Circulation, and Crustal Flow*, J. Geophys. Res. *98*, 6283–6297.

PHIPPS MORGAN, J., and CHEN, Y. J. (1993b), *The Dependence of Ridge-axis Morphology and Geochemistry on Spreading Rate and Crustal Thickness*, Nature *364*, 706–708.

PHIPPS MORGAN, J., and CHEN, Y. J. (1994), *Reply*, J. Geophys. Res. *99*, 1,031–12,032.

PHIPPS MORGAN, J., and FORSYTH, D. W. (1988), *Three-dimensional Flow and Temperature Perturbations due to a Transform Offset: Effects on Oceanic Crustal and Upper Mantle Structure*, J. Geophys. Res. *93*, 2955–2966.

PHIPPS MORGAN, J., PARMENTIER, E. M., and LIN, J. (1987), *Mechanisms for the Origin of Mid-ocean Ridge Axial Topography: Implications for the Thermal and Mechanical Structure of Accretion Plate Boundaries*, J. Geophys. Res. *92*, 12,823–12,836.

PHIPPS MORGAN, J., HARDING, A., ORCUTT, J., KENT, G., and CHEN, Y. J., *An observational and theoretical synthesis of magma geometry and crustal genesis along a Mid-Ocean Spreading Center*, Chapter 7. In *Magmatic Systems* (ed. Ryan, M.) (Academic Press, San Diego 1994) pp. 139–178.

PURDY, G. M., KONG, L. S. L., CHRISTESON, G. L., and SOLOMON, S. (1992), *Relationship between Spreading Rate and the Seismic Structure of Mid-ocean Ridges*, Nature *355*, 815–817.

RABINOWICZ, M., CEULENEER, G., and NICOLAS, A. (1987), *Melt Segregation and Flow in Mantle Diapirs below Spreading Centers: Evidence from the Oman Ophiolite*, J. Geophys. Res. *92*, 3475–3486.

REA, D. K. (1978), *Asymmetric Seafloor Spreading and a Nontransform Axis Offset: The East Pacific Rise 20°S Survey Area*, Geol. Soc. Am. Bull. *89*, 836–844.

REID, I. D., and JACKSON, H. R. (1981), *Oceanic Spreading Rates and Crustal Thickness*, Mar. Geophys. Res. *5*, 165–172.

ROHR, K. M. M., MILKEREIT, B., and YORATH, C. J. (1988), *Asymmetric Deep Crustal Structure across the Juan de Fuca Ridge*, Geology *16*, 533–537.

SANDWELL, D. T., and SMITH, W. H. F. (1994), *New Global Marine Gravity Map/Grid Based on Stacked ERS-1, Geosat and Topex Altimetry (abstract)*, EOS Trans. AGU *75*, 321.

SCHEIRER, D. S., and MACDONALD, K. C. (1993), *Variation in Cross-sectional Area of the Axial Ridge along the East Pacific Rise: Evidence for the Magmatic Budget of a Fast Spreading Center*, J. Geophys. Res. *98*, 7871–7885.

SCOTT, D. R., and STEVENSON, D. J. (1989), *A Self-consistent Model of Melting, Magma Migration and Buoyancy-driven Circulation beneath Mid-ocean Ridges*, J. Geophys. Res. *94*, 2973–2988.

SEMPÉRÉ, J.-C., PURDY, G. M., and SCHOUTEN, H. (1990), *Segmentation of the Mid-Atlantic Ridge between 24°N and 30°40'N*, Nature *344*, 427–429.

SEMPÉRÉ, J.-C., PALMER, J., CHRISTIE, D., PHIPPS MORGAN, J., and SHOR, A. (1991), *Australian-Antarctic Discordance*, Geology *19*, 429–432.

SHAW, P. R. (1992), *Ridge Segmentation, Faulting and Crustal Thickness in the Atlantic Ocean*, Nature *358*, 490–493.

SLEEP, N. H. (1969), *Sensitivity of Heat Flow and Gravity to the Mechanism of Seafloor Spreading*, J. Geophys. Res. *74*, 542–549.

SLEEP, N. H. (1975), *Formation of Oceanic Crust: Some Thermal Constraints*, J. Geophys. Res. *80*, 4037–4042.

SMALL, C., and SANDWELL, D. T. (1989), *An Abrupt Change in Ridge Axis Gravity with Spreading Rate*, J. Geophys. Res. *94*, 17,383–17,392.

SMALL, C., SANDWELL, D., CHEN, Y., and ROYER, J.-Y. (1989), *Transition of Ridge Axis Gravity/Topography with Spreading Rate: The Effect of a Weak Lower Crust? (abstract)*, EOS Trans. AGU *70*, 1301.

SMALLWOOD, J. R., WHITE, R. S., and MINSHULL, T. A. (1995), *The Structure of the Reykjanes Ridge at 61°40'N* (abstract), 1995 RIDGE Theoretical Institute, 0–52.

SMEWING, J. D. (1981), *Mixing Characteristics and Compositional Differences in Mantle-derived Melts beneath Spreading Axes: Evidence from Cyclically Layered Rocks in the Ophiolite of North Oman*, J. Geophys. Res. *86*, 2645–2659.

SOTIN, C., and PARMENTIER, E. M. (1989), *Dynamical Consequences of Compositional and Thermal Density Stratification beneath Spreading Centers*, Geophys. Res. Lett. *16*, 835–838.

SPARKS, D. W., and PARMENTIER, E. M. (1991), *Melt Extraction from the Mantle beneath Spreading Centers*, Earth Planet. Sci. Lett. *105*, 368–377.

SU, W.-J., WOODWARD, R. L., and DZIEWONSKI, A. M. (1992), *Deep Origin of Mid-ocean-ridge Seismic Velocity Anomalies*, Nature *360*, 149–152.

SU, W.-J., WOODWARD, R. L., and DZIEWONSKI, A. M. (1994), *Degree 12 Model of Shear Velocity Heterogeneity in the Mantle*, J. Geophys. Res. *99*, 6945–6980.

TAPONNIER, P., and FRANCHETEAU, J. (1978), *Necking of the Lithosphere and the Mechanics of Slowly-accreting Plate Boundaries*, J. Geophys. Res. *83*, 3955–3970.

TOLSTOY, M., HARDING, A. J., and ORCUTT, J. A. (1993), *Crustal Thickness on the Mid-Atlantic Ridge: Bull's Eye Gravity Anomalies and Focused Accretion*, Science *262*, 726–729.

TOLSTOY, M., HARDING, A. J., ORCUTT, J. A., and PHIPPS MORGAN, J. (1995), *A Seismic Refraction Investigation of the Australian Antarctic Discordance and Neighboring South East Indian Ridges: Preliminary Results (abstract)*, EOS Trans. AGU *76*/Supplement, S275.

TOOMEY, D. R., SOLOMON, S. C., PURDY, G. M., and MURRAY, M. H. (1985), *Microearthquakes beneath the Median Valley of the Mid-Atlantic Ridge near 23°N: Hypocentres and Focal Mechanisms*, J. Geophys. Res. *90*, 5443–5448.

TOOMEY, D. R., SOLOMON, S. C., and PURDY, G. M. (1988), *Microearthquakes beneath Median Valley of Mid-Atlantic Ridge near 23°N: Tomography and Tectonics*, J. Geophys. Res. *93*, 9093–9112.

VERA, E. E., MUTTER, J. C., BUHL, P., ORCUTT, J. A., HARDING, A. J., KAPPUS, M. E., DETRICK, R. S., and BROCHER, T. M. (1990), *The Structure of 0- to 0.2-m.y.-old Oceanic Crust at 9°N on the East Pacific Rise from Expanded Spread Profiles*, J. Geophys. Res. *95*, 15,529–15,556.

WANG, X., and COCHRAN, J. R. (1993), *Gravity Anomalies, Isostasy, and Mantle Flow at the East Pacific Rise Crest*, J. Geophys. Res. *98*, 19,505–19,531.

WANG, X., COCHRAN, J. R., and BARTH, G. A. (1996), *Gravity Anomalies, Crustal Thickness and the Pattern of Mantle Flow at the Fast Spreading East Pacific Rise, 9°–10°N: Evidence for Three-dimensional Upwelling*, J. Geophys. Res., in press.

WEISSEL, J. K., and HAYES, D. E., *Magnetic anomalies in the southeast Indian Ocean. In Antarctic Oceanography II: The Australian-New Zealand Sector, Antarctic Res. Ser. 19* (ed. Hayes, D. E.) (AGU, Washington 1972) pp. 165–196.

WEISSEL, J. K., and HAYES, D. E. (1974), *The Australian-Antarctic Discordance: New Results and Implications*, J. Geophys. Res. *79*, 2579–2587.

WHITE, R. S., MCKENZIE, D., and O'NIONS, R. K. (1992), *Oceanic Crustal Thickness from Seismic Measurements and Rare Earth Element Inversions*, J. Geophys. Res. *97*, 19,683–19,715.

WOODWARD, R. L., and MASTERS, G. (1992), *Upper Mantle Structure from Long-period Differential Traveltimes and Free Oscillation Data*, Geophys. J. Int. *109*, 275–293.

ZHANG, Y. S., and TANIMOTO, T. (1992), *Ridges, Hotspots and their Interaction as Observed in Seismic Velocity Maps*, Nature *355*, 45–49.

(Received March 15, 1995, revised/accepted August 21, 1995)

PAGEOPH, Vol. 146, Nos. 3/4 (1996)

0033–4553/96/040649–11$1.50 + 0.20/0
© 1996 Birkhäuser Verlag, Basel

The Upper Mantle Flow Beneath the North China Platform

Rong-Shan Fu,[1] Jian-hua Huang[1] and Zhe-xun Wei[1]

Abstract — In this paper we establish an upper mantle convection model which is constrained by regional isostatic gravity anomalies. Comparing the computed convection patterns with the tectonic features of the North China Platform we find that there are two positive anomaly centers connected with upward flows. These anomalies belong to the tectonic units of the Shan-Xi geoanticline and the Lu-Xi geoanticline. The centers of downward flows are connected with the tectonic units of the Liao-Ji geosyncline. It is reasonable to suggest that the upward mantle flows push the lithosphere upward and generate the observed positive isostatic gravity anomaly. The downward mantle flows pull the lithosphere down and generate the negative anomaly. However, the use of simple analysis makes it difficult to explain the complex lithospheric dynamics of this region. In order to understand lithospheric structures and tectonic features we must investigate the mechanical properties of the lithosphere and the relationship between the lithosphere and the mantle. These problems are discussed in the last section of this paper.

Key words: Upper mantle, lithosphere, thermal convection.

1. Introduction

The theory of plate tectonics is immensely successful in explaining the global tectonic structure, in reconstructing the continents, and in describing the structure and dynamics of the plate boundary areas. However, many competing hypotheses exist which explain the regional lithospheric structure and dynamics within continental interiors. It is reasonable to think that these regional lithospheric structures and dynamics are connected with the material movements in the upper mantle. Recent research shows that there are about five different styles of convection in the mantle (Fu, 1993). These are: 1) A large-scale convection throughout the entire mantle that is related to global plate tectonics; 2) small-scale convection in the upper mantle that is related to intraplate regional lithospheric tectonics; 3) layered convection somewhere in the mantle; 4) very small-scale convection in the D'' layer; and 5) plumes that are related to very deep dynamical processes. It is easy to understand that global tectonics are related to the large-scale thermal convection in

[1] Department of Earth and Space Sciences, University of Science and Technology of China, Hefei, 230026.

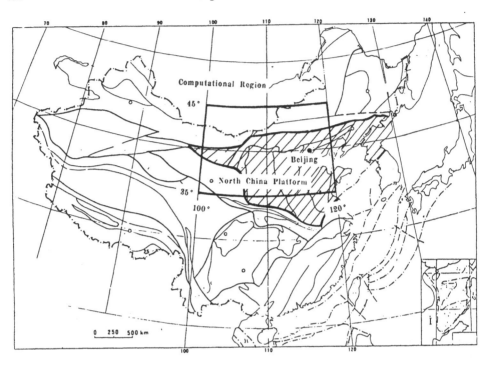

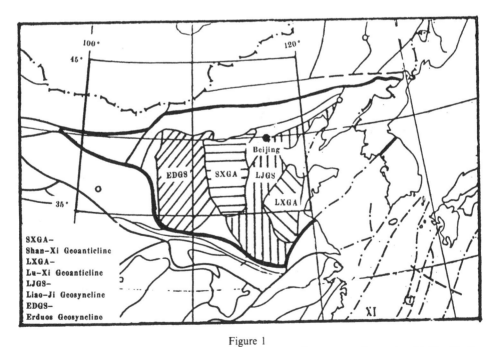

Figure 1

The computational region and the North China Platform. a) The computation region. b) The North China Platform.

the entire mantle, in which the plate divergent boundaries are connected with the upward flows of the mantle convection and the plate convergent boundaries are connected with downward flows that bring old lithosphere into the mantle. However, it is difficult to use this mechanism to explain the intraplate regional lithospheric structure and dynamics. New evidence from seismic tomography (Humphreys et al., 1984), geothermal and short wavelength geoid anomalies (HAXBY and WEISSEL, 1986; BAUDRY and KROENKE, 1991; YAMAJI, 1992), reveal the existence of small-scale convective systems in the upper mantle, with scales ranging from 500 km to 700 km (RICHTER and PARSONS, 1975; PARSONS and McKENZIE, 1978). It is reasonable to suggest that these small-scale convective systems in the upper mantle probably control the regional tectonic structure and the dynamical processes of the lithosphere. Assuming that regional isostatic gravity anomalies or the short wavelength geoid anomalies are derived from density heterogeneities associated with the small-scale convection in the upper mantle. FU et al. (1994) obtained a correlation equation between the regional isostatic gravity anomaly and small-scale convection. They also established a physical-mathematical model to calculate the convection patterns in the upper mantle constrained by regional isostatic gravity anomalies (FU et al., 1995).

The North China Platform extends from longitude 100°E to 130°E and from latitude 35°N to 43°N (Fig. 1b). It includes the northern part of China, the eastern part of Northwest China, the southern part of Northeast China, the Bo-Hai Sea, the northern part of the Huang Hai Sea and the northern part of Korea. After the Yan-Shan orogeny and the Himalaya orogeny, the North China Platform is very active. Large-scale uplift and depression of the crust control the basic tectonic pattern and mineral deposits of this area. Because the North China Platform is distant form the boundary between the Eurasia and the Pacific plates, it is very difficult to imagine that horizontal compressive forces at this convergent plate boundary can produce the observed vertical movements in the lithosphere of the North China Platform. Therefore, mantle flows in the upper mantle may be one of the main driving forces of the lithospheric tectonics in this region.

In this paper we will use the regional isostatic gravity anomaly data to calculate the thermal convection patterns in the upper mantle of the region 100°–120°E, 35°–45°N) and compare these patterns with the regional tectonic features of this area. In order to understand the geodynamical processes of the North China Platform, we will also make some general comments about the driving forces of the regional lithospheric dynamics.

2. Thermal Convection Model of the Upper Mantle

The upper mantle is assumed to be composed of homogeneous, isoviscous liquid. In this case the movement of mantle materials is governed by the following basic equations (with the Boussinesq approximation)

$$\frac{\partial U_i}{\partial t} = -\frac{\partial}{\partial X_i}\left(\frac{\delta p}{\rho_0}\right) + g\alpha\theta\lambda_i + v\nabla^2 U_i$$

$$\frac{\partial T}{\partial t} = -U_j\frac{\partial T}{\partial X_j} + \kappa\nabla^2 T \qquad (1)$$

$$\frac{\partial U_j}{\partial X_j} = 0$$

where U_i and X_i are the vector components of velocity and position respectively, $T = T_0 + \theta$ is the temperature, θ is the temperature perturbation, δp is the perturbation of pressure and g, a, v, κ and ρ_0 are the gravity acceleration, the coefficient of volume expansion, the kinematic viscosity, thermal diffusivity of the liquid and the initial density, respectively. The gravity potential V satisfies the Poisson equation

$$\nabla^2 V = 4\pi G\rho \qquad (2)$$

where G is the universal gravitational constant. For a steady state and at low Rayleigh number (above critical value), we can ignore the nonlinear term $U_j\,\partial\theta/\partial X_j$ in the energy equation. By taking $[L] = h$ (h is the depth of the convection system) and $[T] = h^2/v$ as dimensionless factors, and expanding the poloidal component of velocity and temperature perturbation and the toroidal component of velocity, we have

$$W = W(z)\,\exp\{i(k_x x + k_y y)\}$$

$$\theta = \Theta(z)\,\exp\{i(k_x x + k_y y)\}$$

$$\zeta = Z(z)\,\exp\{i(k_x x + k_y y)\}.$$

Using $a = kh = \sqrt{a_x^2 + a_y^2}$ as a dinensionless wave number and

$$\frac{\partial^2}{\partial x^2} + \frac{\partial^2}{\partial y^2} = -k^2; \quad \nabla^2 = \frac{d^2}{dz^2} - k^2 = \frac{1}{h^2}\left(\frac{\partial^2}{\partial z^2} - a^2\right)$$

we obtain the equations for the poloidal components

$$(D^2 - a^2)^2 W = \frac{g\alpha h^2}{v}\,a^2\theta$$

$$(D^2 - a^2)\theta = -\frac{\beta h^2}{\kappa}\,W \qquad (3)$$

or

$$(D^2 - a^2)^3 = -Ra^2 W \qquad (4)$$

where the differential operator $D = d/dz$ and $R = ga\beta h^4/\kappa v$ is the Rayleigh number (CHANDRASEKHAR, 1961) and β is the initial gradient of temperature T_0. The full solution of equation (4) is

$$W(z) = \sum_{j=1}^{6} A_j \exp\{i(\alpha_j z)\} \tag{5}$$

where coefficients A_j are computed by boundary conditions. The horizontal components of flow are

$$u = \frac{1}{a^2}\frac{\partial^2 W}{\partial z\,\partial x} = \frac{1}{a^2}\frac{\partial W}{\partial z}ia_x\,\exp\{i(a_x x + a_y y)\} \tag{6}$$

$$v = \frac{1}{a^2}\frac{\partial^2 W}{\partial z\,\partial y} = \frac{1}{a^2}\frac{\partial W}{\partial z}ia_y\,\exp\{i(a_x x + a_y y)\}.$$

Usually, following the Bouguer Correction and the Isostatic Correction to the observed gravity data set, the regional isostatic gravity anomalies (their wavelengths are less than 1500 km) are generated by four parts. These are: 1) the density heterogeneities in the upper mantle which are related to the upper mantle thermal convection; 2) boundary deformations between the lithosphere and the mantle (L-M) and between the upper mantle and the lower mantle (UM-LM) which are produced by the upper mantle thermal convection; 3) the elastic or nonelastic vertical deformations of the lithosphere which are produced by horizontal compressions or vertical loads on the surface of the earth; and 4) other dynamic facts. However, in some conditions the last two factors can be ignored and we can consider the first two parts as the main sources which generate the regional isostatic gravity anomalies. Then we have the correlation equation between the regional isostatic gravity anomaly and mantle flow (FU et al., 1994)

$$\left(\frac{d^2}{dz^2} - a^2\frac{d}{dz}\right)W\Big|_{z=H=1}e^{(-a(H_0/H)-1)} - \left(\frac{d^2}{dz^2} - a^2\frac{d}{dz}\right)W\Big|_{z=0}e^{-a(H_0/H)}$$

$$= \frac{gha}{2\pi\mu G}\Delta g(a, z) \tag{7}$$

where μ is the viscosity of the upper mantle, $H = h$ is the distance between the upper and the lower boundaries of the convective system, H_0 is the distance between the surface of the earth and the lower boundary. By using this correlation equation as a constraining condition, we can solve the problem of thermal convection in the upper mantle.

3. Computation and Results

Figure 2 illustrates the computation model, in which viscosity $\mu = 10^{21}$ Pa s, $g = 9.8$ m/s^2, $H = 570$ km, $H_0 = 670$ km and the Rayleigh number $R = 3000$. The computation domain is bounded by 100°E–120°E and 35°N–45°N. We use a FFT

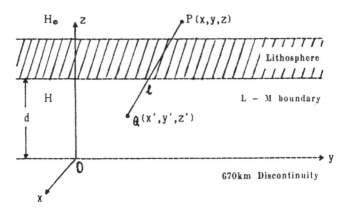

Figure 2
A schematic diagram of the model.

program and a 32×32 horizontal grid to compute the data set and to obtain the patterns of the mantle flows. Figure 3 is the isostatic gravity anomaly contour map of the computed area. We employ the following boundary conditions in our computation models: the vertical component of velocity $W = 0$ for both the upper and the lower boundaries $(Z = 0, Z = 1)$; the temperature perturbations $\theta = 0$ at $Z = 0$; the horizontal components of the velocity at the lower and the upper boundary are equal to zero and W satisfies correlation equation (7).

The isostatic gravity map (Fig. 3) illustrates that there are two large positive anomaly centers. One is located in the region of the Tai-Hang–Lu-Liang mountain

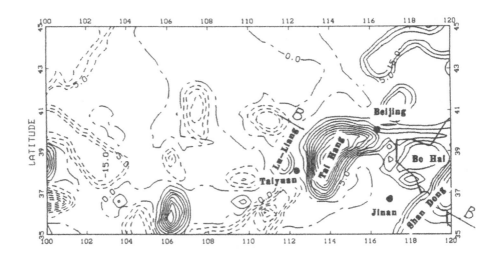

Figure 3
Isostatic gravity anomaly of the computational region.

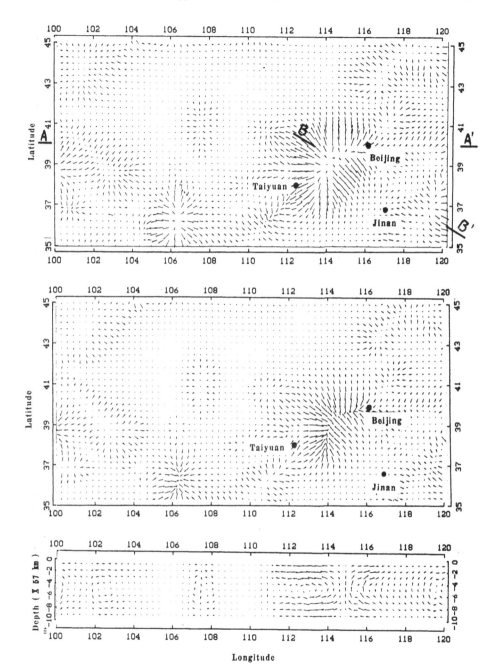

Figure 4
Convection patterns in the upper mantle of the computational region. a) The convection pattern at
157 km depth. b) The convection pattern at 613 km depth. c) The flow pattern of section A–A' (40°N)
in Figure 4(a).

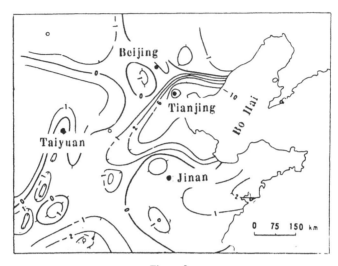

Figure 5
Vertical deformation rate of North China (mm/a) (MA, 1990).

system and the other is located in the region of the western Shan-Dong province and the Bo-Hai Sea. Comparing this map with the computed convection pattern (Fig. 4), we find that these positive anomaly centers are connected directly with upward flows. From the tectonic map shown in Figure 1b, we recognize that either the positive gravity anomaly or the upward flow centers are located in the region of the tectonic units of the Shan-Xi geoanticline and the Lu-Xi geoanticline and the centers of downward flows or negative gravity anomalies are located in the region of tectonic units of Liao-Ji geosyncline and the Erduos geosyncline. This simple result implies that the upward mantle flows push the lithosphere uplift and disturb the original isostatic state of this region to generate the observed positive isostatic gravity anomaly. On the other hand, the downward mantle flows pull the lithosphere depression and generate the observed negative anomaly. This, probably, is the main feature of the lithospheric movements in the North China Platform. This basic feature can fit the recently observed vertical deformation rates (Fig. 5) very well.

4. Lithospheric Dynamics of the North China Platform and Discussions

The North China Platform is one of the oldest cratons in China. In the Mesozoic Era and the Cenozoic Era the North China Platform began a geologically active period. Very strong tectonic movements occurred within the Yan-Shan orogeny and the Himalaya orogeny in this platform. Large uplifts and depressions controlled the main geological structure and mineral deposits of this area, and

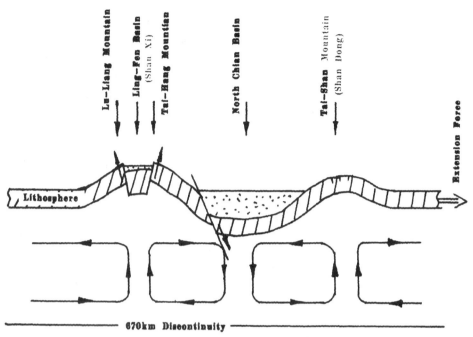

Figure 6
A schematic diagram of the lithospheric dynamics of the North China Platform.

formed the Shan-Xi geoanticline, the Lu-Xi geoanticline, the Liao-Ji geosyncline and the Erduos geosyncline. Geodetic observations in recent decades showed that the recent crustal movements in the North China Platform follow the above tectonic features. The western part of this area (centered in the Tai-Hang and the Lu-Liang mountain system) is uplifting and the eastern part (centered in the North China Sedimentary basin) is subsiding continuously. From the isostatic gravity map we can ascertain that the North China Platform is basically in an isostatic state, and the amplitudes of isostatic gravity anomaly are usually lower than 20 mgal. However, higher isostatic gravity anomalies are observed which are centered in the Tai-Hang mountain and the western part of Shan-Dong province. The anomaly amplitudes are more than 40 mgal and an exceed 60 mgal. The lowest anomaly is located in the southern part of the Lu-Liang mountain and the amplitude is about 40 mgal. On the other hand, the North China Platform is one of the most active regions of neotectonics in China. There are many neomagma connected with the neotectonic structures. Many very strong earthquakes occurred in this region during the past several thousand years. Because the North China Platform is in the eastern part of the Eurasia Plate and it is about one thousand kilometers away from the boundary between the Eurasia–Pacific Plate boundary, the compressive forces between these two plates make it difficult to produce such large vertical movements in this platform. Therefore, we must seek other driving forces to explain its tectonic features. The mantle flow could be regarded as one of the main driving forces to form the tectonic features of the North China Platform. However, we could not ignore the effects of the global tectonics on this region. Some evidence shows that there is also an extension force acting in the lithosphere on the eastern part of the Eurasia Plate (DING 1991). This force could be another driving force of the lithospheric tectonics of this region.

Figure 6 is a schematic diagram of the lithospheric dynamics of the North China Platform. It is implied from the mantle flow patterns (Fig. 3, section B-B') and the tectonic province map (Fig. 1b). In order to emphasize the main structure of this region in the Neozoic era, we place only the extension systems in this diagram. The upwelling materials under the west Shan-Dong province and the Tai-Hang–Lu-Liang mountain push the lithosphere to uplift and form the Lu-Xi geoanticline and the Shan-Xi geoanticline as well as the Ling-Fen extension basin located in the center of the Shan-Xi geoanticline. The downward flows beneath the center of the North China Basin pull the lithosphere downwards to form the Liao-Ji geosyncline. Meanwhile, the large upward and downward movements between the Tai-Hang mountain, the Tai-Shan mountain and the North China Basin, together with the extension forces in the southeast direction, cause the lithosphere to generate thinning, resulting in the fragmentation of the platform to form a series of extensional system in the boundary regions between the North China Basin and the above mountain systems. However, why there exists an upward mantle flow beneath the Bo-Hai Sea, remains to be addressed. This is a very complex problem. Generally, the lithospheric structures not only depend on driving forces, but also depend on the geological evolution of the lithosphere and its thermal mechanical behaviors. Most likely, the Bo-Hai Sea is a new center of mantle upward flows and its lithosphere will undergo thinning to cause the uplift of the mantle. However, the uplift of the Shan-Xi geoanticline superceded the Yan-Shan and the Himalaya orogeny. These geological events occurred within an older craton, where the lithosphere was cold, thick and rigid. Its uplift is entirely produced by the mantle flows.

In summary, the main driving forces of the lithospheric dynamics of the North China Platform are: the vertical push forces of upward mantle flows; the vertical pull forces of downward mantle flows; the horizontal drag forces of the mantle flows acting at the bases of the lithosphere; and the extension forces in the lithosphere caused by the motions of the eastern part of the Eurasia Plate in the southeast direction.

We noted that our computation models are quite simple and based on an assumption that the regional isostatic gravity anomalies may be derived from the density heterogeneities and the boundary deformations related to small-scale mantle convection in the upper mantle. However, we could not ignore the effects of other geodynamical processes to the deformations of the lithosphere, such as the vertical loading on the surface of the earth and horizontal compressional forces. Therefore we must consider these factors in future models. Conversely, as we have mentioned, the lithospheric tectonics not only depend on the driving forces, but also depend on the geological evolution of the lithosphere and its thermal mechanical behaviors. We also understand that our model is limited by its assumption of linearity and steady state. This system could not fully describe the structures, the movements and the evolutions of the real North China Platform. To control the geological evolutions and geodynamical processes of this region, a more complex nonlinear and time-dependent model is necessary.

Acknowledgment

This research work is supported by the National Science Foundation of China and the Geodynamics and Geodesy Laboratory of Wuhan of Science Academe of China.

REFERENCES

BAUDRY, N., and KROENKE, L. (1991), Intermediate Wavelength (400–600 km), South Pacific Geoidal Undulations: Their Relationship of Linear Volcanic Chains, Earth Planet. Sci. Lett. 102, 430–443.

CHANDRASEKHAR, U., Hydrodynamic and Hydromagnetic Stability (Clarendon Press, Oxford 1961) 652.

DING, G. Y., Introduction of the Lithospheric Geodynamics of China (Seismological Press 1991) (in Chinese with English Abstract).

FU, R. S. (1993), Mantle Thermal Dynamic Model, Progress in Geophysics 8, 13–26 (in Chinese with English Abstract).

FU, R. S., HUANG, J. H., and LIU, W. Z. (1994), Correlation Equation between Regional Isostatic Gravity Anomalies and Small-scale Mantle Convection, Acta Geophysica Sinica 37, 638–646 (in Chinese with English Abstract).

HAXBY, W. F., and WEISSEL, J. K. (1986), Evidence for Small-scale Mantle Convection from SEASAT Altimeter Data, J. Geophys. Res. 91, 3507–3520.

HUANG, J. H., CHANG, X. H., FU, R. S., and LIU, W. Z. (1995), Small-scale Convection in the Upper Mantle Constrained on the Regional Isostatic Gravity Anomalies, Acta Geophysica Sinica, Supp. II, in press (in Chinese with English Abstract).

HUMPHREYS, E., CLAYTON, R. W., and HAGER, B. H. (1984), A Tomography Image of Mantle Structure beneath Southern California, Geophys. Res. Lett. 11, 625–627.

LU, Z. J., Regional Tectonic Structure of China (USTC Press 1990) (in Chinese).

MA, X. Y., Lithospheric Geodynamics ATLAS of China (China Map Press 1990) (in Chinese with English Abstract).

PARSONS, B., and MCKENZIE, D. P. (1978), Mantle Convection and the Thermal Structure of the Plates, J. Geophys. Res. 83, 4485–4496.

RICHTER, F. M., and PARSONS, B. (1975), On the Interaction of Two Scale Convection in the Mantle, J. Geophys. Res. 80, 2529–2541.

RICHTER, F. M., and MCKENZIE, D. P. (1978), Simple Plate Models and Mantle Convection, J. Geophys. Res. 44, 441–471.

YAMAJI, A. (1992), Periodic Hotspots Distribution and Small-scale Convection in the Upper Mantle, Earth and Planet. Sci. Lett. 109, 107–116.

ZHANG, Z. M., Tectonic Structure of Oil Deposits in China (Oil Industry Press 1982) (in Chinese).

(Received October 20, 1994, revised May 30, 1995, accepted June 5, 1995)

PAGEOPH, Vol. 146, Nos. 3/4 (1996)

0033-4553/96/040661-15$1.50 + 0.20/0

Toward a Physical Understanding of Earthquake Scaling Relations

Z.-M. Yin[1,2] and G. C. Rogers[1,2]

Abstract —In seismological literature, there exist two competing theories (the so-called W model and L model) treating earthquake scaling relations between mean slip and rupture dimension and between seismic moment and rupture dimension. The core of arguments differentiating the two theories is whether the mean slip should scale with the rupture width or with the rupture length for large earthquakes. In this paper, we apply the elastic theory of dislocation to clarify the controversy. Several static dislocation models are used to simulate strike-slip earthquakes. Our results show that the mean slip scales linearly with the rupture width for small earthquakes with a rupture length smaller than the thickness of the seismogenic layer. However, for large earthquakes with a rupture length larger than the thickness of the seismogenic layer, our models show a more complicated scaling relation between mean slip and rupture dimension. When the rupture length is smaller than a cross-over length, the mean slip scales nearly linearly with the rupture length. When the rupture length is larger than a cross-over length, the mean slip approaches asymptotically a constant value and scales approximately with the rupture width. The cross-over length is a function of the rupture width and is about 75 km for earthquakes with a saturated rupture width of 15 km. We compare our theoretical predictions with observed source parameters of some large strike-slip earthquakes, and they match up well. Our results also suggest that when large earthquakes have a fixed aspect ratio of rupture length to rupture width (which seems to be the case for most subduction earthquakes) the mean slip scales with the rupture dimension in the same way as small earthquakes.

Key words: Earthquake scaling, seismic moment, mean slip, rupture dimension.

Introduction

The scaling relation between mean slip and rupture dimension or between seismic moment and rupture dimension is a fundamental problem in earthquake seismology. Dynamic crack models (KNOPOFF, 1958; MADARIAGA, 1976) predict that the mean slip scales linearly with rupture width for small earthquakes with rupture length $L \leq W_0/\sin(\theta)$, where W_0 and θ are the thickness of the seismogenic layer and the dip of the fault plane, respectively. Because the seismic moment (M_0) is the product of the shear modulus (μ), the mean slip ($\bar{u}$) and the rupture area (A)

$$M_0 = \mu \bar{u} A, \tag{1}$$

[1] Pacific Geoscience Centre, Geological Survey of Canada, Sidney, B.C., Canada V8L 4B2.
[2] School of Earth and Ocean Sciences, University of Victoria, Victoria, B.C., Canada.

assuming a constant average stress drop and $L \approx W$ leads to a scaling relation of $M_0 \propto L^3$ for small earthquakes. This theoretical prediction is in agreement with earthquake observations (HANKS, 1977). However, a controversy arises for large earthquakes with rupture length $L > W_0/\sin(\theta)$. Because the thickness of the seismogenic layer is limited to about 15–25 km and the rupture width must saturate at this depth, the rupture grows along the strike direction only. Therefore, one may expect a change in scaling relation from small earthquakes to large earthquakes. On the one hand, dynamic crack models (KNOPOFF, 1958; MADARIAGA, 1976) show that $\bar{u} \propto W$ when $L \gg W$, which predicts a scaling relation of $M_0 \propto L$ (W model). On the other hand, based on the observations that the mean slip of surface ruptures increases with the rupture length, SCHOLZ (1982a) proposed a model (L model) that predicts a scaling relation of $M_0 \propto L^2$. More recently, ROMANOWICZ (1992) found that a change in the scaling relation for strike-slip faults occurs at a rupture length of about 60–70 km (or a seismic moment of 0.6–0.8 × 10^{20} N-m) rather than at the expected length of 15–25 km identical to the saturated rupture width. Romanow-icz's statistical results pose a new problem in the scaling relations because a rupture length of 60–70 km is not an obvious characteristic length. To date, no consensus has been reached on whether the scaling relation for large earthquakes should be $M_0 \propto L^2$ or $M_0 \propto L$, and whether the change should occur at $L \approx 15$–25 km or at $L \approx 60$–70 km (see the discussion among SCHOLZ (1982a,b), ROMANOWICZ (1992), ROMANOWICZ and RUNDLE (1993), and SCHOLZ (1994)). Because statistical results from small earthquakes are often used to evaluate the seismic risk of large earthquakes, a reappraisal of the earthquake scaling relations is of not only theoretical, but also practical importance.

In this paper, we use the elastic theory of dislocation to study the scaling relation between mean slip and rupture dimension for both small and large earthquakes. To our knowledge, no thorough discussion of this subject has been made. Since the elastic theory of dislocation was first introduced to the field of earthquake seismology by STEKETEE (1958), it has been widely applied to calculation of the coseismic radiation fields (e.g., CHINNERY, 1961, 1963; PRESS, 1965; IWASAKI and SATO, 1979) and to inversion of the distribution of the slip on the rupture surface from geodetic measurements (e.g., DU et al., 1992; DU and AYDIN, 1993; SEGALL and DU, 1993). Recently, the displacement field of the Landers earthquake, mapped by radar interferometry, revealed a remarkable match of that calculated using a dislocation model (MASSONNET et al., 1993), which demonstrates that earthquake rupture is well represented by a dislocation model. Consequently, it is anticipated that the elastic theory of dislocation may help clarify the problem of earthquake scaling relations. In other words, scaling relations predicted by the elastic theory of dislocation should not contradict the earthquake observations. We use a number of dislocation models to simulate strike-slip earthquakes. They are (1) a rectangular rupture with uniform slip and (2) a circular rupture with non-uniform slip for small earthquakes, and (3) a rectangular rupture with uniform slip and (4) an elliptical rupture with nonuniform slip for large earthquakes. The theoretical results are compared with those obtained from observations. We also discuss the effect of average stress drop on the controversial statistics of the correlation between mean slip and rupture length. Although the analysis here is limited to strike-slip earthquakes, it can be readily extended to normal, thrust and oblique-slip earthquakes.

Small Earthquakes

Let x_i be a Cartesian coordinate system with the origin at the surface of the earth, and with x_1 parallel to the strike direction, x_2 perpendicular to the (vertical) fault plane, and x_3 downward (Fig. 1). Assuming a Poissonian medium, i.e., $\lambda = \mu$ (λ and μ are Lamé's constants), the displacement field $u_i(x_1, x_2, x_3)$ due to horizontal displacement $\Delta u_1(y_1, y_2, y_3)$ along the x_1 direction and across a surface A in an isotropic medium is given by (STEKETEE, 1958; OKADA, 1985)

$$u_i(x_1, x_2, x_3) = \frac{\mu}{F} \int \int_A \Delta u_1 \left(\frac{\partial u_i^1}{\partial y_2} + \frac{\partial u_i^2}{\partial y_1} \right) dA \qquad (2)$$

where μ is the shear modulus, and u_i^1 and u_i^2 are the ith component of displacement at (x_1, x_2, x_3) due to point double-couple forces of magnitude F acting in the x_1 and x_2 direction at (y_1, y_2, y_3), respectively (note that y_1, y_2, and y_3 denote the coordinates of a dislocation across the fault, and x_1, x_2, and x_3 denote the coordinates of the displacement field produced by the dislocation). The expressions for u_i^1 and u_i^2 are listed in PRESS (1965).

Integrating equation (2) and allowing y_2 be zero, we obtain the expressions for the three components of $u_i(x_1, x_2, x_3)$ at any depth for a rectangular fault with

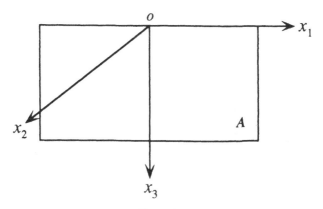

Figure 1
The directions of the Cartesian coordinate system x_i with respect to a vertically oriented fault (A). The origin (o) is on the surface of the earth.

length L, width W and uniform slip Δu_1. Then, differentiating $u_i(x_1, x_2, x_3)$ with respect to x_1 and x_2, we derive the expression for coseismic shear stress σ_{12} on the rupture surface, which is identical to the stress drop $\Delta\sigma$. Because $\Delta\sigma$ is continuous across the rupture, let x_2 be zero, then the stress drop $\Delta\sigma(x_1, x_3)$ on the rupture surface can be expressed as

$$\Delta\sigma(x_1, x_3) = \frac{\mu \Delta u_1}{12\pi} \, f(x_1, x_3, L_1, L_2, W_1, W_2) \tag{3}$$

where L_1, L_2, W_1, and W_2 are the coordinates of the four corners of the rectangular rupture in the x_i-coordinate system (the rupture length and the width are, respectively, $L = L_2 - L_1$ and $W = W_2 - W_1$), f is a polynomial function of x_1, x_3, L_1, L_2, W_1, and W_2, whose expressions are listed in the Appendix. A similar expression for σ_{12} (or $\Delta\sigma$) can be found in IWASAKI and SATO (1979).

Equation (3) can be used to directly calculate the stress drop $\Delta\sigma(x_1, x_3)$ for a rectangular rupture with uniform slip. For a rupture with nonuniform slip, a numerical method must be employed to calculate $\Delta\sigma(x_1, x_3)$, which will be addressed later. Once $\Delta\sigma(x_1, x_3)$ is obtained, the average stress drop $\overline{\Delta\sigma}$ can be calculated by numerical integration. Thereafter, the relation between mean slip and rupture dimension can be found. We devise two dislocation models. The first model is a rectangular rupture with uniform slip Δu_1. In this case, Δu_1 is identical to the mean slip $\bar{u}$. The aspect ratio $L/W = 1$ is assumed. As shown in Figure 2, the size of the rupture increases until $L = W_0$ and all the ruptures nucleate at the middle depth of the seismogenic layer. For each rupture we calculate its average stress drop $\overline{\Delta\sigma}$, using $\mu = 3 \times 10^4$ MPa. Because $\bar{u}$ has a linear relation with $\overline{\Delta\sigma}$, we plot $\bar{u}/\overline{\Delta\sigma}$ versus W rather than plot $\bar{u}$ directly versus W (Fig. 3). One can see from Figure 3 that, given a constant average stress drop, the mean slip scales linearly with W (or L), except that when W is close to W_0 (i.e., the upper edge of the rupture approaches the free surface) the curve deviates from linearity. This result reveals

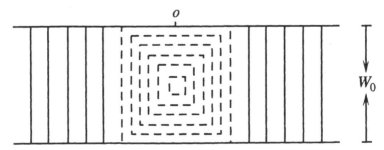

Figure 2
A schematic diagram of the rectangular rupture model. The dashed squares denote small earthquakes and the solid rectangles denote large earthquakes. W_0 is the thickness of the seismogenic layer and o is the origin of the coordinate system x_i.

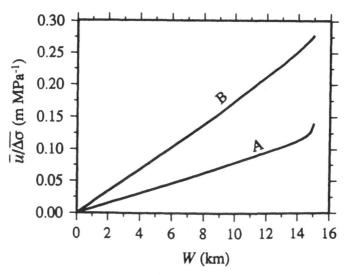

Figure 3

The mean slip per unit stress drop ($\bar{u}/\overline{\Delta\sigma}$) versus the rupture width (W), calculated for small earthquakes using a rectangular rupture model with uniform slip (A) and a more realistic circular rupture model with a nonuniform slip (B). The shear modulus $\mu = 3 \times 10^4$ MPa.

that $M_0 \propto L^3$ for small earthquakes, which confirms the results obtained both from dynamic crack models and from earthquake observations.

In order to study the effect of nonuniform slip, the second model is devised to be a circular rupture with nonuniform slip. The model is similar to the previous one shown in Figure 2, but with replacement of the square rupture surfaces with circles. The distribution of nonuniform slip is assumed to be in an elliptical form with the aspect ratio $W/L = 1$, that is

$$u_1(y_1, y_3) = \Delta u_1 \left[1 - \frac{y_1^2}{\left(\frac{w}{2}\right)^2} - \frac{\left(y_3 - \frac{w_0}{2}\right)^2}{\left(\frac{w}{2}\right)^2} \right] \tag{4}$$

where Δu_1 is a constant. The slip attains its maximum value (equal to Δu_1) at the center and decreases to zero at the edge of the rupture. In this case, the mean slip $\bar{u}$ is equal to $\Delta u_1/2$. Because no analytical solutions are available for such a distribution of nonuniform slip, we use a numerical method to calculate the stress drop $\Delta\sigma(x_1, x_3)$. We divide the rupture area into n^2 rectangular small areas, each of which has a size of $W/n \times W/n$, and regard the slip obtained from equation (4) as uniformly distributed within each of the small areas. Next, we use equation (3) to calculate the proportion of stress drop $\Delta\sigma(x_1, x_3)$ contributed by each of the rectangular small areas. Finally, we obtain the stress drop $\Delta\sigma(x_1, x_3)$ and the average stress drop $\overline{\Delta\sigma}$ by numerical integration. Figure 3 shows that the nonuniform slip does not change the scaling relation between mean slip and rupture dimension, that is, the mean slip still scales linearly with the rupture width (or length). However, the nonuniform slip affects the proportionality constant (or the slope) in the scaling relation. The proportionality constant for the nonuniform slip model is larger than that for the uniform slip model. It is interesting to notice that the proportionality constant for the circular nonuniform slip model is much closer to the observed one than that for the rectangular uniform slip model. The proportionality constant in the relation between $\bar{u}/\Delta\sigma$ and W is 1.8×10^{-5} for the nonuniform slip model and 0.8×10^{-5} for the uniform slip model, whereas the observed proportionality constant is about 2×10^{-5} for small earthquakes, calculated from HANKS' (1977) data and using $\mu = 3 \times 10^4$ MPa. This result is not surprising, because the model of circular rupture with nonuniform slip is more realistic than the model of rectangular rupture with uniform slip. Figure 3 also shows that, even if the average stress drop, the shear modulus and the rupture dimension are the same, different slip distributions may give rise to slightly different mean slips.

Large Earthquakes

For large strike-slip earthquakes the rupture grows only as L, because W is limited to the thickness of the brittle part of the crust of 15–25 km. Consequently, a large seismic fault can be regarded as a long and narrow rupture. We also devise two models to simulate large earthquakes. First, we consider a rectangular rupture with uniform slip, in which the rupture width is set to be constant ($W = W_0 = 15$ km) and the length is increased from 15 km to 300 km, the range of which covers nearly all large earthquakes (see Fig. 2). To eliminate the effect of variation in average stress drop, we plot $\bar{u}/\Delta\sigma$ versus L. As shown in Figure 4, large earthquakes show a more complicated scaling relation between the mean slip and the rupture dimension, which can be explained by neither the W model nor SCHOLZ's (1982a) L model alone. When L is small, $\bar{u}$ increases nearly linearly with L. As L increases, the curve becomes flat and approaches asymptotically a constant value. For the first order of approximation, this scaling relation can be interpreted in terms of a combination of the W and L models. If L is small, $\bar{u}$ scales approximately with L and if L is large, $\bar{u}$ scales approximately with W. It is noteworthy that the change in the scaling relation from scaling with L to scaling with W occurs at a rupture length of about 75 km. This result not only coincides with, but also provides a theoretical explanation for ROMANOWICZ's (1992) statistical result that a cross-over length occurs at 60–70 km (equivalent to a cross-over seismic moment of 0.6–0.8×10^{20} N-m). The above result suggests that the thickness of the seismogenic layer (W_0) is not the characteristic length in the $\bar{u} - L$ scaling relation, but the characteristic length is about 4–5 times W_0. However, our theoretical modeling predicts a scaling relation of $M_0 \propto L^2$ for large earthquakes

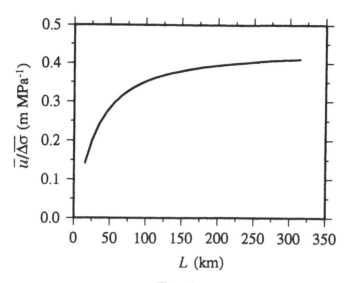

Figure 4

The mean slip per unit stress drop ($\bar{u}/\Delta\sigma$) versus the rupture length (L), calculated for large earthquakes using the model of rectangular rupture with uniform slip. $W = W_0 = 15$ km and $\mu = 3 \times 10^4$ MPa.

with the rupture length smaller than the cross-over length, which is in agreement with SCHOLZ's (1982a) but not ROMANOWICZ's (1992) statistical result.

Next, we consider a more realistic model of an elliptical rupture with nonuniform slip. In this model, the rupture width is constant and equal to the thickness of the seismogenic layer; three examples $W = W_0 = 15, 20, 25$ km are considered. The rupture length is increased from $L = W$ until 300 km. As shown in Figure 5, all the ruptures are composed of two half ellipses. The lower half ellipse has a constant short axis equal to W_0 and the upper half ellipse has a fixed ratio $(b - W_0)/(a - W_0) = 1/5$ so as to allow the rupture to break to the free surface, where $a(a = L)$ and b are the long and the short axes of the upper half ellipse, respectively.

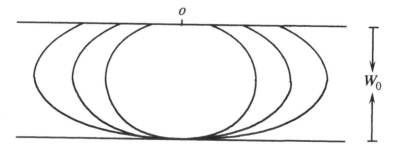

Figure 5

A schematic diagram of the elliptical rupture model for large earthquakes. W_0 and o denote the thickness of the seismogenic layer and the origin of the coordinate system x_i.

The slip distribution is assumed to follow an elliptical function, that is

$$u_1(y_1, y_3) = \Delta u_1 \left[1 - \frac{y_1^2}{\left(\frac{a}{2}\right)^2} - \frac{\left(y_3 - \frac{W_0}{2}\right)^2}{\left(\frac{b}{2}\right)^2} \right], \quad y_3 \leq \frac{W_0}{2}$$

$$u_1(y_1, y_3) = \Delta u_1 \left[1 - \frac{y_1^2}{\left(\frac{L}{2}\right)^2} - \frac{\left(y_3 - \frac{W_0}{2}\right)^2}{\left(\frac{W_0}{2}\right)^2} \right], \quad y_3 > \frac{W_0}{2}$$

(5)

where Δu_1 is a constant. We use the numerical method as described above to calculate the stress drop $\Delta\sigma(x_1, x_3)$ on the rupture surface and consequently the average stress drop $\overline{\Delta\sigma}$. The results are plotted in Figure 6. Comparing Figure 6 with Figure 4, one may find that the nonuniform slip model gives rise to a similar scaling relation between the mean slip and the rupture length, that is, when L is small $\bar{u}$ increases nearly linearly with L and when L is large $\bar{u}$ becomes asymptotically constant. However, the cross-over length at which the scaling relation changes drastically is more obvious in this case. The cross-over length increases from about 75 km to 125 km as the thickness of the seismogenic layer increases from 15 km to 25 km, and the asymptotical constant is different for different rupture width and

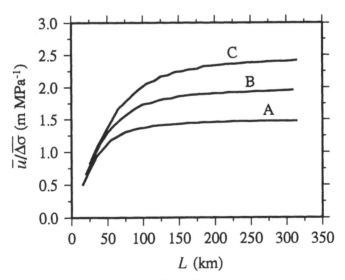

Figure 6

The mean slip per unit stress drop ($\bar{u}/\overline{\Delta\sigma}$) versus the rupture length (L), calculated for large earthquakes with different saturated rupture widths $W_0 = 15$ km (A), $W_0 = 20$ km (B), and $W_0 = 25$ km (C), using the model of elliptical rupture with nonuniform slip. $\mu = 3 \times 10^4$ MPa.

is larger when the rupture is deeper. Like the results for small earthquakes, the nonuniform slip model yields a larger mean slip than the uniform slip model for a given constant average stress drop and rupture dimension.

We also consider a uniform slip rectangular rupture model with a fixed aspect ratio W/L. In this model, the rupture width increases with the length according to a fixed aspect ratio. Three cases $W/L = 1/2, 1/3$ and $1/4$ are studied. As shown in Figure 7, we obtain a linear scaling relation between mean slip and rupture length, which is expected from the scaling relation for small earthquakes. Different aspect ratios lead to different proportionality constants. The proportionality constant increases with the aspect ratio W/L. Because different proportionality constants in the $\bar{u} - L$ relation affect only the intercept but not the slope of the $\mathrm{Log}(M_0) - \mathrm{Log}(L)$ relation, we may predict a scaling relation of $M_0 \propto L^3$ for large earthquakes maintaining a fixed aspect ratio W/L. This result, although obtained from the strike-slip dislocation model, has implications for interplate thrust earthquakes. For subduction zone earthquakes, the rupture can propagate substantially deeper and there is no obvious saturation in rupture width. It has been documented that, over a wide range of the seismic moment, $W/L \approx 1/2$ for circum-Pacific subduction zone earthquakes (ABE, 1975; PURCARU and BERCKHEMER, 1982). Therefore, one may predict that $M_0 \propto L^3$ for most subduction zone earthquakes, which is indistinguishable from the scaling relation for small earthquakes. This theoretical prediction is in agreement with earthquake observations (ROMANOWICZ, 1992).

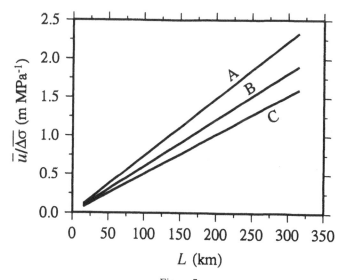

Figure 7
The mean slip per unit stress drop ($\bar{u}/\Delta\sigma$) versus the rupture length (L), calculated for large earthquakes with a fixed aspect ratio $W/L = 1/2$ (A), $W/L = 1/3$ (B), and $W/L = 1/4$ (C), using the rectangular rupture model with uniform slip. $\mu = 3 \times 10^4$ MPa.

Comparison between Theoretical Modeling and Earthquake Observations

Many attempts have been made to formulate the scaling relations between mean slip and rupture dimension and/or between seismic moment and rupture dimension by means of earthquakes statistics (e.g., WYSS and BRUNE, 1968; KANAMORI and ANDERSON, 1975; SCHOLZ, 1982a; SHIMAZAKI, 1986; ROMANOWICZ, 1992). A common approach is to plot the logarithmic seismic moment versus the logarithmic rupture length. However, since among the two scaling relations the mean slip versus the rupture dimension is fundamental, an alternative and more direct approach is to plot the mean slip versus the rupture length. Here, we adopt the latter approach.

As addressed above, many factors may affect the scaling relation between mean slip and rupture length, such as average stress drop, elastic properties of the media (μ and λ), rupture width, faulting regimes, and heterogeneous distribution of the slip. Among them, the average stress drop, the rupture width and the faulting regimes are vital. As noticed by previous workers (SCHOLZ, 1982a; ROMANOWICZ, 1992), in order to develop meaningful statistics, earthquakes must be classified into different groups according to their faulting regime and tectonic setting, and the effect of rupture width must be taken into account. The effect of average stress drop is more difficult to evaluate. On the one hand, the average stress drop is statistically constant, independent of rupture size over a broad range (e.g., AKI, 1972; THATCHER and HANKS, 1973; KANAMORI and ANDERSON, 1975; HANKS, 1977). On the other hand, it has been observed that the average stress drop varies considerably (a range of about $0.03-30$ MPa (SCHOLZ, 1990, p. 182)). In other words, the mean value of average stress drop is constant, but its variance is enormous. Since the mean slip has a linear relation with the average stress drop, an amount of variation in the average stress drop will cause the mean slip to vary proportionally. Statistical methods that treat all earthquakes as having a constant stress drop are inappropriate because the average stress drop is not constant for individual earthquakes. If the average stress drop can be determined independently, then using the normalized parameters $\bar{u}/\Delta\sigma$ and $M_0/\Delta\sigma$ (i.e., mean slip per unit stress drop and seismic moment per unit stress drop) is better than using $\bar{u}$ and M_0. Here, we plot $\bar{u}/\Delta\sigma$ versus L to reduce the effect of variation in stress drop.

The data listed in Table 1 derive from SCHOLZ (1982a), from which the sources of original references can be found. The earthquakes are all large strike-slip earthquakes with rupture widths ranging from $10-15$ km. Figure 8 shows the variation of $\bar{u}/\Delta\sigma$ with rupture length for the earthquakes listed in Table 1, compared with the theoretical curves calculated using the same elliptical rupture model as shown in Figure 6 except for $\mu = 2.8 \times 10^4$ MPa. The theoretical curves agree with the general trend of variation of $\bar{u}/\Delta\sigma$ with L, and each individual curve matches up well with the corresponding earthquakes that have the same rupture width. It is noteworthy that, although the rectangular uniform slip model leads to a similar form of the scaling relation for large earthquakes (see Fig. 4), the mean

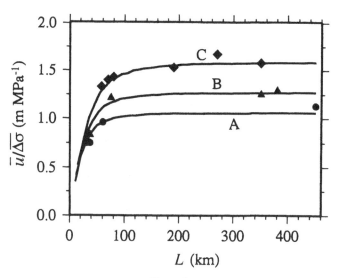

Figure 8

Comparison of the scaling relation between the mean slip per unit stress drop ($\bar{u}/\overline{\Delta\sigma}$) and the rupture length (L) calculated using the same dislocation model as for Figure 6 (curves) with that obtained from the earthquakes listed in Table 1 (symbols). Circles denote the earthquakes with rupture width $W = 10$ km, triangles $W = 12$ km, and diamonds denote $W = 15$ km. The curves A, B and C denote the rupture width $W = 10$ km, 12 km and 15 km, respectively. $\mu = 2.8 \times 10^4$ MPa.

Table 1

List of source parameters of large strike-slip earthquakes (from SCHOLZ (1982a))

No.	Date	Location	M_0 (10^{20} N-m)	L (km)	W (km)	$\overline{\Delta u}$ (m)	$\overline{\Delta\sigma}$ (MPa)
1	07/10/1958	SE Alaska	4.3	350	12	3.25	2.6
2	01/09/1857	S. California	7.0	380	12	4.65	3.6
3	04/18/1906	San Francisco	4.0	450	10	4.50	4.4
4	05/19/1940	Imperial Valley, Calif.	0.23	60	10	1.25	1.3
5	01/27/1966	Parkfield, Calif.	0.03	37	10	0.30	0.4
6	04/09/1968	Borrego Mountain, Calif.	0.08	37	12	0.25	0.3
7	10/15/1979	Imperial Valley, Calif.	0.03	30	10	0.30	0.4
8	02/04/1976	Guatemala	2.6	270	15	1.50	0.9
9	10/16/1974	Gibbs Fault Zone	0.45	75	12	1.70	1.4
10	12/26/1939	Ercincan, Turkey	4.5	350	15	2.85	1.8
11	12/20/1942	Erbaa Niksar, Turkey	0.35	70	15	1.12	0.8
12	02/01/1944	Gerede-Bolu, Turkey	2.4	190	15	2.75	1.8
13	03/18/1953	Gönen-Yenice, Turkey	0.73	58	15	2.80	2.1
14	07/22/1967	Mudurnu, Turkey	0.36	80	15	1.00	0.7

slip predicted by that model is too small to fit the observations. This results in an independent confirmation that the slip distribution on a rupture surface can be approximately modeled by an elliptical function.

Discussion

Both the theoretical modeling and the earthquake observations consistently suggest that seismic rupture can be represented by a well devised dislocation model grounded in the elastic theory of dislocation. Hypothetical rupture models, such as that proposed by SCHOLZ (1982a), are not necessary to explain the earthquake scaling relations. The observed linear correlation of the mean slip with rupture length is predicted by the elastic dislocation models rather than contradict the dislocation models. This is because most of those earthquakes used in statistics have rupture lengths that are smaller than the cross-over length. However, our analysis suggests that the mean slip will saturate for very large earthquakes (unless the rupture breaks deeper and/or the average stress increases with rupture length). In earthquake statistics, the average stress drop is often assumed to be constant. This assumption must be used with care because the average stress drop, although it has a constant mean value, varies within a wide range of 0.03–30 MPa (i.e., three orders of magnitude). For instance, it is inadequate to make the inference that increasing $\bar{u}$ with L must mean that $\bar{u}$ scales with L for very large earthquakes with a rupture length larger than the cross-over length. Because the proportionality constant in the $\bar{u}/\Delta\sigma - L$ linear scaling relation is small (about a few meters per 100 km) and the variation in stress drop can be very large, a small amount of stress drop increase is enough to contribute to a significant increase in $\bar{u}$.

The static parameters, the seismic moment and the mean slip, are not the only parameters that are relevant to the rupture dimension. Such dynamic parameters as the rise time, the peak velocity and the peak acceleration are also important. Based on the W and L models, SCHOLZ (1982b) found a huge discrepancy between the scaling relations of the dynamic parameters with rupture length predicted by the W model and that predicted by the L model. He noted in his paper that "...the possibility remains that some intermediate model may be appropriate, in which case the scaling of strong ground motions will be intermediate to the extremes discussed here...". Detailed discussion of the scaling of dynamic ground motions is beyond the scope of this paper. Nevertheless, our dislocation models may predict the intermediate results. For instance, assuming the rise time $t_r \approx W/(2v_r)$, where v_r is the velocity of rupture propagation (KASAHARA, 1981, p. 105; SCHOLZ, 1982b), we predict a range of the rise time of about 1.7–4.2 s for large continental earthquakes with a rupture width ranging from 10–25 km. Given a slip-time function, our models predict that the peak velocity in the very near field close to the fault zone increases with the rupture length and does not saturate until the rupture exceeds the cross-over length. These theoretical predictions are found to be in principle in agreement with the observations (e.g., KASAHARA, 1981, pp. 220–221).

Conclusions

This paper has dealt with the earthquake scaling problem of whether the mean slip should scale with the rupture width (the W model) or scale with the rupture length (the L model) for large earthquakes whose rupture width is limited by the thickness of the seismogenic layer. On the basis of the elastic theory of dislocation, we have devised several static dislocation models which have different rupture shapes and different slip distributions to simulate strike-slip earthquakes. For each of the models we have calculated the mean slip and the average stress drop, and have found their relations with rupture dimension. The conclusions can be drawn as follows.

(1) For small earthquakes with rupture length equal to or smaller than the thickness of the seismogenic layer, our models predict a linear scaling relation between the mean slip and the rupture width (Fig. 3).

(2) For large earthquakes with a rupture length larger than the thickness of the seismogenic, our models predict a more complicated scaling relation between mean slip and rupture length (Figs. 4 and 6). When the rupture length is smaller than a cross-over length, the mean slip scales nearly linearly with the rupture length. When the rupture length is larger than the cross-over length the mean slip approaches asymptotically a constant value and scales approximately with rupture width. The cross-over length, which is a function of rupture width is about 75 km for $W_0 = 15$ km and can reach 125 km for $W_0 = 25$ km. This result suggests that, in the scaling relation of the mean slip with the rupture dimension, the characteristic length is the cross-over length rather than the thickness of the seismogenic layer, which is in agreement with ROMANOWICZ's (1992) statistical result.

(3) The scaling relation between seismic moment and rupture dimension can be derived from the mean slip-rupture dimension relation (see equation (1)). For small earthquakes, the seismic moment scales with the cube of rupture width. For large earthquakes with a rupture length smaller than the cross-over length, the seismic moment scales approximately with the square of rupture length. For very large earthquakes with a rupture length larger than the cross-over length, the seismic moment scales approximately with the rupture length.

(4) Our theoretical predictions have been compared with observed source parameters of large strike-slip earthquakes available in the literature and they match up well (Fig. 8).

(5) The heterogeneous distribution of slip does not affect the scaling relation between the mean slip and the rupture dimension, but it does affect the parameters in the scaling relation. The mean slips predicted by the uniform slip model are too

small to fit the observations. Only the mean slips predicted by the nonuniform slip model fit the observations well (Figs. 3, 4, 6, and 8). This result is an independent confirmation that the distribution of coseismic slip on the fault surface can be modeled approximately by an elliptical function.

(6) Our results show that, when large earthquakes have a fixed aspect ratio of the rupture length to the rupture width (such as appears to be the case for most subduction earthquakes), the mean slip scales with the rupture dimension in the same way as that for small earthquakes (Fig. 7).

Acknowledgements

This work has been supported by a Natural Sciences and Engineering Council of Canada postdoctoral fellowship to Yin and by the Geological Survey of Canada at the Pacific Geoscience Centre. We have benefited from a discussion with K. Wang.

Appendix

Here, we use CHINNERY's (1963) notation $\|$ to represent the integral substitution

$$g(y_1, y_3) \| = g(L_2, W_2) - g(L_2, W_1) - g(L_1, W_2) + g(L_1, W_1) \qquad (A1)$$

where L_1, L_2, W_1, and W_2 are the coordinates of the four corners of the rectangular rupture, and the rupture length is $L = L_2 - L_1$ and the width is $W = W_2 - W_1$, respectively. Thus, the function f in equation (3) can be expressed as

$$f = \left[-\frac{z_1}{R(R + z_3)} - \frac{2z_3}{Rz_1} - \frac{3R}{z_1 z_3} + \frac{z_1}{2Q(Q + s)} + \frac{3Q}{sz_1} + \frac{6s^2}{z_1^3} - \frac{3Qs}{z_1^3} \right.$$
$$\left. + \frac{3s^2 - 8x_3^2}{2Qsz_1} + \frac{4(s^2 - x_3^2)}{Qsz_1} - \frac{Q(3s^2 - 8x_3^2)}{sz_1^3} + \frac{8z_1 x_3(s - x_3)}{Q^3 s} \right] \| \qquad (A2)$$

where

$$\left\{ \begin{array}{l} z_1 = x_1 - y_1 \\ z_3 = x_3 - y_3 \\ s = x_3 + y_3 \\ R^2 = z_1^2 + z_3^2 \\ Q^2 = z_1^2 + s^2. \end{array} \right. \qquad (A3)$$

With appropriate substitution of L_1, L_2, W_1, and W_2, equations (3) and (A2) allow us to calculate the stress drop on the rupture surface for vertically oriented strike-slip faults with any sizes.

REFERENCES

ABE, K. (1975), *Reliable Estimation of the Seismic Moment of Large Earthquakes*, Phys. Earth Planet. Int. *23*, 381–390.

AKI, K. (1972), *Earthquake Mechanics*, Tectonophysics *13*, 423–446.

CHINNERY, M. A. (1961), *The Deformation of the Ground Around Surface Faults*, Bull. Seismol. Soc. Am. *50*, 336–372.

CHINNERY, M. A. (1963), *The Stress Changes that Accompany Strike-slip Faulting*, Bull. Seismol. Soc. Am. *53*, 921–932.

DU, Y., AYDIN, A., and SEGALL, P. (1992), *Comparison of Various Inversion Techniques as Applied to the Determination of a Geophysical Deformation Model for the 1983 Borah Peak Earthquake*, Bull. Seismol. Soc. Am. *82*, 1840–1866.

DU, Y., and AYDIN, A. (1993), *Stress Transfer During Three Sequential Moderate Earthquakes Along the Central Calaveras Fault, California*, J. Geophys. Res. *98*, 9947–9962.

HANKS, T. C. (1977), *Earthquake Stress-drops, Ambient Tectonic Stresses, and Stresses that Drive Plates*, Pure and Appl. Geophys. *115*, 441–458.

IWASAKI, T., and SATO, R. (1979), *Train Field in a Semi-infinity Medium due to an Inclined Rectangular Fault*, J. Phys. Earth *27*, 285–314.

KANAMORI, H., and ANDERSON, D. L. (1975), *Theoretical Basis of Some Empirical Relations in Seismology*, Bull. Seismol. Soc. Am. *65*, 1073–1095.

KASAHARA, K., *Earthquake Mechanics* (Cambridge University Press, Cambridge 1981).

KNOPOFF, L. (1958), *Energy Release in Earthquakes*, Geophys. J. R. A. S. *1*, 44–52.

MADARIAGA, R. (1976), *Dynamics of an Expanding Circular Fault*, Bull. Seismol. Soc. Am. *66*, 639–666.

MASSONNET, D., ROSSI, M., CARMONA, C., ADRAGNA, F., PELTZER, G., FEIGL, L., and RABAUTE, T. (1993), *The Displacement Field of the Landers Earthquake Mapped by Radar Interferometry*, Nature *364*, 138–142.

OKADA, Y. (1985), *Surface Deformation due to Shear and Tensile Faults in a Half-space*, Bull. Seismol. Soc. Am. *75*, 1135–1154.

PRESS, F. (1965), *Displacements, Strains and Tilts at Teleseismic Distance*, J. Geophys. Res. *70*, 2395–2412.

PURCARU, G., and BERCKHEMER, H. (1982), *Quantitative Relations of Seismic Source Parameters and a Classification of Earthquakes*, Tectonophysics *84*, 57–128.

ROMANOWICZ, B. (1992), *Strike-slip Earthquakes on Quasi-vertical Transcurrent Faults: Inferences for General Scaling Relations*, Geophys. Res. Lett. *19*, 481–484.

ROMANOWICZ, B., and RUNDLE, J. B. (1993), *On Scaling Relation for Large Earthquakes*, Bull. Seismol. Soc. Am. *83*, 1294–1297.

SCHOLZ, C. H. (1982a), *Scaling Laws for Large Earthquakes: Consequences for Physical Models*, Bull. Seismol. Soc. Am. *72*, 1–14.

SCHOLZ, C. H. (1982b), *Scaling Relations for Strong Ground Motion in Large Earthquakes*, Bull. Seismol. Soc. Am. *72*, 1903–1909.

SCHOLZ, C. H., *The Mechanics of Earthquakes and Faulting* (Cambridge University Press, Cambridge 1990).

SCHOLZ, C. H. (1994), *A Reapprasial of Large Earthquake Scaling*, Bull. Seismol. Soc. Am. *84*, 215–218.

SEGALL, P., and DU, Y. (1993), *How Similar Were the 1934 and 1966 Parkfield Earthquakes?* J. Geophys. Res. *98*, 4527–4538.

SHIMAZAKI, K., *Small and large earthquakes: The effects of the thickness of the seismogenic layer and the free surface. In Earthquake Source Mechanics* (eds. Das, S., Boatwright, J., and Scholz, C.) (American Geophysical Union, Washington, D.C. 1986), AGU Geophys. Mono. *37*, 209–216.

STEKETEE, J. A. (1958), *On Volterra's Dislocation in a Semi-infinite Elastic Medium*, Can. J. Phys. *36*, 192–205.

THATCHER, W., and HANKS, T. C. (1973), *Source Parameters of Southern California Earthquakes*, J. Geophys. Res. *78*, 8547–8576.

WYSS, M., and BRUNE, J. N. (1968), *Seismic Moment, Stress and Source Dimensions for Earthquakes in the California-Nevada Region*, J. Geophys. Res. *73*, 4681–4694.

(Received September 7, 1994, revised April 6, 1995, accepted May 20, 1995)

PAGEOPH, Vol. 146, Nos. 3/4 (1996)

0033–4553/96/040677–12$1.50 + 0.20/0

Reflection of Waves from the Boundary of a Random Elastic Semi-infinite Medium

RABINDRA KUMAR BHATTACHARYYA[1]

Abstract — In this paper the smooth perturbation technique is employed to investigate the problem of reflection of waves incident on the plane boundary of a semi-infinite elastic medium with randomly varying inhomogeneities. Amplitude ratios have been obtained for various types of incident and reflected waves. It has been shown that an incident SH or SV type of wave gives rise to reflected SH, P and SV waves, the main components being SH and P, SV in the respective cases. The reflected amplitudes have been calculated depending upon the randomness of the medium to the square of the small quantity ε, where ε measures the deviation of the medium from homogeneity. An incident P-type wave produces mainly a P component and also a weak SH component to the order of ε^2. The reflected amplitudes obtainable for elastic media are also altered by terms of the same order. The direction of the reflected wave is influenced by randomness in some cases.

Key words: Smooth perturbation, elastic waves, reflection, random medium.

Introduction

KARAL and KELLER (1964) applied the theory of smooth perturbation to investigate problems of wave propagation in random elastic, electromagnetic and other media. CHOW (1973) adopted the same technique in examining the problem of wave propagation in a random thermoelastic medium. BHATTACHARYYA (1986) studied problems of wave propagation in a random magneto-thermo-viscoelastic medium as also in a random micropolar elastic medium.

In the present paper the same procedure has been adopted to investigate the problem of reflection of waves incident on the plane boundary of a random elastic and semi-infinite medium. The Lamé parameters and density are assumed to vary as random functions of position. If the displacement

$$\vec{u}(\vec{x}, t) = \vec{u}_0(\vec{x}) \, e^{-i\omega t} \tag{1}$$

[1] Department of Mathematics, Brahmananda Keshab Chandra College, 111/2. Barrackpore Trunk Road, Bon-Hooghly, Calcutta 700035, India.

is such that the mean displacement $\langle \vec{u}_0(\vec{x}) \rangle$, given by

$$\langle \vec{u}_0(\vec{x}) \rangle = \vec{A} \, e^{i\vec{k} \cdot \vec{x}} \tag{2}$$

represents a plane wave incident on the boundary, then $\langle \vec{u}_0(\vec{x}) \rangle$ satisfies the integro-differential field equation (upto terms of order ε^2):

$$L_0 \langle \vec{u}_0(\vec{x}) \rangle - \varepsilon^2 \langle L_1 L_0^{-1} L_1' \rangle \langle \vec{u}_0(\vec{x}) \rangle = 0, \tag{3}$$

where

$$L_0 = (\lambda_0 + \mu_0)\vec{\nabla}(\vec{\nabla} \cdot) + \mu_0 \nabla^2 + \omega^2 \rho_0 \tag{4}$$

$$L_1 = (\lambda_1 + \mu_1)\vec{\nabla}(\vec{\nabla} \cdot) + \mu_1 \nabla^2 + \nabla \lambda_1 \cdot \vec{\nabla} + \nabla \mu_1 \times (\vec{\nabla} x)$$

$$+ 2\nabla \mu_1 \cdot \vec{\nabla} + \omega^2 \rho_1 \tag{5}$$

with

$$\rho(\vec{x}) = \rho_0 + \varepsilon \rho_1(\vec{x}), \quad \lambda(\vec{x}) = \lambda_0 + \varepsilon \lambda_1(\vec{x})$$

$$\mu(\vec{x}) = \mu_0 + \varepsilon \mu_1(\vec{x})$$

and

$$L_0 G_{lj}(\vec{x}, \vec{x}') = \delta(\vec{x}, \vec{x}')\delta_{lj}, \tag{6}$$

G_{lj} being the Green's matrix for the deterministic elastic field (KARAL and KELLER, 1964).

It has been demonstrated by KARAL and KELLER (1964) that only two types of waves with wave numbers $k = k_d$ and $k = k_l$ can propagate in the medium, where

$$k_d = k_s - \varepsilon^2 P, \quad k_s = \sqrt{\rho_0 \omega^2 / \mu_0}, \quad P = D_1 / 2\mu_0 k_s \tag{7}$$

and

$$k_l = k_c - \varepsilon^2 Q, \quad k_c = \sqrt{\rho_0 \omega^2 / (\lambda_0 + 2\mu_0)}, \quad Q = (D_1 + D_2)/2(\lambda_0 + 2\mu_0)k_c. \tag{8}$$

The expressions for D_1 and D_2 are given by relations (57) and (58) in KARAL and KELLER (1964), D_1, D_2 being functions of various auto- and cross-correlation functions of the parameters λ, μ, ρ depicting the medium.

In the sequel we have been concerned with these two types of waves incident on the plane boundary of the random elastic half-space, giving rise to various types of reflected waves. Several reflected amplitudes to the order of ε^2 are available in addition to the main amplitudes for elastic media (MIKLOWITZ, 1980). The presence of these additional terms are due to the randomness of the medium. Moreover, the main components have themselves been modified by the addition of terms of the order of ε^2. In computing the various amplitude ratios, it has however been considered necessary to make the mean stresses instead of the usual stresses vanish at the boundary. This is due to the fact that the stresses have been assumed to vary with different probabilities $p(\alpha)$ over the entire medium (KELLER, 1964). This step is also appropriate in as much as the mean displacement vector and not the usual displacement vector appears in the field equation (3).

Problem. Let the plane $x_3 = 0$ represent the boundary of the random elastic half-space. For an incident k_d-type wave the mean displacement is assumed to be of the form

$$\langle \vec{u}_0(\vec{x}) \rangle = \vec{A}^{(1)} e^{ik_d \hat{n} \cdot \vec{x}} + \vec{A}^{(2)} e^{ik_d \vec{l} \cdot \vec{x}} + \vec{B}^{(2)} e^{ik_l \hat{m} \cdot \vec{x}} \tag{9}$$

where $\vec{A}^{(1)}$, $\vec{A}^{(2)}$, and $\vec{B}^{(2)}$ are the amplitudes of the incident k_d-type, reflected k_d-type and reflected k_l-type waves, respectively and

$$\hat{n} = (l_1, m_1, n_1), \quad \vec{l} = (l_1, m_2, n_2), \quad \hat{m} = (l_3, m_3, n_3)$$

are the corresponding unit wave normals. Then the wave form (9) must satisfy the boundary stress conditions

$$\langle \overset{v}{\vec{T}} \rangle = 0 \quad \text{at} \quad x_3 = 0 \tag{10}$$

where the mean stress $\langle \overset{v}{\vec{T}} \rangle$ is represented by

$$\langle \overset{v}{\vec{T}} \rangle = \lambda_0 (\vec{\nabla} \cdot \langle \vec{u}_0(\vec{x}) \rangle) \vec{v} + \mu_0 (\vec{v} \cdot \vec{\nabla}) \langle \vec{u}_0(\vec{x}) \rangle + \mu_c \vec{\nabla} (\langle \vec{u}_0(\vec{x}) \rangle \cdot \vec{v})$$
$$+ \varepsilon^2 [\langle \lambda_1(\vec{x}) (\vec{\nabla} \cdot \{L_0^{-1} L_1 \langle \vec{u}_0(\vec{x}) \rangle\}) \vec{v} \rangle + \langle \mu_1(\vec{x}) (\vec{v} \cdot \vec{\nabla}) (L_0^{-1} L_1 \langle \vec{u}_0(\vec{x}) \rangle) \rangle$$
$$+ \langle \mu_1(\vec{x}) \vec{\nabla} \{(L_0^{-1} L_1 \langle \vec{u}_0(\vec{x}) \rangle) \cdot \vec{v}\} \rangle] \tag{11}$$

and

$$\vec{v} = (0 \quad 0 \quad 1). \tag{12}$$

Then substitution of (9) in (10) yields the following three equations

$$(i\mu_0 k_d n_1 + \varepsilon^2 a_{11}^{(1)}) A_1^{(1)} + \varepsilon^2 a_{12}^{(1)} A_2^{(1)} + (i\mu_0 k_d l_1 + \varepsilon^2 a_{13}^{(1)}) A_3^{(1)}$$
$$+ (-i\mu_0 k_d n_1 + \varepsilon^2 a_{11}^{(2)}) A_1^{(2)} + \varepsilon^2 a_{12}^{(2)} A_2^{(2)} + (i\mu_0 k_d l_1 + \varepsilon^2 a_{13}^{(2)}) A_3^{(2)}$$
$$+ (i\mu_0 k_l n_3 + \varepsilon^2 a_{11}^{(3)}) B_1^{(2)} + \varepsilon^2 a_{12}^{(3)} B_2^{(2)} + (i\mu_0 k_l l_3 + \varepsilon^2 a_{13}^{(3)}) B_3^{(2)} = 0 \tag{13}$$

$$\varepsilon^2 b_{11}^{(1)} A_1^{(1)} + (i\mu_0 k_d n_1 + \varepsilon^2 b_{12}^{(1)}) A_2^{(1)} + (i\mu_0 k_d m_1 + \varepsilon^2 b_{13}^{(1)}) A_3^{(1)}$$
$$+ \varepsilon^2 b_{11}^{(2)} A_1^{(2)} + (-i\mu_0 k_d n_1 + \varepsilon^2 b_{12}^{(2)}) A_2^{(2)} + (i\mu_0 k_d m_1 + \varepsilon^2 b_{13}^{(2)}) A_3^{(2)}$$
$$+ \varepsilon^2 b_{11}^{(3)} B_1^{(2)} + (i\mu_0 k_l n_3 + \varepsilon^2 b_{12}^{(3)}) B_2^{(2)} + (i\mu_0 k_l m_3 + \varepsilon^2 b_{13}^{(3)}) B_3^{(2)} = 0 \tag{14}$$

$$(i\lambda_0 k_d l_1 + \varepsilon^2 c_{11}^{(1)}) A_1^{(1)} + (i\lambda_0 k_d m_1 + \varepsilon^2 c_{12}^{(1)}) A_2^{(1)}$$
$$+ \{i(\lambda_0 + 2\mu_0) k_d n_1 + \varepsilon^2 c_{13}^{(1)}\} A_3^{(1)}$$
$$+ (i\lambda_0 k_d l_1 + \varepsilon^2 c_{11}^{(2)}) A_1^{(2)} + (i\lambda_0 k_d m_1 + \varepsilon^2 c_{12}^{(2)}) A_2^{(2)}$$
$$+ \{-i(\lambda_0 + 2\mu_0) k_d n_1 + \varepsilon^2 c_{13}^{(2)}\} A_3^{(2)}$$
$$+ (i\lambda_0 k_l l_1 + \varepsilon^2 c_{11}^{(3)}) B_1^{(2)} + (i\lambda_0 k_l m_1 + \varepsilon^2 c_{12}^{(3)}) B_2^{(2)}$$
$$+ \{i(\lambda_0 + 2\mu_0) k_l n_3 + \varepsilon^2 c_{13}^{(3)}\} B_3^{(2)} = 0, \tag{15}$$

where, obeying the law of reflection

$$(l_2, m_2, n_2) = (l_1, m_1, -n_1), \tag{15a}$$

and

$$l_3 = \frac{k_s}{k_c} \left\{ 1 + \varepsilon^2 \frac{D_2}{2\rho_0 \omega^2} \right\} l_1 \tag{15b}$$

$$m_3 = \frac{k_s}{k_c} \left\{ 1 + \varepsilon^2 \frac{D_2}{2\rho_0 \omega^2} \right\} m_1 \tag{15c}$$

$$n_3 = \left[1 - (1 - n_1^2) \frac{k_s^2}{k_c^2} \right]^{1/2} \left[1 - \varepsilon^2 \frac{D_2 k_s^2 (1 - n_1^2)}{2\rho_0 \omega^2 k_c^2 \left\{ 1 - (1 - n_1^2) \frac{k_s^2}{k_c^2} \right\}} \right]. \tag{15d}$$

Hence

$$\vec{v} \times \vec{l} = \vec{v} \times \vec{m} \quad \text{and} \quad \vec{v} \times \vec{m} = \frac{k_s}{k_c} \left(1 + \varepsilon^2 \frac{D_2}{2\rho_0 \omega^2} \right) \vec{v} \times \vec{n}.$$

This illustrates that the reflected wave normal is also in the plane of $\vec{v}$ and $\vec{n}$. The relations (15b)–(15d) indicate the dependence of the direction of reflected k_l-type waves on the randomness of the medium. The 27 coefficients $a_{11}^{(1)}, \ldots, b_{11}^{(1)}, \ldots,$ $c_{11}^{(1)}, \ldots,$ are integrals involving various correlation functions of the forms

$$\langle \lambda_1(\vec{x}) \lambda_1'(\vec{x}') \rangle, \quad \langle \lambda_1(\vec{x}) \mu_1'(\vec{x}') \rangle, \quad \langle \lambda_1(\vec{x}) \rho_1'(\vec{x}') \rangle, \ldots, \text{etc.}$$

The integrands are lengthy expressions and only a few of them have been included in Appendix 1.

Next the substitution of (9) in the field equation (3) leads to the following relations between the components of incident or reflected amplitudes:

$$A_1^{(1)} = -\frac{1}{l_1} [m_1 A_2^{(1)} + n_1 A_3^{(1)}] \tag{22}$$

$$A_1^{(2)} = -\frac{1}{l_1} [m_1 A_2^{(2)} - n_1 A_3^{(2)}]. \tag{23}$$

Therefore

$$\vec{A}^{(1)} \cdot \vec{n} = 0 \quad \text{and} \quad \vec{A}^{(2)} \cdot \vec{n} = 0.$$

By (22) and (23), the first and second terms in $\langle \vec{u}_0(\vec{x}) \rangle$ represent transverse waves. Also,

$$B_1^{(2)} = g_1 M, \quad B_2^{(2)} = g_2 M, \quad B_3^{(2)} = g_3 M.$$

$$g_1 = l_3 n_3 [(\lambda_0 + \mu_0)k_i^2 + \varepsilon^2 D_2(k_i)]$$

$$g_2 = m_3 n_3 [(\lambda_0 + \mu_0)k_i^2 + \varepsilon^2 D_2(k_i)]$$

$$g_3 = -[\mu_0 k_i^2 - \omega^2 \rho_0^* + \varepsilon^2 D_1(k_i) + (l_3^2 + m_3^2)$$

$$\times \{(\lambda_0 + \mu_0)k_i^2 + \varepsilon^2 D_2(k_i)\}]. \tag{24}$$

Clearly, $A_1^{(1)}$, $A_2^{(1)}$, $A_3^{(1)}$ being the known amplitudes of the incident wave, there remain only three unknowns $A_2^{(2)}$, $A_3^{(2)}$, and M, connected by the three equations (13)–(15). By (15b)–(15d), it has been verified that the incident wave-normal (l_1, m_1, n_1), the reflected wave-normal $(l_1, m_1, -n_1)$ and the normal to the boundary $(0, 0, 1)$ lie in one plane, i.e., $\hat{n} \cdot (\vec{v} \times \vec{l}) = 0$. Similarly, (l_1, m_1, n_1), (l_3, m_3, n_3), $(0, 0, 1)$ are coplanar, i.e.,

$$\hat{n} \cdot (\vec{v} \times \hat{m}) = 0.$$

Again restricting the incident and reflected waves to propagate in the plane $x_2 = 0$, one at once finds $g_2 = 0$, and hence

$$B_2^{(2)} = 0. \tag{25}$$

Also equations (13)–(15), by virtue of (22)–(24) and the conditions

$$m_1 = m_2 = m_3 = 0$$

reduce to

$$\varepsilon^2 l_1 a_{12}^{(2)} A_2^{(2)} + [i\mu_0 k_d(l_1^2 - n_1^2) + \varepsilon^2(a_{11}^{(2)} n_1 + a_{13}^{(2)} l_1)]A_3^{(2)}$$

$$+ [i\mu_0 k_i(n_3 g_1 + l_3 g_3) + \varepsilon^2(a_{11}^{(3)} g_1 + a_{13}^{(3)} g_3)]l_1 M$$

$$+ [\varepsilon^2 l_1 a_{12}^{(1)} A_2^{(1)} + \{i\mu_0 k_d(l_1^2 - n_1^2) - \varepsilon^2(a_{11}^{(1)} n_1 - a_{13}^{(1)} l_1)\}A_3^{(1)}] = 0 \tag{26}$$

$$[-i\mu_0 k_d l_1 n_1 + \varepsilon^2 b_{12}^{(2)} l_1]A_2^{(2)} + \varepsilon^2(l_1 b_{13}^{(2)} + n_1 b_{11}^{(2)})A_3^{(2)}$$

$$+ \varepsilon^2(b_{11}^{(3)} g_1 + b_{13}^{(3)} g_3)l_1 M + [\{i\mu_0 k_d l_1 n_1 + \varepsilon^2 b_{12}^{(1)} l_1\}A_2^{(1)}$$

$$+ \varepsilon^2(b_{13}^{(1)} l_1 - b_{11}^{(1)} n_1)A_3^{(1)}] = 0 \tag{27}$$

$$\varepsilon^2 l_1 c_{12}^{(2)} A_2^{(2)} + [-2i\mu_0 k_d l_1 n_1 + \varepsilon^2(c_{11}^{(2)} n_1 + c_{13}^{(2)} l_1)]A_3^{(2)}$$

$$+ [i\lambda_0 k_i(l_3 g_1 + n_3 g_3) + 2i\mu_0 k_i n_3 g_3 + \varepsilon^2\{c_{11}^{(3)} g_1 + c_{13}^{(3)} g_3\}]l_1 M$$

$$+ [\varepsilon^2 c_{12}^{(1)} l_1 A_2^{(1)} + \{2i\mu_0 k_d l_1 n_1 + \varepsilon^2(l_1 c_{13}^{(1)} - c_{11}^{(1)} n_1)\}A_3^{(1)}] = 0. \tag{28}$$

Solving these equations for $A_2^{(2)}$, $A_3^{(2)}$, and M results in the expressions of the reflected amplitudes in terms of the incident amplitudes in the following manner

$$A_2^{(2)} = \frac{1}{r_0}\left[p_0 A_2^{(1)} + \varepsilon^2\left\{p_0' A_2^{(1)} - p_0 \frac{r_0'}{r_0} A_2^{(1)} + q_0 A_3^{(1)}\right\}\right] \tag{29}$$

$$A_3^{(2)} = \frac{1}{r_0} m_0 A_3^{(1)} + \varepsilon^2 \left\{ s_0 A_2^{(1)} + m_0' A_3^{(1)} - m_0 \frac{r_0'}{r_0} A_3^{(1)} \right\} \right] \tag{30}$$

$$A_1^{(2)} = \frac{n_1}{l_1} A_3^{(2)} \tag{31}$$

$$B_1^{(2)} = \frac{g_1}{r_0} \left[\varepsilon^2 t_0 A_2^{(1)} + n_0 A_3^{(1)} + \varepsilon^2 n_0' A_3^{(1)} - \varepsilon^2 n_0 \frac{r_0'}{r_0} A_3^{(1)} \right] \tag{32}$$

$$B_2^{(2)} = 0 \tag{33}$$

$$B_3^{(2)} = \frac{g_3}{g_1} B_1^{(2)} \tag{34}$$

where $p_0, r_0, \ldots$, are defined in Appendix 2.

Two possible types of incidence that may take place are considered below:

Type I. The incident k_d-type waves may propagate with only amplitude $A_2^{(1)}$ (that is *SH*-type) such that

$$A_1^{(1)} = A_3^{(1)} = 0. \tag{35}$$

In this case relations $(29)-(34)$ reduce to

$$A_2^{(2)} = \frac{p_0}{r_0} A_2^{(1)} + 0(\varepsilon^2)$$

$$A_3^{(2)} = 0(\varepsilon^2)$$

$$A_1^{(2)} = 0(\varepsilon^2)$$

$$B_1^{(2)} = 0(\varepsilon^2)$$

$$B_2^{(2)} = 0$$

$$B_3^{(2)} = 0(\varepsilon^2). \tag{36}$$

Thus the *SH* incident wave with amplitude $A_2^{(1)}$ gives rise to a reflected k_d and reflected k_l waves. Only $A_2^{(2)}$ is dominant since

$$A_2^{(2)} = \frac{p_0}{r_0} A_2^{(1)} + 0(\varepsilon^2),$$

indicating that the reflected k_d amplitude $A_2^{(2)}$ depends upon $A_2^{(1)}$. But $A_3^{(2)}$, $B_3^{(2)}$, $A_1^{(2)}$ represent small quantities of order ε^2. Moreover no transverse k_l-type wave is possible in this case (since $B_2^{(2)} = 0$). (Fig. 1).

Type II. The incident k_d-type waves with amplitudes $A_1^{(1)}$, $A_3^{(1)}$ only (that is *SV* type) may propagate with $A_2^{(1)} = 0$. Also in this case reflected k_d- and k_l-type waves are available such that

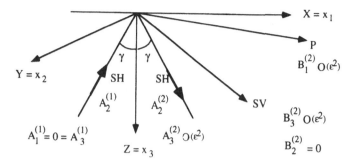

Figure 1
Incident *SH* wave. Reflected *SH* main, reflected *P* wave weak, reflected *SV* weak.

$$A_2^{(2)} = 0(\varepsilon^2)$$

$$A_3^{(2)} = \frac{m_0}{r_0} A_3^{(1)} + 0(\varepsilon^2)$$

$$A_1^{(2)} = \frac{n_1}{l_1} \cdot \frac{m_0}{r_0} A_3^{(1)} + 0(\varepsilon^2)$$

$$B_1^{(2)} = \frac{g_1 n_0}{r_0} A_3^{(1)} + 0(\varepsilon^2)$$

$$B_2^{(2)} = 0$$

$$B_3^{(2)} = \frac{g_3 n_0}{r_0} A_3^{(1)} + 0(\varepsilon^2). \tag{37}$$

Also in this case no transverse k_l-type reflection is possible as $B_2^{(2)} = 0$.

The k_d type of reflected wave also does not arise in the case of the classical elastic medium (Fig. 2).

Incident k_l-type waves. The mean displacement in the case of incident k_l-type waves is taken in the form

$$\langle \vec{u}_0(\vec{x}) \rangle = \vec{E}^{(1)} e^{ik_l \vec{n}' \cdot \vec{x}} + \vec{E}^{(2)} e^{ik_l \vec{l}' \cdot \vec{x}} + \vec{F}^{(2)} e^{ik_d \vec{m}' \cdot \vec{x}} \tag{38}$$

where $\vec{E}^{(1)}$, $\vec{E}^{(2)}$, $\vec{F}^{(2)}$ respectively stand for incident k_l type, and reflected k_l, k_d-type amplitudes. Also

$$(\vec{n}', \vec{l}', \vec{m}') = (l_1', m_1', n_1'), \quad (l_2', m_2', n_2'), \quad (l_3', m_3', n_3') \tag{39}$$

are the unit normals for the incident k_l, reflected k_l and reflected k_d waves. Assuming $m_1' = m_2' = m_3' = 0$, it can be shown as before that

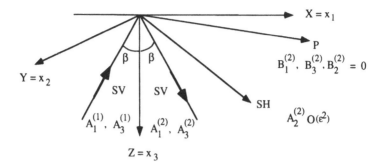

Figure 2
Incident SV wave. Reflected SH weak, reflected SV main, reflected P main.

$$E_1^{(1)} = h_1 N, \quad E_2^{(1)} = h_2 N = 0, \quad E_3^{(1)} = -h_3 N$$

$$E_1^{(2)} = h_1 N', \quad E_2^{(2)} = h_2 N' = 0, \quad E_3^{(2)} = h_3 N'$$

$$F_1^{(2)} = -\frac{n_3'}{l_3'} F_3^{(2)}$$

$$h_1 = l_1' n_1' [(\lambda_0 + \mu_0)k_l^2 + \varepsilon^2 D_2(k_l)]$$

$$h_3 = \mu_0 k_l^2 - \omega^2 \rho_0 + \varepsilon^2 D_1(k_l) + l_1'^2 \{(\lambda_0 + \mu_0)k_l^2 + \varepsilon^2 D_2(k_l)\} \qquad (40)$$

and N, $N' = $ constants.

In this case also $E_1^{(1)}$, $E_2^{(1)}$, $E_3^{(1)}$ (and hence N) being known, there remain only three unknowns N', $F_2^{(2)}$, $F_3^{(2)}$. The relations which express reflected amplitudes in terms of incident amplitudes are deduced as before

$$E_1^{(2)} = \frac{1}{r_0}\left(\bar{q}_0 + \varepsilon^2 \bar{q}_0' - \varepsilon^2 \bar{q}_0' \frac{\bar{r}_0'}{\bar{r}_0}\right) F_1^{(1)}$$

$$E_2^{(2)} = 0$$

$$E_3^{(2)} = \frac{h_3}{h_1 r_0}\left(\bar{q}_0 + \varepsilon^2 \bar{q}_0' - \varepsilon^2 \bar{q}_0' \frac{r_0'}{\bar{r}_0}\right) E_1^{(1)}$$

$$F_2^{(2)} = \varepsilon^2 \frac{\bar{p}_0}{\bar{r}_0 h_1} E_1^{(1)}$$

$$F_3^{(2)} = \frac{1}{\bar{r}_0 h_1}\left(\bar{s}_0 + \varepsilon^2 \bar{s}_0' - \varepsilon^2 \frac{\bar{s}_0 \bar{r}_0'}{\bar{r}_0}\right) E_1^{(1)}$$

$$F_1^{(2)} = -\frac{n_3'}{l_3'} F_3^{(2)}. \qquad (41)$$

The quantities $\bar{p}_0, \bar{r}_0, \ldots$, etc. are defined in Appendix 3.

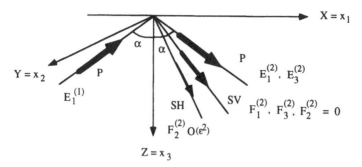

Figure 3
Incident P wave. Reflected P main, reflected SV main, reflected SH weak.

The relations (41) may be rewritten to display the main amplitude ratios

$$E_1^{(2)} = \frac{\bar{q}_0}{\bar{r}_0} E_1^{(1)} + 0(\varepsilon^2)$$

$$E_2^{(2)} = 0$$

$$E_3^{(2)} = \frac{h_3 \bar{q}_0}{h_1 \bar{r}_0} E_1^{(1)} + 0(\varepsilon^2)$$

$$F_2^{(2)} = 0(\varepsilon^2)$$

$$F_3^{(2)} = \frac{\bar{s}_0}{\bar{r}_0 h_1} E_1^{(1)} + 0(\varepsilon^2)$$

$$F_1^{(2)} = -\frac{\bar{s}_0 n_3'}{\bar{r}_0 h_1} E_1^{(1)} + 0(\varepsilon^2). \tag{42}$$

It is observed that $E_2^{(1)} = 0$, $E_2^{(2)} = 0$, illustrating that only $E_1^{(1)}$ and $E_3^{(1)}$ amplitudes can propagate. Reflected k_l-type waves are available with amplitudes $E_1^{(2)}$ and $E_3^{(2)}$. Reflected k_d-type waves propagate with amplitudes $F_1^{(2)}$ and $F_3^{(2)}$ but the amplitude $F_2^{(2)}$ is only of the order of ε^2 (Fig. 3).

Discussion and Conclusion

The results obtained in (15a)–(15d), (36), (37), (42) etc., are certainly valid as long as the effect of random inhomogeneity is sufficiently smaller than the corresponding nonrandom terms. In these cases ε^2 or ε clearly measures the order of random variation of inhomogeneity which is always small. Thus the terms involving ε^2 in (15a)–(15d) are much smaller than the corresponding nonrandom terms. In other words, the relations are valid only when ε^2 order terms are smaller than the nonrandom terms.

The phenomena of decoupling obtainable in elastic wave reflections of (P, SV) and SH waves no longer hold for random media. This fact is discernible from the results obtained in equations (36), (37), and (42).

The effects of random inhomogeneity are presented qualitatively and also quantitatively by terms involving D_2 in the expressions (15a)–(15d). These relations were deduced by application of the law of reflection.

Acknowledgement

The author expresses his sincere thanks to Professor Dr. M. Mitra, Presidency College, Calcutta, for his kind advice in the preparation of this paper.

Appendix 1

The coefficients a_{11}, $l = 1, 2, 3$, are the coefficients of $A_l^{(1)}$ respectively in

$$
f_{11} = f_{11}(\vec{A}^{(1)}, k_d, \vec{n})
$$

$$
= \int \left\{ \frac{\partial G}{\partial z} \left[R_1(-k_d \vec{n} \cdot \vec{A}^{(1)}) k_d l_1 + R_2(-k_d^2) A_1^{(1)} \right. \right.
$$

$$
+ ik_d(\vec{n} \cdot \vec{A}^{(1)}) \frac{\partial R_3}{\partial x'} + ik_d \{ (\vec{\nabla}' R_2 \cdot \vec{A}^{(1)}) l_1 - (\vec{\nabla}' R_2 \cdot \vec{n}) A_1^{(1)} \}
$$

$$
\left. + 2ik_d(\vec{n} \cdot \vec{\nabla}' R_2) A_1^{(1)} + \omega^2 R_4 A_1^{(1)} \right] + \frac{\partial G}{\partial x} \left[-k_d^2 n_1 (\vec{n} \cdot \vec{A}^{(1)}) R_1 \right.
$$

$$
- k_d^2 R_2 A_3^{(1)} + ik_d(\vec{n} \cdot \vec{A}^{(1)}) \frac{\partial R_3}{\partial z'} - ik_d(n_1 A_1^{(1)})
$$

$$
- l_1 A_3^{(1)}) \frac{\partial R_2}{\partial x'} - ik_d(m_1 A_3^{(1)} - n_1 A_2^{(1)}) \frac{\partial R_2}{\partial y'}
$$

$$
\left. \left. + 2ik_d(\vec{n} \cdot \vec{\nabla}' R_2) A_3^{(1)} + \omega^2 R_4 A_3^{(1)} \right] \right\} e^{-ik_d \vec{n} \cdot (\vec{x} - \vec{x}')} \, d\vec{x}',
$$

$R_1 = \langle \mu_1 \lambda_1' + \mu_1 \mu_1' \rangle$, $R_2 = \langle \mu_1 \mu_1' \rangle$, $R_3 = \langle \mu_1 \lambda_1' \rangle$, and $R_4 = \langle \mu_1 \rho_1' \rangle$. The other coefficients are similarly defined. Illustration:

$$
a_{11}^{(1)} = \int \left\{ \frac{\partial G}{\partial z} \left[-R_1 k_d^2 l_1^2 - R_2 k_d^2 + ik_d l_1 \frac{\partial R_3}{\partial x'} \right. \right.
$$

$$
\left. + ik_d \left\{ -m_1 \frac{\partial R_2}{\partial y'} - \frac{\partial R_2}{\partial z'} n_1 \right\} + 2ik_d(\vec{n} \cdot \vec{\nabla}' R_2) + \omega^2 R_4 \right]
$$

$$
\left. + \frac{\partial G}{\partial x} \left[-k_d^2 l_1 n_1 R_1 + ik_d l_1 \frac{\partial R_3}{\partial z'} - ik_d n_1 \frac{\partial R_2}{\partial x'} \right] \right\} e^{-ik_d \vec{n} \cdot (\vec{x} - \vec{x}')} \, d\vec{x}'
$$

Appendix 2

Solutions of equations (26)–(28) are given by Cramer's Rule

$$\frac{A_2^{(2)}}{\Delta_1} = \frac{A_2^{(3)}}{\Delta_2} = \frac{M}{\Delta_3} = \frac{1}{\Delta}$$

where Δ, Δ_1, Δ_2, Δ_3 can, on expansion, be arranged in the forms

$$\Delta = r_0 + \varepsilon^2 r_0'$$

$$\Delta_1 = p_0 A_2^{(1)} + \varepsilon^2 p_0' A_2^{(1)} + \varepsilon^2 q_0 A_3^{(1)}$$

$$\Delta_2 = \varepsilon^2 s_0 A_2^{(1)} + m_0 A_3^{(1)} + \varepsilon^2 m_0' A_3^{(1)}$$

$$\Delta_3 = \varepsilon^2 t_0 A_2^{(1)} + n_0 A_3^{(1)} + \varepsilon^2 n_0' A_3^{(1)}.$$

The expressions for $r_0, p_0, p_0', \ldots$, may be recorded explicitly, if necessary, such as,

$$r_0 = -i\mu_0^2 k_d^2 l_1^2 n_1 k_l [\lambda_0 (l_1^2 - n_1^2)^2 (l_3 g_1 + n_3 g_3) + 2\mu_0 n_1 (n_3 g_1 + l_3 g_3)].$$

Appendix 3

We replace k_d by k_l and k_l by k_d and $(\hat{n}, \vec{l}, \hat{m})$ by $(\hat{n}', \vec{l}', \hat{m}')$, use relation (40) in the equations (13)–(15) and solve them for N', $F_2^{(2)}$, $F_3^{(2)}$ by Cramer's Rule:

$$\frac{F_2^{(2)}}{\bar{\Delta}_1} = \frac{F_3^{(2)}}{\bar{\Delta}_2} = \frac{N}{\bar{\Delta}_3} = \frac{1}{\bar{\Delta}}$$

where

$$\bar{\Delta} = \bar{r}_0 + \varepsilon^2 r_0'$$

$$\bar{\Delta}_1 = \varepsilon^2 \bar{p}_0 N$$

$$\bar{\Delta}_2 = \bar{s}_0 N + \varepsilon^2 s_0' N$$

$$\bar{\Delta}_3 = \bar{q}_0 N + \varepsilon^2 \bar{q}_0' N.$$

REFERENCES

BHATTACHARYYA, R. K. (1986), *On Wave Propagation in a Random Magneto-thermo-viscoelastic Medium*, Indian J. Pure and Appl. Math. *17* (5), 705–725.
BHATTACHARYYA, R. K., *Wave Propagation in a Random Micropolar Elastic Medium* (to be published).
CHOW, P. L. (1973), *Thermoelastic Wave Propagation in a Random Medium and Some Related Problems*, Int. J. Eng. Sci. *11*, 953.

KARAL, F. C., and KELLER, J. B. (1964), *Elastic, Electromagnetic and Other Waves in a Random Medium*, J. Math. Phys. *5* (4), 537–547.

KELLER, J. B. (1964), *Stochastic Equations and Wave Propagation in Random Media*, Proc. Sympos. Appl. Math. *16*, Am. Math. Soc., Providence, R.I. 145–170.

MIKLOWITZ, J., *The Theory of Elastic Waves and Waveguides* (North-Holland Publishing Co., Amsterdam 1980).

(Received October 31, 1994, revised August 17, 1995, accepted August 27, 1995)

PAGEOPH, Vol. 146, Nos. 3/4 (1996)

0033–4553/96/040689–08$1.50 + 0.20/0

An Effective Approach to Determine the Dynamic Source Parameters

Xiaofei Chen[1] and Keiiti Aki[1]

Abstract—In this study, we present a new and effective method to determine the dynamic source parameters (i.e., stress drop and strength distribution). We first assume that the kinematic source parameters, i.e., the slip and rupture time distributions on the fault plane, are known from the previous source inversion studies. Then, using the seismic source representation theorem we determine the dynamic stress field on a fault plane from known kinematic parameters. Finally, we determine the strength of the fault defined as the peak stress just before the rupture. We have tested the validity of this method by using an illustrative two-dimensional analytical example. To assess the applicability of this method, we have applied it to study the 1979 Imperial Valley earthquake, and obtained consistent results with those of Miyatake's (1992) and Quin's (1990). Compared with previous methods, this new method is simple, straightforward and accurate, and needs much less calculation. Therefore, it is expected to be useful in exploring the seismic source process.

Key words: Seismic source parameters, dynamic stress drop.

Introduction

Reconstruction of seismic source parameters plays an important role in studying the seismic source process, strong motion simulation and earthquake prediction, etc. particularly the reconstruction of dynamic source parameters (e.g., dynamic stress drop and strength) is very useful in understanding the complex faulting process. So far, most of the source inversion methods dealt with the kinematic source parameters (e.g., Hartzell and Heaton, 1983; Olson and Apsel, 1982), only few of them study the dynamic source parameters (Quin, 1990; Miyatake, 1992). Quin proposed a method to reconstruct the dynamic source parameters from the estimated kinematic source parameters. He used a trial and error method to fit both the distributions of slip and slip velocity obtained from the previous kinematic inversion studies, and each trial was carried out by solving Das' boundary integral equation (Das and Aki, 1977). He attempted more than 200 models for the 1979

---

[1] Department of Earth Sciences, University of Southern California, Los Angeles, California 90089-0740, U.S.A.

Imperial Valley earthquake. Obviously, Quin's method is too time-consuming for solving a practical problem.

MIYATAKE (1992) developed a more efficient method than Quin's to reconstruct the dynamic source parameters from the results of kinematic source inversion. First, he calculated the static stress drop distribution on the fault by solving a static mixed boundary value problem using finite-difference algorithm from the known static slip distribution. Secondly, he constructed an approximate dynamic stress-drop distribution from the calculated static stress-drop distribution, then used this dynamic stress drop and rupture time distributions to calculate the dynamic stress distribution by using dynamic finite-difference algorithm. Finally, he calculated the strength excess distribution which was defined as the peak shear stress just before the rupture at each grid point on the fault plane. As discussed above, Miyatake's approach involves finite-difference calculations for both static and dynamic mixed boundary value problems. Those finite-difference calculations are usually expensive in both computing time and storage. We can see that these dynamic source inversion methods are more complicated than kinematic source inversion, therefore, they are usually difficult to be applied to the comparative study of large data sets. In this article, we shall present a new and effective method to determine the dynamic seismic source parameters from the results of the kinematic source inversion studies. In the following, we shall first present the theory of this approach, then test its validity by an illustrative two-dimensional analytic example. Finally, we shall apply this approach to investigate the 1979 Imperial Valley earthquake, and discuss its applicability.

Theory

As QUIN (1990) and MIYATAKE (1992) did in their studies, we assume that the kinematic source parameters, i.e., the slip and rupture time distributions on the fault plane, are known from the previous source inversion studies. There exist many effective algorithms to determine the kinematic source parameters by inversion of seismic wave-form data (e.g., HARTZELL and HEATON 1983; OLSON and APSEL, 1982). To relate the stress field to the slip function, we start with the representation theorem (see, e.g., AKI and RICHARDS, 1980). The displacement field at point x can be written as,

$$u_i(\vec{x}, t) = \iint_\Sigma [u_p(\vec{\xi}, t)] * c_{pqlk} v_q G_{il,k}(\vec{x}, \vec{\xi}; t) \, d\Sigma(\vec{\xi}), \qquad (1)$$

where, * denotes time-domain convolution, Σ is the fault plane, $[u]$ is displacement discontinuity, c_{pqlk} is elastic constant, v_q is the normal vector of the fault plane and $G_{il}(\vec{x}, \vec{\xi}; t)$ is Green's function. The corresponding stress field on fault plane can be written as,

$$\sigma_{nm}(\vec{x}, t) = \int\!\!\int_{\Sigma} [u_p(\vec{\xi}, t)] * c_{nmij} c_{pqlk} v_q G_{il,kj}(\vec{x}, \vec{\xi}; t) \, d\Sigma(\vec{\xi}). \tag{2}$$

This equation is valid for any observational point including those on the fault plane, though the displacement crossing fault plane is discontinuous. We see that equation (2) can directly be used for computing the stress distribution on a fault plane from the known slip distribution. Then, the strength distribution of fault can be derived according to its definition, i.e., the values of peak stress just before the rupture for each point on the fault plane (see Fig. 1). If the Green's function and elastic coefficients are known, we can immediately obtain the stress field on the fault plane from the known slip distribution, without processing any sophisticated simulation. It is noted that although the Green's function in equation (2) has a singularity at $\vec{x} = \vec{\xi}$, this singularity is integrable over any symmetric finite area. To avoid the singularity, we used the following discrete formula derived from equation (2) to calculate the dynamic stress field,

$$\bar{\sigma}_{nm}(\vec{x}_N, t) = c_{nmij} c_{pqlk} v_q \sum_M [\bar{u}_p(\vec{x}_M, t)] * \bar{G}_{il,kj}(\vec{x}_N, \vec{x}_M; t), \tag{3}$$

where, $\{\vec{x}_N, \vec{x}_M\}$ are the grid points of the fault plane, and

$$[\bar{u}_p(\vec{x}_M, t)] = \frac{1}{\Delta S_M} \int\!\!\int_{\Delta S_M} [u_p(\vec{x}, t)] \, d\Sigma(\vec{x}),$$

$$\bar{G}_{il,kj}(\vec{x}_N, \vec{x}_M; t) = \int\!\!\int_{\Delta S_M} G_{il,kj}(\vec{x}_N, \vec{x}; t) \, d\Sigma(\vec{x}),$$

and ΔS_M is area element corresponding to grid point $\vec{x}_M$.

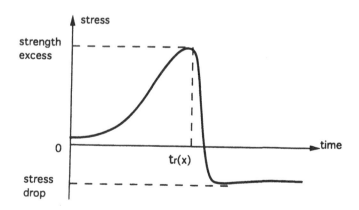

Figure 1
The definition of strength excess of a point on fault plane. The strength excess of point x is defined as the peak shear stress just before the rupture ($t < t_r(\mathbf{x})$) at this point.

An Analytical Example

To check the validity of equation (2), we consider a two-dimensional anti-plane static crack problem which is described by the following equations

$$\frac{\partial^2 u(x, y)}{\partial x^2} + \frac{\partial^2 u(x, y)}{\partial y^2} = 0$$

$$u(x, y) = 0, \qquad \text{when} \quad y = 0 \quad \text{and} \quad |x| > 1$$

$$\frac{\partial u(x, y)}{\partial y} = -\frac{\sigma_0}{\mu}, \qquad \text{when} \quad y = 0 \quad \text{and} \quad |x| < 1.$$

The solution of this mixed boundary value problem had been found as (KNOPOFF, 1958; or see KASAHARA, 1981),

$$u(x, 0) = \begin{cases} (\sigma_0/\mu)\sqrt{1 - x^2}, & \text{for} \quad |x| < 1 \\ 0, & \text{for} \quad |x| > 1 \end{cases} \tag{4a}$$

and

$$\tau(x, 0) = \begin{cases} -\sigma_0 & , & \text{for} \quad |x| < 1 \\ \sigma_0[|x|/(x^2 - 1)^{1/2} - 1], & \text{for} \quad |x| > 1. \end{cases} \tag{4b}$$

We now use this analytical solution to verify the validity of equation (2). In this case, equation (2) becomes,

$$\tau(x, 0) = \mu^2 \int_{-1}^{+1} [u(\xi, 0)]\{\partial^2 G(x, y; \xi, 0)/\partial y^2\} \,|_{y=0} \, d\xi, \tag{5}$$

where, $[u] = 2u$ and $G(x, y; \xi, \eta) = (4\pi\mu)^{-1} \ln[(x - \xi)^2 + (y - \eta)^2]$. Substituting the expressions of Green's function and slip function, $[u] = 2u$, into equation (5), we find

$$\tau(x) = \frac{\sigma_0}{\pi} \int_{-1}^{+1} \{\sqrt{1 - \xi^2}/(x - \xi)^2\} \, d\xi$$

$$= -\sigma_0 + \begin{cases} \dfrac{\sigma_0|x|}{\pi\sqrt{1 - x^2}} \ln\left[\left|\dfrac{\sqrt{1 - \xi^2} + \sqrt{1 - x^2}}{\xi - x}\right| - x\sqrt{1 - x^2}\right]_{\xi = -1}^{\xi = +1}, & |x| < 1 \\[4mm] \dfrac{\sigma_0|x|}{\pi\sqrt{x^2 - 1}} \sin^{-1}\left[\dfrac{1 - x^2 - x(\xi - x)}{\xi - x}\right]_{\xi = -1}^{\xi = +1}, & |x| > 1 \end{cases}$$

$$= -\sigma_0 + \begin{cases} 0, & |x| < 1 \\[3mm] \sigma_0 \dfrac{|x|}{\sqrt{x^2 - 1}}, & |x| > 1. \end{cases}$$

This result is identical with the analytic solution (4b), though there is a singularity at point $\xi = x$, confirming the validity of equation (2).

Application to 1979 Imperial Valley, California Earthquake

We apply our method to study the dynamic rupture process of the 1979 Imperial Valley earthquake. This event has been investigated by many seismologists (e.g., ARCHULETA, 1984; QUIN, 1990). The kinematic source parameters used here are those estimated by ARCHULETA (1984) using a wave-form inversion technique shown in Figure 2. We used these kinematic parameters to determine the corresponding dynamic parameters. First, we divided the fault into 45 by 15 grids with grid size of 1 km^2, and calculated the averaged values of the Green's function and the slip distribution over each grid, so that our integral formula is reduced to a summation. Then, we calculated the dynamic stress field distribution using the summation formula. Finally, we derived the strength excess of each grid from the dynamic stress distribution. Figures 3(a) and 3(b) show the final results of dynamic stress drop and strength distributions calculated by using our method. We find that

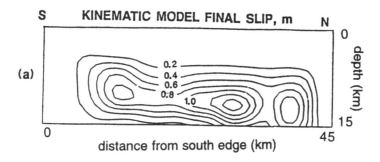

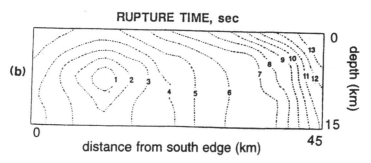

Figure 2
The kinematic source parameters obtained by inversion study (ARCHULETA, 1984; QUIN, 1990). (a) The final slip distribution; (b) the rupture time distribution, the rupture started from the point with zero strength excess.

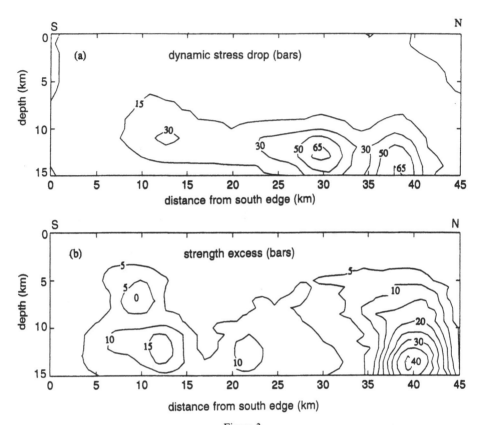

Figure 3
The dynamic source parameters calculated by our new method based on equation (2). (a) Dynamic stress drop distribution; (b) distribution of strength excess.

there are three strong areas located at the northern edge, at the center of the fault, and below the starting point of the rupture; whereas the weakest place is the starting point of the rupture. The stress-drop distribution also has three peaks located near the northern edge, at the center part and near the part of initial rupture. Figures 4(a) and 4(b) show the same parameters obtained by MIYATAKE (1992) using a finite-difference algorithm. If we consider that the dynamic stress drop is about 80% of the static stress drop, we find our results agree with those of Miyatake's, confirming the validity of our method.

Conclusions

In this study, we have presented an effective method for calculating the dynamic seismic source parameters from the results of the previous kinematic source inversion studies. To test its validity, we have applied it to a two-dimensional crack

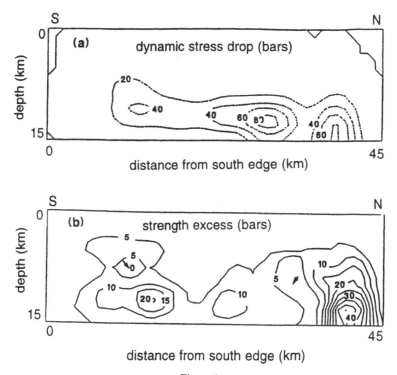

Figure 4
The dynamic source parameters calculated by MIYATAKE's finite-difference method (1992). (a) Static stress drop distribution; (b) distribution of strength excess (after MIYATAKE, 1992).

problem; and obtained an exact solution. To show its applicability, we have investigated the dynamic source parameters of the 1979 Imperial Valley earthquake using our new method. We found that our new method is simple, straightforward, and accurate, therefore it can be used for a comparative study of large data sets.

Acknowledgements

We are grateful to Professor Ren Wang for his critical review on this manuscript. This work is partially supported by the U.S. National Science Foundation under grant EAR-9218923, and by the Southern California Earthquake Center.

REFERENCES

AKI, K., and RICHARDS, P. G., Quantitative Seismology: Theory and Methods (W. H. Freeman, San Francisco, 1980).

ARCHULETA, R. J. (1984), A Faulting Model for the 1979 Imperial Valley Earthquake, J. Geophys. Res. 89, 4559–4585.

DAS, S., and AKI, K. (1977), *A Numerical Study of Two-dimensional Spontaneous Rupture Propagation*, Geophys. J.R. Astr. Soc. *50*, 643–668.

HARTZELL, S., and HEATON, T. (1983), *Inversion of Strong Ground Motion and Teleseismic Wave Form Data for the Fault Rupture History of the 1979 Imperial Valley, California Earthquake*, Bull. Seismol. Soc. Am. *73*, 1553–1583.

KASAHARA, K., *Earthquake Mechanics* (Cambridge University Press, New York 1981).

KNOPOFF, L. (1958), *Energy Release in Earthquake*, Geophys. J. MNRAS *1*, 44–52.

MIYATAKE, T. (1992), *Reconstruction of Dynamic Rupture Process of an Earthquake*, Geophys. Res. Lett. *19*, 349–352.

OLSON, A. H., and APSEL, R. (1982), *Finite Faults and Inverse Theory with Applications to the 1979 Imperial Valley Earthquake*, Bull. Seismol. Soc. Am. *72*, 1969–2002.

QUIN, H. (1990), *Dynamic Stress Drop and Rupture Dynamics of the October 15, 1979 Imperial Valley, California, Earthquake*, Tectonophysics *175*, 83–117.

(Received September 7, 1994, revised May 31, 1995, accepted June 20, 1995)

PAGEOPH, Vol. 146, Nos. 3/4 (1996)

0033–4553/96/040697–19$1.50 + 0.20/0

Source Process of the 1990 Gonghe, China, Earthquake and Tectonic Stress Field in the Northeastern Qinghai-Xizang (Tibetan) Plateau*

Y. T. Chen,[1] L. S. Xu,[1] X. Li,[1] and M. Zhao[1]

Abstract — The $M_s = 6.9$ Gonghe, China, earthquake of April 26, 1990 is the largest earthquake to have been documented historically as well as recorded instrumentally in the northeastern Qinghai-Xizang (Tibetan) plateau. The source process of this earthquake and the tectonic stress field in the northeastern Qinghai-Xizang plateau are investigated using geodetic and seismic data. The leveling data are used to invert the focal mechanism, the shape of the slipped region and the slip distribution on the fault plane. It is obtained through inversion of the leveling data that this earthquake was caused by a mainly reverse dip-slipping buried fault with strike 102°, dip 46° to SSW, rake 86° and a seismic moment of 9.4×10^{18} Nm. The stress drop, strain and energy released for this earthquake are estimated to be 4.9 MPa, 7.4×10^{-5} and 7.0×10^{14} J, respectively. The slip distributes in a region slightly deep from NWW to SEE, with two nuclei, i.e., knots with highly concentrated slip, located in a shallower depth in the NWW and a deeper depth in the SEE, respectively.

Broadband body waves data recorded by the China Digital Seismograph Network (CDSN) for the Gonghe earthquake are used to retrieve the source process of the earthquakes. It is found through moment-tensor inversion that the $M_s = 6.9$ main shock is a complex rupture process dominated by shear faulting with scalar seismic moment of the best double-couple of 9.4×10^{18} Nm, which is identical to the seismic moment determined from leveling data. The moment rate tensor functions reveal that this earthquake consists of three consecutive events. The first event, with a scalar seismic moment of 4.7×10^{18} Nm, occurred between 0–12 s, and has a focal mechanism similar to that inverted from leveling data. The second event, with a smaller seismic moment of 2.1×10^{18} Nm, occurred between 12–31 s, and has a variable focal mechanism. The third event, with a scalar seismic moment of 2.5×10^{18} Nm, occurred between 31–41 s, and has a focal mechanism similar to that inverted from leveling data. The strike of the 1990 Gonghe earthquake, and the significantly reverse dip-slip with minor left-lateral strike-slip motion suggest that the pressure axis of the tectonic stress field in the northeastern Qinghai-Xizang plateau is close to horizontal and oriented NNE to SSW, consistent with the relative collision motion between the Indian and Eurasian plates. The predominant thrust mechanism and the complexity in the tempo-spatial rupture process of the Gonghe earthquake, as revealed by the geodetic and seismic data, is generally consistent with the overall distribution of isoseismals, aftershock seismicity and the geometry of intersecting faults structure in the Gonghe basin of the northeastern Qinghai-Xizang plateau.

Key words: Qinghai-Xizang (Tibetan) plateau, source process, moment tensor, tectonic stress field.

Introduction

On April 26, 1990, a destructive earthquake occurred in Gonghe, Qinghai Province of western China (Fig. 1). The epicentral location of the earthquake is

* Contribution No. 96 B0006, Institute of Geophysics, State Seismological Bureau, Beijing, China.
[1] Institute of Geophysics, State Seismological Bureau, Beijing 100081, China.

latitude $\phi = 35.986°N$, longitude $\lambda = 100.245°E$. The focal depth, $h = 8.1$ km. The origin time, $O = 09$ h 37 min 15 s UTC (17 h 37 min 15 s BTC). The magnitude, $M_s = 6.9$. The earthquake occurred beneath the Gonghe basin of the northeastern part of Qinghai-Xizang (Tibetan) plateau near the south margin of the Gonghe basin. This was the largest earthquake to have occurred in the northeastern part of the Qinghai-Xizang plateau, and the first significant earthquake in this region to be recorded by the modern digital broadband seismograph network. No earthquake larger than 6.0 was documented historically (780 B.C. to 1900 A.D.), nor recorded instrumentally (1900 to 1989). The microearthquake seismicity in this region, as compared to the overall level of seismicity in the Qinghai Province of China, is very low in the last 20 years since 1970 (Fig. 1). The maximum intensity in this region, given by the seismic zoning map of the Qinghai Province of China, as assessed largely on the basis of surficial geologic observations and historic seismicity, is grade IV, a numeral even lower than that in the regions surrounding the Gonghe basin (ZENG, 1990, 1991, 1995).

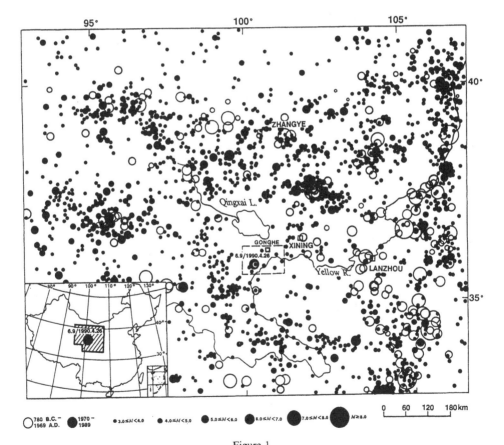

Figure 1

Map showing the epicenters of the $M_s = 6.9$ Gonghe, China, earthquake of April 26, 1990 and seismicity from 780 B.C. to 1989 A.D. in the northeastern Qinghai-Xizang (Tibetan) plateau.

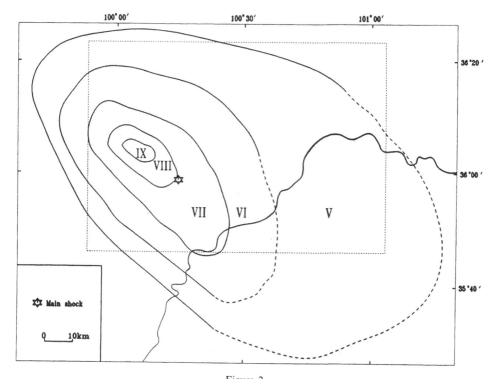

Figure 2
Isoseismals of the 1990 Gonghe earthquake. Dotted rectangle shows the area depicted in Figure 3.

The earthquake caused 126 casualties, 2,049 injuries, the collapse of 21,200 houses, and damage to 66,800 houses. During the earthquake, buildings in the meizoseismal region of intensity grade IX totally collapsed. The earthquake was felt in the entire Qinghai Province (~93° to 108°E, 33° to 41°N), as far east as Lanzhou (about 300 km to the east), and as far north as Zhangye (about 360 km to the north) (Fig. 2).

The focal mechanism and seismogenic structure of the Gonghe earthquake have been studied by ZENG (1990) and TU (1990) using mainly field geologic data, and routinely determined by PERSON (1991), and DZIEWONSKI et al. (1991) using seismic data, among others. While considerable effort has been expended in the studies of the focal mechanism and seismogenic structure of the earthquake, the answer regarding these questions remains controversial.

The relatively low level of historical as well as present seismic activity prevents our understanding of the tectonic stress field in the Gonghe basin and its neighbouring region, using the seismological method. Using P-wave first motion data from a large number of small earthquakes, XU et al. (1992) found that the pressure axis of the tectonic stress field in the northeastern Qinghai-Xizang plateau is close to horizontal and trends NE-SW to NNE-SSW. Nevertheless, as XU et al. (1992)

noted, the data available for the determination of this region are relatively rarer and less reliable.

To understand the focal mechanism and seismogenic structure of the Gonghe earthquake and the tectonic stress field in the northeastern Qinghai-Xizang plateau, we study the source process of the Gonghe earthquake, using geodetic and seismic data. We invert leveling data obtained before and after the occurrence of the Gonghe earthquake to determine the focal mechanism, fault shape, and spatial distribution of variable slip on the fault. We also invert broadband body waves data obtained from the China Digital Seismograph Network (CDSN) to determine the moment release history, using the moment-tensor inversion technique. We then apply the empirical Green's function (EGF) technique to seismic waves recorded at stations of the CDSN to invert the records for the tempo-spatial rupture process of this earthquake. Finally, we compare the results obtained with geological, meizo-seismal and aftershock data obtained by other investigators.

Focal Mechanism from Leveling Data

GONG and GUO (1992) and GONG *et al.* (1993) at the No. 2 Crustal Deformation Monitoring Center, State Seismological Bureau, Xi'an, China, have conducted

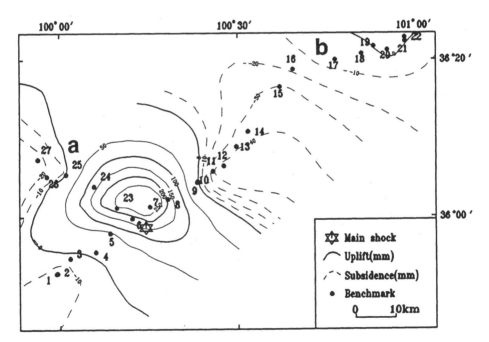

Figure 3
Vertical displacement distribution of the $M_s = 6.9$ Gonghe earthquake of April 26, 1990.

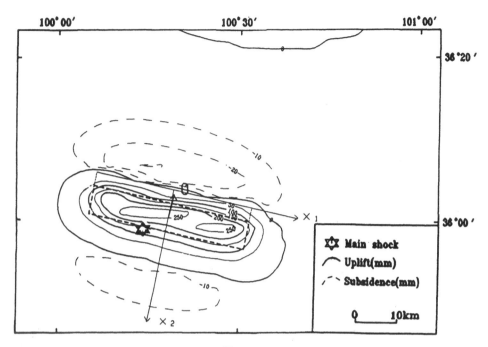

Figure 4
Theoretical vertical displacement distribution contours for the single, rectangular uniform slip model.

a leveling survey in an area of 100 km × 60 km which enclosed the meizoseismal regions, in 1978 to 1979. Following the main shock, in May to June, 1990, they carried out leveling remeasurements along the same leveling route, and obtained valuable geodetic data of this earthquake (Fig. 3). These data provide a good opportunity to study the focal mechanism and seismogenic structure of this earthquake. In this article, the leveling data obtained by GONG and GUO (1992) and GONG et al. (1993) are used to invert the focal mechanism of the Gonghe earthquake.

There are 27 benchmarks in total, distributed in two leveling routes of 162.6 km in total length. The leveling route (a) runs northwestward of the epicenter. The leveling route (b) runs southwestward from benchmark No. 22, crossing the epicentral area, and ending about 25 km southwest of the epicenter of the main shock. Relative to the benchmark No. 22 the largest vertical deformation located in an uplift region of 30 km-wide, is 358 mm. The random survey error σ_1 is 0.43 mm/km$^{1/2}$, and the root-mean-squared errors of the observation data σ is 3.9 mm. Considering that the ground deformation in the northeastern part of the epicentral area is substantially influenced by the tectonic movement of the Nanshan mountain, which is close to the benchmark No. 22, in this study we exclude the data of the benchmarks Nos. 16 to 22 from inversion, and take benchmark No. 15 as a reference point.

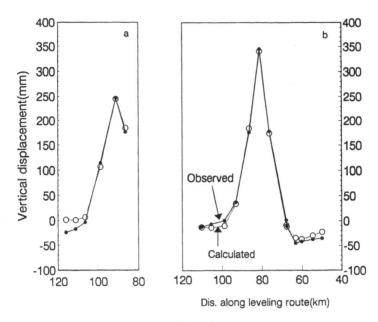

Figure 5

Comparison of theoretical (open circles with dotted lines) and observed (solid circles with solid lines) vertical displacements for the single, rectangular uniform slip model. Left part (a) and right part (b) refer to leveling routes shown in Figure 3, respectively.

In the first step of the leveling data inversion, we model the observed coseismic deformation by a single rectangular fault buried in an isotropic, homogeneous and perfect elastic half-space, with uniform oblique slip. We define fault length $2L$ along the fault-strike direction α, width W along perpendicular direction to the strike, with upper edge d and lower edge D, dip angle θ, strike-slip ΔU_s and dip-slip ΔU_d. In the Cartesian coordinate system, as shown in Figure 4, the i-th component of the displacement, u_i ($i = 1, 2, 3$) at an observation point (x_1, x_2, x_3), due to this fault, is given by MANSINHA and SMYLIE (1971) for the Poisson medium, and is given by CHEN *et al.* (1975) for the case of unequal Lamé constants.

If we denote the geographical coordinates of the origin of the source coordinates system as $(s_1, s_2, 0)$, the vertical component of the displacement at a given point on the ground surface $(X_1, X_2, 0)$ can be expressed as follows (CHEN *et al.*, 1979):

$$u_3 = u_3(\alpha, \theta, L, d, D, \Delta U_s, \Delta U_d, s_1, s_2; X_1, X_2). \tag{1}$$

The displacement at a given point on the ground surface is a nonlinear function of the source parameters ($\alpha, \theta, L, d, D, \Delta U_s, \Delta U_d, s_1, s_2$). To invert the data of the vertical displacement on the ground surface for the source parameters, the function u_3 was linearized by making Taylor series expansion of the function u_3 about an initial model, and neglecting the higher order terms. Starting from initial model parameters, the estimation of the model parameters which satisfy convergence criteria described in CHEN *et al.* (1979), was obtained as $\alpha = 102°$, $\theta = 46°$, $2L = 2 \times 20$ km, $d = 5$ km, $D = 14$ km, $\Delta U_s = 5$ cm, $\Delta U_d = -79$ cm, $s_1 = 93.0$ km, and $s_2 = 34.5$ km. For the single rectangular uniform slip model, this final model is the best fit to the observed leveling data. The root-mean-squared residual r for the final model is 14.2 mm.

Figure 4 shows the projection of the inverted fault model on the ground surface and the contour of the theoretical vertical displacement due to the inverted fault model. Figure 5 shows the fitness of the theoretical vertical displacement due to the inverted fault model and the observation data. The overall focal mechanism we obtained for this earthquake from leveling data indicates that this earthquake is mainly reverse faulting with a small left-lateral component on a dipping 46° to SSW striking 102° fault plane.

We also calculated seismic moment, stress drop, strain drop and energy released by the Gonghe earthquake, using the inverted fault parameters. Using a rigidity of 3.0×10^{11} dyn/cm^2, the seismic moment M_0 is estimated to be 9.4×10^{18} Nm. The stress drop $\Delta\sigma = 4.9$ MPa. The strain drop $\Delta\varepsilon = 7.4 \times 10^{-5}$, and the energy released $\Delta E = 7.0 \times 10^{14}$ J. It can be seen that the stress drop of the Gonghe earthquake is low (4.9 MPa) as compared to the usual intraplate earthquake (normally several ten MPa) (KANAMORI and ANDERSON, 1975).

While the single rectangular fault model inverted from leveling data accounts well for the observed vertical deformation, it is worthwhile to note that the misfit for this model is still several times larger than the root-mean-squared errors of the observation data ($\sigma = 3.9$ mm). These discrepancies are probably caused by the inadequacy of the single rectangular dislocation model and one would therefore not expect the oversimplified model to fit the observation data quite well.

In the second step of the leveling data inversion, following WARD and BARRIENTOS (1986), we use a variable slip fault model in which we initially fix the focal mechanism, i.e., fault strike (α), dip angles (θ), depth and width parameters (d and D) and rake (λ), and allow the slip amount to change on the fault plane. The fault plane is then divided into subfaults of equal size, and point dislocation sources are distributed uniformly across each of the subfaults. At individual observation points, displacement is computed by summing the contribution of each point source:

$$u_i = K_{ij}D_j, \quad i = 1, 2, \ldots, n, \tag{2}$$

where u_i is the displacement at the i-th observation point, $D_j, j = 1, 2, \ldots, m$ is the slip of the j-th subfault, and K_{ij} the displacement at the i-th observation point from the j-th subfault of unit slip.

Equation (2) represents a linear relation between the displacements at observation points on the ground surface and the slips of the subfaults. The theoretical and observed displacements describe an underdetermined system of linear equations. We solve for D_i using a gradient inversion scheme proposed by WARD and BARRIENTOS (1986), which invokes a positivity constraint on the solutions. The fault

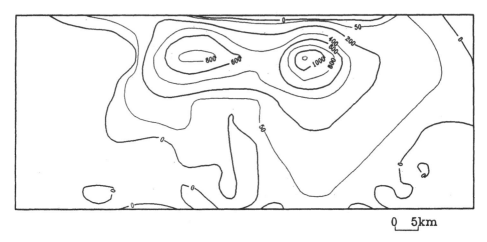

Figure 6
Contoured distribution of variable slip on a fault plane. The figure in the contour shows the slip in mm.

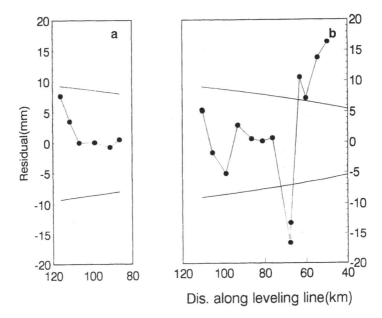

Dis. along leveling line(km)

Figure 7
Residuals of the observed and theoretical vertical displacements for the variable slip models. Left part
(a) and right part (b) refer to leveling routes shown in Figure 3, respectively. Ordinate represents
residuals in mm. Abscissa represent distance along leveling route. Dashed lines represent twice standard
deviation along the leveling route.

plane of $100 \, \text{km} \times 40 \, \text{km}$ is divided into 20×8 subfaults with dimensions of
$5 \, \text{km} \times 5 \, \text{km}$. The fault strike α, dip angles δ and rake λ are taken from the
inversion result for the single rectangular fault model, i.e., $\alpha = 102°$, $\delta = 46°$, and
$\lambda = 86°$.

The inversion yields a variable slip distribution on the fault plane (Fig. 6) and an estimation of the seismic moment of 9.8×10^{18} Nm, which is very close to the estimation of the seismic moment from the single, rectangular fault model of uniform slip. The slip distributes in a region slightly deep from NWW to SEE with two nuclei, i.e., knots with highly concentrated slip (equal to and larger than 60 cm), located in a shallower depth (4 to 12 km) at the NWW and, about 25 km away, in a deeper depth (5 to 16 km) at the SEE, respectively. At most observation points the theoretical vertical displacements for the variable slip fault model fit the observed vertical displacements quite well. Figure 7 illustrates the residuals at individual observation points versus distance along the leveling routes. Dashed lines represent an envelope of two times standard deviation. The standard deviation at distance l along the leveling route, $\sigma(l)$, is expressed as (WARD and BARRIENTOS, 1986)

$$\sigma(l) = \sigma_1 \sqrt{l} \tag{3}$$

where σ_1 is the random survey error. It can be seen that the residuals at most observation points are within one standard deviation uncertainty. The root-mean-squared residuals for the variable slip models is 7.8 mm, which is twice the root-mean-squared errors of 3.9 mm.

Source Process from Moment-tensor Inversion

Moment-tensor inversion technique is used to study the source process of the Gonghe earthquake (DZIEWONSKI et al., 1991, among others).

If the dimension of the source is considerably smaller than the dominant wavelength, the relationship between the seismic displacement $u_i(\mathbf{r}, t)$ and the seismic moment $M_{jk}(\mathbf{0}, t)$ is linear:

$$u_i(\mathbf{r}, t) = G_{ij,k}(\mathbf{r}, t; \mathbf{0}, 0) * M_{jk}(\mathbf{0}, t) \tag{4}$$

where $G_{ij,k}(\mathbf{r}, t; \mathbf{0}, 0)$ is the partial derivative with respect to source coordinates x'_k, of the Green's function $G_{ij}(\mathbf{r}, t; \mathbf{0}, 0)$. The Green's function $G_{ij}(\mathbf{r}, t; \mathbf{0}, 0)$ denotes the i-th component of displacement at the position $\mathbf{r}$ and at the time t due to a unit impulse applied at the origin $\mathbf{0}$ and at the time 0 and in the j-th direction.

In the frequency domain, equation (4) can be written

$$u_i(\mathbf{r}, \omega) = G_{ij,k}(\mathbf{r}, 0; \omega) \cdot M_{jk}(\mathbf{0}, \omega). \tag{5}$$

In this study the generalized reflection-transmission coefficient matrix method (KENNETT, 1979, 1983) and the discrete wave slowness integration method are used to synthesize the partial derivatives of the Green's function $G_{ij,k}$. In our calculation, the wave slowness interval is taken to be 0.09 to 0.32 s/km and the frequency interval, 0.01 and 0.5 Hz. We use the broadband P and S body wave phases of the

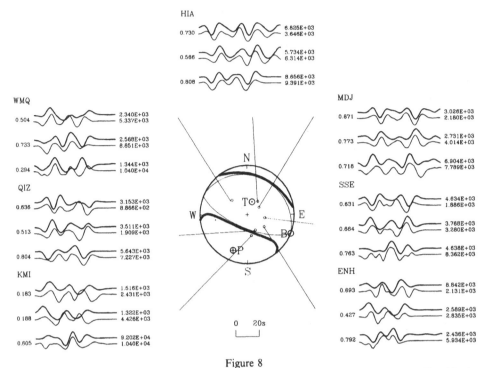

Figure 8

Focal mechanism of the $M_s = 6.9$ Gonghe earthquake of April 26, 1990 inverted from broadband body waves data, and observed and synthetic seismograms at seven stations of the CDSN. For detail refer to text.

CDSN to invert the moment tensor of the Gonghe earthquake. The original seismograms were preprocessed for N-E-U to Z-R-T and filtered in the passband between 0.05 and 0.2 Hz.

Figure 8 shows the observed (thick lines) and synthetic (thin lines) broadband displacement wave forms at seven stations of the CDSN, for vertical (upper traces), radial (middle traces) and tangential (lower traces) components, respectively, and the lower hemisphere projection of the moment-tensor solution. In the lower hemisphere projection of the focal sphere, P nodals for the moment tensor and the best double-couple are represented by thick and thin lines, respectively. The best double-couple solutions for the Gonghe earthquake have one nodal plane (N.P.1) with strike 113°, dip 68°, rake 89°, and another nodal plane (N.P.2) with strike 294°, dip 22°, rake 91°. Considering that the nodal plane 1 is so close to the fault plane inverted from the leveling data, we prefer the nodal plane 1 to the fault plane. The focal mechanism of the Gonghe earthquake obtained from moment-tensor inversion is a mainly reverse dip slip fault with strike 113°, dip 68° to SSW and rake 89°. The pressure axis P is close to horizontal (plunge 23°) and oriented NNE to SSW (azimuth 23°). The plunge and the azimuth of the tension axis T are 67° and

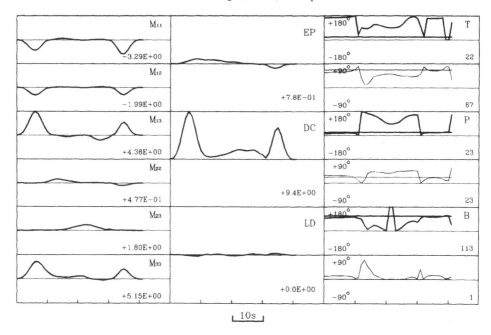

Figure 9

Moment rate tensor of the $M_s = 6.9$ Gonghe earthquake of April 26, 1990 inverted from broadband body waves data. For detail refer to text.

22°, respectively. The null axis B is almost horizontal (plunge 1°) and oriented NWW to SEE (azimuth 113°).

Figure 9 shows the moment rate tensor (left part), isotropic component (EP), best double-couple (DC) and compensated linear vector dipole (LD) (central part), and the azimuth (thick lines) and plunge (thin lines) of the T, P, and B axes (right part), of the Gonghe earthquake, versus time. The figures in the lower right are the integrations of individual quantities. It is evident that the Gonghe earthquake is a rupture process dominated by shear faulting. The scalar sesmic moment of the best double-couple is 9.4×10^{18} Nm, which is identical to that determined from leveling data. The moment of the isotropic part is only 7.8×10^{17} Nm, about 8.3% of that of the best double-couple, and the moment of the compensated linear vector dipole is vanishing.

The central part of Figure 9 clearly depicts that the Gonghe earthquake consists of three consecutive events. The first event which occurred between 0–12 s, with a scalar seismic moment of 4.7×10^{18} Nm, has a focal mechanism of strike 96°, dip 73°, and rake 77°. The second event which occurred between 12–31 s with a smaller seismic moment of 2.1×10^{18} Nm, has a variable focal mechanism with average strike 107°, dip 8°, and rake 169°. The third event which occurred between 31–41 s, with a seismic moment of 2.5×10^{18} Nm, has a focal mechanism of strike 100°, dip 67°, and rake 75°. As the right part of Figure 9 indicated, the principal axes in the time intervals for the first and the third events are stable and almost identical. The principal axes in the time interval for the second event (12–31 s) are different from that for the first and the third events, and evidently change with time.

Tempo-spatial Rupture Process from Inversion of Source Time Functions

The time domain inversion technique proposed by HARTZELL and IIDA (1990) and DREGER (1994) are used to image the rupture process of the Gonghe earthquake. The source time functions (STF) retrieved from the recordings at a station is the summation of the STFs of subfaults,

$$S_i(t) = \sum_{j=1}^{N} m_j(t - \tau_{ij}) \tag{6}$$

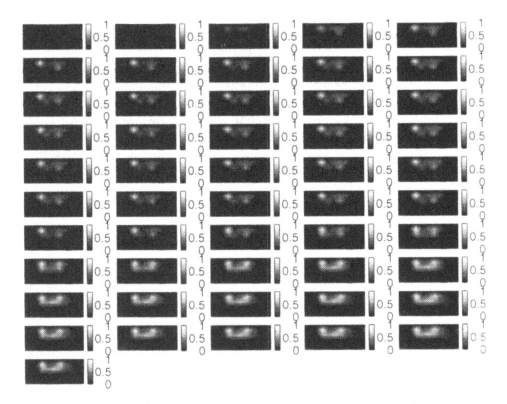

Figure 10
Snapshots showing tempo-spatial rupture process of the 1990 Gonghe earthquake. From top to bottom and from left to right, these pictures indicate variable slip on the fault plane of 88 km long and 40 km wide between 0–41 s at time interval of 1 s.

where S_i is the STF observed at the i-th station, N is the number of subfaults, m_j is the STF of the j-th subfault, and τ_{ij}, the time delay of the j-th subfault to the reference point, determined by

$$\tau_{ij} = \frac{R_{ij}}{V} \tag{7}$$

where R_{ij} is the distance of the i-th station to the j-th subfault, and V, the wave velocity at the earthquake source.

Equation (6) can be rewritten as convolution of the STF of the j-th subfault with Dirac δ-function:

$$S_i(t) = \sum_{j=1}^{N} \delta(t - \tau_{ij}) * m_j(t). \tag{8}$$

Equation (8) represents a linear relation between the STF observed at a station and the STFs of subfaults.

Solving the $m_j(t)$ is an underdetermined problem. In order to obtain a stable solution, some appropriate constraints should be imposed. In our study, the positivity constraint, which physically means that no backward slip is allowed to take place, is used in the inversion.

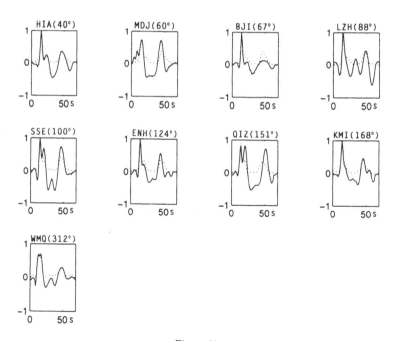

Figure 11

Comparison of theoretical (dotted lines) and observed (solid lines) source time functions of the Gonghe earthquake at nine stations of the CDSN. Figure enclosed in parenthesis represents the azimuth of the station.

In the study of Gonghe earthquake, the fault plane of 88 km long and 40 km wide is divided into 11×5 subfaults with dimensions of 8 km × 8 km. The STFs retrieved from Love waves are used to invert for the tempo-spatial rupture process of the Gonghe earthquake. In the inversion, the wave velocity is taken to be 6.2 km/s. The inversion yields a result with final slip distribution similar to the static variable slip distribution inverted from leveling data. The slip distribution on the fault plane inverted from the STFs (Fig. 10) also exhibits two slip concentrated regions about 25 km distant in the NWW to SEE direction. Using a total seismic moment of 9.4×10^{18} Nm and rigidity of 3.0×10^{11} dyn/cm^2, the maximum slip on the fault is estimated to be about 50 cm, which is very close to that (about 60 cm) obtained using leveling data.

The advantage of the technique described above is that not only the static, final slip distribution could be inverted but also the tempo-spatial variable slip on the fault plane could be obtained. The snapshots depicted in Figure 10 demonstrate that the rupture of the Gonghe earthquake is complex, both temporally and spatially. Two noticeable events, the first event and the third event as revealed by moment-tensor inversion, do not simply respectively correspond to the nuclei at the NWW and at the SEE ends of the final, variable slip model inverted from leveling data (Fig. 6). The tempo-spatial variable slip from STF inversion, shows that both nuclei were involved in the entire rupture process. At the onset of the earthquake, rupture initiated at the NWW end of the fault plane, and then expanded mainly toward the SEE end of the fault plane and triggered rupture at the SEE end of the fault plane. During the time period of the first event, rupture involved most of the fault plane but mainly concentrated at the NWW end of the fault plane, and during the time period of the third event, rupture also involved most of the fault plane but mainly concentrated at the SEE end of the fault plane.

We also calculate the theoretical STFs at different stations from the inverted tempo-spatial slip distribution (Fig. 11). In general, the theoretical STFs at individual stations are in good agreement with the observed STFs.

Conclusions and Discussion

Table 1 summarizes the focal mechanisms from geodetic and seismic data. Comparing the fault-plane solutions from geodetic and seismic data, it is concluded that the seismogenic structure of the 1990 Gonghe earthquake is a buried fault with strike NWW to SEE ($\alpha = 102°$ to $113°$). The *P* axis of the earthquake shown is close to horizontal (plunge $23°$) and oriented NNE to SSW (azimuth $23°$). The seismogenic fault of the Gonghe earthquake is a mainly reverse dip-slip fault with a minor left-lateral strike component. The scalar seismic moments determined from geodetic data, and broadband data range from 9.4×10^{18} to 9.8×10^{18} Nm. By using rigidity $\mu = 3.0 \times 10^{11}$ dyne/cm^2 the stress drop, $\Delta\sigma$, is calculated to be 4.9 to

Table 1

Focal mechanisms from geodetic and seismic data

Data	strike /(°)	dip /(°)	rake /(°)	M_0 /10^{18} Nm	Remarks
Leveling (1)	102	46	86	9.4	a single rectangular uniform-slip fault
Leveling (2)	102	46	86	9.8	a variable-slip fault with two nuclei
Broadband body					one source
waves (1)	113	68	89	9.4	total rupture duration 41 s
Broadband body					three events
waves (2)	96	73	77	4.7	occurred between 0–12 s
	107	8	169	2.1	occurred between 12–31 s
	100	67	75	2.5	occurred between 31–41 s

5.1 MPa. The slip distributes in a region slightly deep from NWW to SEE, with two nuclei located in a shallower depth in the NWW and a deeper depth in the SEE, respectively. The earthquake consists of three consecutive events. The first event

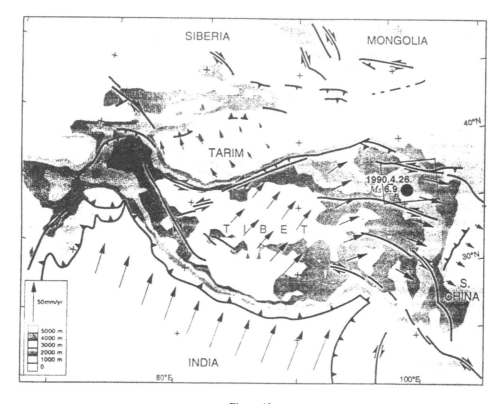

Figure 12
Simplified map of tectonics of Qinghai-Xizang (Tibetan) plateau after Avouac and Tapponnier (1993). Inner solid rectangle represents the area shown in Figure 13.

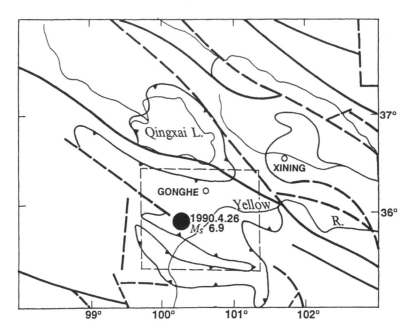

Figure 13
Map of the northeastern Qinghai-Xizang (Tibetan) plateau showing the epicentral location of the $M_s = 6.9$ Gonghe earthquake of April 26, 1990 and the faults (heavy solid lines), inferred faults (heavy dashed lines) and basins after INSTITUTE OF GEOLOGY, STATE SEISMOLOGICAL BUREAU (1981). Inner thin dash rectangle represents the area shown in Figure 14.

which occurred between 0–12 s, with a scalar seismic moment of 4.7×10^{18} Nm, has a focal mechanism with strike 96°, dip 73°, and rake 77°, which is similar to that inverted from leveling data. The second event which occurred between 12–31 s, with a smaller scalar seismic moment of 2.1×10^{18} Nm, has a variable focal mechanism with average strike 107°, dip 8°, and rake 169°. The third event which occurred between 31–41 s, with a scalar seismic moment of 2.5×10^{18} Nm, has a focal mechanism almost identical with that of the first event. Both slip concentrated nuclei were involved in the entire rupture process, but during the first 12 s, rupture mainly occurred at the NWW end, and during the last 10 s, mainly at the SEE end, of the fault plane.

The 1990 Gonghe earthqiake occurred in the south margin of the Gonghe basin (Figs. 12 and 13). The Gonghe, Qinghai Lake, Xining-Minghe and Qaidam basins are known as large Cenozoic basins (Ma, 1987). The thickness of the Quaternary accumulation beneath these basins is more than 1000 m. The thickness of the Quarternary accumulation beneath the Qaidam and Gonghe basins is 2800 m and 1200 m, respectively. The block of Qaidam and Gonghe basins, separated by the left-lateral strike-slip faults such as Qilianshan, Haiyuan, Altyn Tagh and Kunlun fault zones is a region of strong differential tectonic movement. As AVOUAC and TAPPONNIER (1993) noted, this curved block rotates clockwise with respect to the

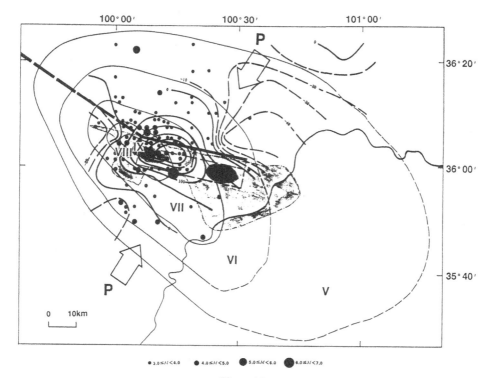

Figure 14

Map showing the pressure axis in the northeastern Qinghai-Xizang (Tibetan) plateau, vertical ground deformation, surface projection of the uniform slip and variable slip model inverted from leveling data, epicenters of aftershocks and isoseismals, of the Gonghe earthquake.

Siberia block. Also, ZENG *et al.* (1993) and ZENG and SUN (1993) proposed a new model of the continental collision from the Indian to Eurasian plates and an eastward transfer of the crustal material underneath the Qinghai-Xizang plateau. The 1990 Gonghe earthquake strike, and the significantly reverse dip-slip with minor left-lateral strike-slip motion obtained in this study, confirm that the pressure axis of the tectonic stress field in the northeastern Qinghai-Xizang plateau is close to horizontal and oriented NNE to SSW, consistent with the relative collision motion between the Indian and Eurasian plates. This result gives support to the new model proposed by ZENG *et al.* (1993) and ZENG and SUN (1993) of the eastward transfer of the crustal material beneath the Qinghai-Xizang plateau, and to the kinematic model proposed by AVOUAC and TAPPONNIER (1993), of the clockwise rotation of the Qaidam basin-Gonghe basin block with respect to the Siberian block as a whole.

Figure 14 summarizes the pressure axis of the tectonic stress field in the northeastern Qinghai-Xizang plateau, isoseismals, aftershock activity and the projection of the inverted fault model on the ground surface. Comparing these results, it is believed that the occurrence of the Gonghe earthquake is mostly related to a buried fault striking, NW to SE, near the south margin of the Gonghe basin, inferred by the INSTITUTE OF GEOLOGY, STATE SEISMOLOGICAL BUREAU (1981). The predominantly thrust mechanism and the complexity in the tempo-spatial rupture process of the Gonghe earthquake, as revealed by the geodetic and seismic data, is generally consistent with the overall distribution of isoseismals, aftershock seismicity and intersecting faults structure in the Gonghe basin of the northeastern Qinghai-Xizang plateau.

REFERENCES

AVOUAC, J. P., and TAPPONNIER, P. (1993), *Kinematic Model of Active Deformation in Central Asia*, Geophys. Res. Lett. *20*, 895–898.
CHEN, Y. T., LIN, B. H., LIN, Z. Y., and LI, Z. Y. (1975), *The Focal Mechanism of the 1966 Hsingtai Earthquake as Inferred from the Ground Deformation Observations*, Acta Geophysica Sinica *18*, 164–182 (in Chinese with English abstract).
CHEN, Y. T., LIN, B. H., WANG, X. H., HUANG, L. R., and LIU; M. L. (1979), *A Dislocation Model of the Tangshan Earthquake of 1976 from the Inversion of Geodetic Data*, Acta Geophysica Sinica *22*, 201–217 (in Chinese with English abstract).
DREGER, D. S. (1994), *Investigation of the Rupture Process of the 28th June 1992 Landers Earthquake Utilizing TERRA Scope*, Bull. Seismol. Soc. Am. *84*, 713–724.
DZIEWONSKI, A. M., EKSTRÖM, G., WOODHOUSE, J. H., and ZWART, G. (1991), *Centroid-moment Tensor Solutions for April-June in 1990*, Phys. Earth Planet. Inter. *66*, 133–143.
GONG, S. W., and GUO, F. Y. (1992), *Vertical Ground Deformation in the Earthquake of Gonghe, Qinghai Province*, Acta Seismologica Sinica (Chinese edition) *14* (Supplement), 725–727 (in Chinese).
GONG, S. W., WANG, Q. L., and LIN, J. H. (1993), *Study of Dislocation Model and Evolution Characteristics of Vertical Displacement Field of Gonghe $M_S = 6.9$ Earthquake*, Acta Seismologica Sinica (English edition) *6* (3), 641–648.
HARTZELL, S., and IIDA, M. (1990), *Source Complexity of the 1987 Whittier Narrows, California, Earthquake from the Inversion of Strong Motion Records*, J. Geophys. Res. *98* (B12), 22123–22134.
INSTITUTE OF GEOLOGY, STATE SEISMOLOGICAL BUREAU (ed.), *Seismotectonic Map of Asia and Europe* (China Cartography Press, Beijing 1981) (in Chinese).
KANAMORI, H., and ANDERSON, D. L. (1975), *Theoretical Basis of Some Empirical Relations in Seismology*, Bull Seismol. Soc. Am. *65*, 1073–1095.
KENNETT, B. L. N., *Seismic Wave Propagation in Stratified Media* (Cambridge University Press, 1983), 342 pp.
MANSINHA, L., and SMYLIE, D. E. (1971), *The Displacement Fields of Inclined Faults*, Bull. Seismol. Soc. Am. *61*, 1433–1400.
MA, X. Y. (ed.), *Outlines of Lithospheric Dynamics of China* (Seismological Press, Beijing, 1987) (in Chinese).
PERSON, W. J. (1991), *Seismological Notes—March-April 1990*, Bull. Seismol. Soc. Am. *81*, 297–302.
TU, D. L. (1990), *Geological Structure Background of Gonghe Earthquake $M_S = 6.9$, on April 26, 1990*, Plateau Earthq. Res. *2* (3), 15–20 (in Chinese with English abstract).
VELASCO, A. A., AMMON, C. J., and LAY, T. (1994), *Empirical Green Function Deconvolution of Broadband Surface Waves: Rupture Directivity of the 1992 Landers, California ($M_w = 7.3$) Earthquake*, Bull. Seismol. Soc. Am. *84* (3), 735–750.
WARD, S. N., and BARRIENTOS, S. (1986), *An Inversion for Slip Distribution and Fault Shape from Geodetic Observations of the 1983, Borah Peak, Idaho, Earthquake*, J. Geophys. Res. *91*, 4909–4919.
XU, Z. H., WANG, S. Y., HUANG, Y. R., and GAO, A. J. (1992), *Tectonic Stress Field of China Inferred from a Large Number of Small Earthquakes*, J. Geophys. Res. *97* (B8), 11867–11877.

ZHAO, M., CHEN, Y. T., GONG, S. W., and WANG, Q. L. (1993), *Inversion of focal mechanism of the Gonghe, China, Earthquake of April 26, 1990, using leveling data.* In *Continental Earthquakes* (eds. Ding, G. Y. and Chen, Z. L.) IASPEI Publication Series for the IDNDR *3*, 246–252.

ZENG, Q. S. (1990), *Survey of the Earthquake $M_s 6.9$ between Gonghe and Xinghai on April 26, 1990,* Plateau Earthq. Res. *2* (3), 3–12 (in Chinese with English abstract).

ZENG, Q. S. (1991), *Seismicity and Earthquake Disaster of Qinghai Province,* Plateau Earthq. Res. *3* (1), 1–11 (in Chinese with English abstract).

ZENG, Q. S. (1995), *Earthquake Resistance and Disaster Reduction and Short-impending Prediction of Qinghai Province,* Plateau Earthq. Res. *7* (1), 42–51 (in Chinese with English abstract).

ZENG, R. S., ZHU, J. S., ZHOU, B., DING, Z. F., HE, Z. Q., ZHU, L. R., LUO, X., and SUN, W. G. (1993), *Three-dimensional Seismic Velocity Structure of the Tibetan Plateau and its Eastern Neighboring Areas with Implications to the Model of Collision between Continents,* Acta Seismologica Sinica (English edition) *6* (2), 251–260.

ZENG, R. S., and SUN, W. G. (1993), *Seismicity and Focal Mechanism in Tibetan Plateau and its Implications to Lithospheric Flow,* Acta Seismologica Sinica (English edition) *6* (2), 261–287.

(Received December 10, 1995, revised/accepted December 30, 1995)

PAGEOPH, Vol. 146, Nos. 3/4 (1996)

0033–4553/96/040717–24$1.50 + 0.20/0

Aseismic Fault Slip and Block Deformation in North China

Lanbo Liu,[1,2] Alan T. Linde,[1] I. Selwyn Sacks,[1] and Shihai He[2]

Abstract — In North China, the tectonic fault-block system enables us to use the Discontinuous Deformation Analysis (DDA) method to simulate the long-term cross-fault survey and other geodetic data related to aseismic tectonic deformation. By the simulation we have found that: (1) Slips on faults with different orientation are generally in agreement with the ENE-WSW tectonic stress field, but the slip pattern of faulting can vary from nearly orthogonal, to pure shear along the strike of the faults, this pattern cannot be explained by simple geometric relation between the strike of the fault and the direction of the tectonic shortening. This phenomenon has been observed at many sites of cross-fault geodetic surveys, and might be caused by the interactions between different blocks and faults. (2) According to the DDA model, if the average aseismic slip rate along major active faults is at the order of several tenths of millimeter per year as observed by the cross-fault geodetic surveys, the typical strain rate inside a block is at the order of 10^{-8} year^{-1} or less, so that the rate of 10^{-6} year^{-1}, as reported by observations in smaller areas, cannot be the representative deformation rate in this region. (3) Between the slips caused by regional compression and block rotation, there is a possibility that the sense of slip caused by rigid body rotation in two adjacent blocks is opposite to the slip caused by the tectonic compression. But the magnitude of slip resulting from the tectonic compression is much larger than that due to the block rotation. Thus, in general, the slip pattern on faults as a whole agrees with the sense of tectonic compression in this region. That is to say, the slip caused by regional compression dominates the entire slip budget. (4) Based on (3), some observed slips in contradiction to ENE tectonic stress field may be caused by more localized sources, and have no tectonic significance.

Key words: Discontinuous Deformation Analysis (DDA), North China, fault-block system, aseismic fault slip, block rotation.

Introduction

The collision between the Indian and Eurasian plates is the dominant factor in the neotectonic movement in the Eurasian continent. Besides the uplift of the Tibet Plateau and crustal thickening close to the collision zone (Fig. 1), this strong collision also caused a relatively higher deformation rate in intraplate areas around the Tibetan Plateau. In farther regions of continental China, relatively high deformation rates also exist and are represented by high seismicity. For example, in

[1] Department of Terrestrial Magnetism, Carnegie Institution of Washington, Washington, D.C. 20015–1305, U.S.A.
[2] Center for Analysis and Prediction, State Seismological Bureau, Beijing, 100036, P. R. China.

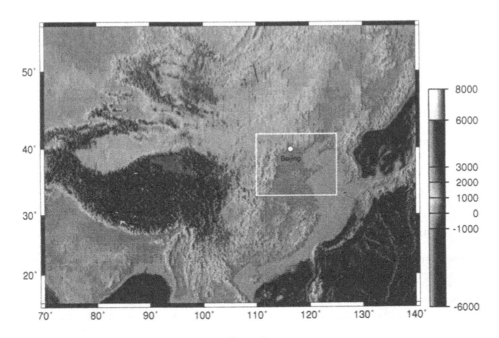

Figure 1
The relative position and tectonic environment of northern China in the Eurasian Continent.

North China, the average strain rate may reach the order of 10^{-8} year^{-1} (MA *et al.*, 1989; F. HUANG *et al.*, 1991). In contrast, the tectonic strain rate in the intraplate eastern North America is estimated on the order of $10^{-11} \sim 10^{-9}$ year^{-1} for most regions (ANDERSON, 1986; MORGAN, 1987). Only in some substantially smaller areas where there are intraplate seismic zones (such as the New Madrid seismic zone on the borders of Missouri, Arkansas, and Tennessee, or Charleston in South Carolina) the strain rates may reach the order of 10^{-7} year^{-1} (LIU and ZOBACK, 1992).

Intraplate seismicity in China has been highly active and rather episodic since Holocene time. For instance, from 1966 to 1976, 15 earthquakes with magnitudes greater than $M \sim 7$ occurred in continental China, nine of them in heavily populated eastern China (east of longitude 98°E). Seismicity is closely related to the tectonic framework of continental China. Generally speaking, seismicity is more active in western China than in eastern China since western China is closer to the collision zone between the Indian and the Eurasian plates. However, the northern part of east China (North China, area in the box in Fig. 1) has experienced 15 $M_s > 5.8$ earthquakes during that period, with many more smaller events since 1966 (Fig. 2). The period of 1966–1978 is usually referred to as the third seismic active episode in this century in continental China. (The first and second active episodes are 1909–1923 and 1929–1952, respectively. See MA *et al.*, 1990; ZHANG

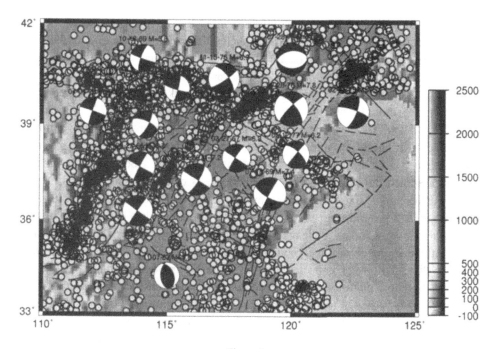

Figure 2
Northern China map with faults, seismicity, and focal mechanisms of major earthquakes (1966–1993).

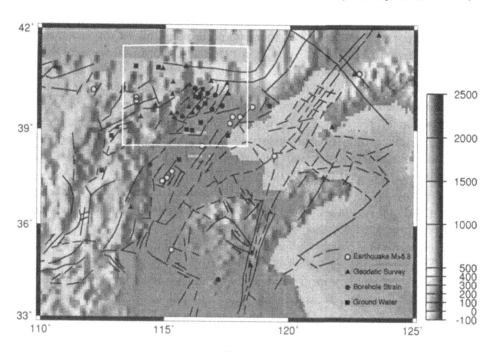

Figure 3
Station locations of the northern China earthquake monitoring network.

and MA, 1993). This active episode is represented by active seismicities in North China and southwest China (Yunnan and Sichuan Provinces) (MA *et al.*, 1990). In general, higher seismicity does indicate that the eastern China continent is undergoing a higher deformation rate in the latest several decades.

In North China various kinds of geophysical networks have been operated after the Xingtai earthquake of 1966 ($M_s = 6.8$, March 8, and $M_s = 7.2$, March 22, 1966) (Figs. 2 and 3). These networks consist of seismicity monitoring, conventional triangulation, trilateration, and leveling surveys, cross-fault geodetic surveys (short baseline length and short leveling), groundwater wells and a borehole dilatometer network installed during 1985–1987 under the aegis of the United Nations Development Program. Data of geodetic surveys, cross-fault continuous monitoring, strainmeter, and groundwater for the period of 1966–1993 have been collected with the assistance of Chinese colleagues in the State Seismological Bureau (SSB). Among these data, cross-fault geodetic observations such as short baseline and short leveling data were collected by the State Seismological Bureau (SSB) of China. A borehole dilatometer (SACKS *et al.*, 1971) network was installed by SSB during 1985–1987 under the aegis of the Carnegie Institution (LIU *et al.*, 1986). These observations provide a large data base for earthquake studies in intraplate continental regions. A unique data set of various pre-, co-, and post-seismic geophysical and geodetic observations from this region, especially data of cross-fault geodetic surveys and the borehole strain, have been acquired with the assistance of Chinese colleagues in the State Seismological Bureau (SSB) of China. During this period 15 $M_s > 5.8$ earthquakes occurred in this region, including the tragic Tangshan earthquake ($M_s = 7.8$) in 1976 that claimed at least 240,000 lives. The 1989 Datong earthquake (M 5.8, see ZHU *et al.*, 1993) is the largest event in North China subsequent to the 1976 great Tangshan earthquake (M 7.8) and has drawn great attention in the seismological community in terms of the evaluation and reduction of the seismic hazard in the metropolitan areas of Beijing and Tianjin. Various geophysical precursory anomalies such as crustal deformation, groundwater, geomagnetism, and geoelectricity were reported for the Datong earthquake (ZHANG and MA, 1993).

Understanding the seismogenic process is one of the most challenging topics in earthquake studies. From late 1970 to early 1980, numerical modelings have been conducted for the nucleation-occurrence process of earthquakes in China, especially in North China. WANG *et al.* (1980) used a finite-element method to model the earthquake sequence from 1966 to 1976. SONG *et al.* (1990) combined finite-element modeling, earthquake focal mechanism studies, and crustal deformation measurements in an attempt to understand the migration of earthquake events. More recently, J. Ma (in GAO and MA, 1993) has reported the possible occurrence of block rotations in this region, based upon the Quaternary basin distributions and fault slips abstracted from geological observations.

The rate of long-term aseismic (interseismic) deformation controls how often large earthquakes can occur in a region. Investigation of the deformation behavior in the aseismic period is important in understanding the seismogenic cycle. In this paper we will use the Discontinuous Deformation Analysis (DDA) method (SHI and GOODMAN, 1988, 1989; SHI, 1993), which has gained wide use in rock mechanics and geotechnical engineering, to model the rates of stress accumulation, block deformation, and fault slips associated with the interseismic period in northern China. By this modeling, we try to shed light on the distribution of deformation in different tectonic blocks and the deformation budget between block interior and block boundary, i.e., between the blocks themselves and the faults which divide the area into blocks. We use mapped faults as the boundaries of different blocks.

In this paper we first describe different data sets and the tectonic background of North China because they provide the bases for modeling. Then we discuss the selection of the geometric and physical parameters used in the DDA model. Finally we model the stress-strain variation, block deformation, and fault slip associated with tectonic shortening, investigate the interaction among different faults and blocks in the fault-block system in northern China during an interseismic period and compare this with the estimates of deformation rates in different blocks and aseismic slip across faults. Since the model is linear-elastic, for modeling the stress and strain accumulation and fault slips in the interseismic period, we used the shortening of the entire system between points at two opposite borders in the maximum compression direction as the control of the temporal process, and rated the shortening to coincide with the strain rate inferred from geological and tectonic information in this region.

Discontinuous Deformation Analysis (DDA)

In the fault-block tectonic environment such as North China, an efficient numerical modeling technique is essential for the study of earthquake processes and crustal geodynamics. Discontinuous deformation analysis (DDA) is a powerful tool for analyzing the force displacement interactions of a fault-block system (SHI and GOODMAN, 1988, 1989; SHI, 1993). This method can compute block translation, rotation and deformation simultaneously. The DDA method has been used in rock mechanics and geotechnical engineering with great success. This study is the first in which DDA has been applied to a tectonic problem and we compare and contrast its capacity with more commonly used techniques.

There are notable advantages of the DDA technique over the finite-element method (FEM) for the fault-block system. DDA and FEM approach the real jointed or blocky rock masses from two extremes. DDA does not imply continuity at block boundaries, i.e., it is fundamentally discontinuous. In contrast, FEM approaches the deformation behavior of real rock mass by assuming the rock mass to be a continuum. Thus the displacement of real rock mass should be larger than the displacements from FEM computation and less than that from DDA (SHI, 1993). However, high tension strength and cohesion can be applied to the joint lines or fault lines between blocks when joints or faults do not completely separate the rock mass into distinct blocks. DDA solves a finite-element type of mesh where all the elements are real isolated blocks, bounded by preexisting discontinuities. While the elements or blocks used by the DDA method can be of any convex or concave shape, or even multi-connected polygons with holes, the finite-element method encompasses elements of standard shape. In the DDA method, generally the number of unknowns is the sum of the degrees of freedom of all blocks; while in the finite-element method, the number of unknowns is the sum of the degrees of all nodes. From a theoretical point of view, DDA may be considered as a generalization of the finite-element method.

DDA is also different in nature from the widely used distinct element method (DEM). DDA is a displacement method, where the unknowns in the equilibrium equations are displacements; DEM is a force method which attempts to adjust the contact forces by using fictitious forces and artificial damping to control numerical instability. In DDA, on the other hand, the displacements and strains of the blocks in each time step are computed by solving implicitly a system of simultaneous equilibrium equations. Since this system of equations is derived from minimizing the total potential energy of the block system, its solution satisfies equilibrium at all times (SHI, 1993). Furthermore, a complete kinematic theory governs how blocks interact in DDA. Because DDA is inherently stable, it does not require artificial forces for stability, and could therefore model real-time dynamic behavior accurately. DDA is an implicit method, which is fast, accurate, and has a correct energy consumption. However, there is no dynamic modeling involved in this study. This paper only deals with static problems.

Unlike the tectonic environment along major passive plate boundaries where subduction of one plate beneath another is the principal motion, in the intracontinental fault-block system, tectonic deformation is mainly expressed as fault slips and block compression and shear deformation, but without penetration of one block into another. In the DDA model, opening and closing as well as slipping along any faults is allowed, but no intrusion. It assumes each block has constant stress and constant strain throughout the entire block. For this reason, to obtain detailed information about deformation properties over space, smaller blocks within a larger block can be used to redefine the distribution of mechanical parameters. High tensile strength and inherent shear strength (cohesion) can be assigned to the faults or joints between these small blocks to prevent any slip or opening from occurring along them.

We have modified the DDA code to handle larger scale geological problems compared with the much smaller geotechnical problems treated previously with this technique. This was necessary because tectonic deformations are typically several orders of magnitude smaller than those for the smaller scale geotechnical problems.

Seismotectonic Background of North China

Our study region (North China) is 110°E to 125°E in longitude and 33°N to 42°N in latitude. This region encompasses the major portion of the North China fault-block region (Fig. 2, also see ZHANG et al., 1980; MA et al., 1989). The time window under study is from 1966 to 1993, which includes a seismic active episode and a quiescent episode (ZHANG and MA, 1993; MA et al., 1990). In the west, the Erdos block is a relatively strong and cold geological body. According to MA et al. (1990), it moves in the direction of ENE at a rate of 2–4 mm/year under the effect of the India-Tibet intracontinental collision. If we assume the east edge of North China is not moving and the total length of North China is about 600–700 km, then the total strain rate caused by pushing of the Erdos block in the west is about 0.5×10^{-8} year^{-1}. If we assume shortening is also caused by pushing of the Pacific plate from the east, then the strain rate may reach about 1×10^{-8} year^{-1}. We will discuss another constraint of total strain rate in the following section.

1. Seismicity Since 1966

From 1966 to 1994, fifteen earthquakes with magnitudes greater than 5.8 have occurred in this region. The source parameters for these events are listed in Table 1 in which parameters for earthquakes before 1983 are taken from S. WANG et al. (1991) and the last two are from the Bulletin of the International Seismological Center (BISC).

In March 1966, the Xingtai earthquakes marked the start of the third active episode (1966–1977) of this century in North China (ZHANG and MA, 1993; MA et al., 1990). From 1966 to 1977, 12 $M_s > 6$ earthquakes occurred, five of them exceeding a magnitude of 7. The main feature of the seismicity in this time period is that large earthquakes migrated from southwest to northeast (Fig. 2). This active episode ended with the tragic Tangshan earthquake of 1976 and its large after-shocks in 1977. Tangshan earthquake is the largest event in the active episode from 1966 to 1977 in North China. From 1978 to 1994 no earthquake exceeding a magnitude of 6 has occurred in this region, thus this period may be regarded as a quiescent episode.

2. Tectonic Stress Field

The stress field of the North China fault-block system experienced several significant changes during the course of geological time frame. In pre-Mesozoic

Table 1

Earthquake (M_s > 5.8) Parameters in North China (1966–1993)

λ (°)	ϕ (°)	h (km)	T_0 mm-dd-yy	M (M_s)	M_0 (Nm)	Nodal plane 1 strike	dip	rake	Nodal plane 2 strike	dip	rake
114.92	37.35	10	03-08-66	6.8	6.6E18	215	88	−178	125	88	−2
115.10	37.50	9	03-22-66	6.7	2.3E18	208	86	−178	118	88	−4
115.10	37.50	15	03-22-66	7.2	1.8E19	32	77	−180	302	90	−13
115.30	37.70	15	03-26-66	6.2	1.4E18	30	74	−157	293	68	−17
116.50	38.50	30	03-27-67	6.3	1.7E18	213	82	−178	123	88	−8
119.40	38.20	35	07-18-69	7.4	4.0E19	209	87	−160	118	70	−3
122.80	40.70	16	02-04-75	7.3	3.0E19	288	78	−18	22	72	−167
112.10	40.20	18	04-06-76	6.2	5.7E17	289	89	4	199	86	179
118.00	39.40	22	07-28-76	7.8	5.8E19	225	80	157	319	67	11
118.50	39.70	22	07-28-76	7.1	3.0E19	260	50	−99	94	41	−79
117.70	39.40	17	11-15-76	6.9	2.6E18	322	67	−10	56	81	−157
117.60	39.20	19	05-12-77	6.2	3.8E17	311	69	2	220	88	159
115.30	35.20	15	11-07-83	5.9	4.2E17	176	49	104	335	43	74
113.83	39.97	9	10-18-89	5.8	3.3E17	200	75	175	109	85	−15
113.83	39.97	10	03-25-91	5.8	1.6E17	106	82	7	15	83	172

time, the main feature is north-south compression, the stress regime is reverse to strike-slip faulting. From the Mesozoic time, the stress field started to change direction to a NE compression, and the stress regime gradually changed to pure strike-slip faulting. The stress direction continued to change into an ENE compression during the Cenozoic, and at present the stress regime has changed into strike-slip to normal faulting.

Various studies (XU *et al.*, 1992; SONG *et al.*, 1990) concluded that the modern regional tectonic stress in North China is in the strike-slip to normal faulting stress regime (i.e., $\sigma_1 = \sigma_v > \sigma_3$) with the principal horizontal stress oriented in an ENE direction ($\sim$N80°E) (XU *et al.*, 1992; MA *et al.*, 1989). The stress regime of strike-slip to normal faulting implies that the maximum horizontal compressive stress is approximately equal to the vertical stress (i.e., the overburden $\sigma_v = \rho g z$), with the maximum and minimum principal stresses in the horizontal plane. Besides the regional stress field with rather uniform orientation, secondary sources of stress field also exist. Topographic effect is among the most important secondary sources which may give rise to flexure in the lithosphere and perturbation of the local stress field. For instance, the orientation of the horizontal principal stress in the Shanxi Uplift block (west of 112°E in latitude and between 35°N and 40°N in latitude in Figs. 2 and 3) of the North China fault-block system is more northward than for other blocks (ZHANG *et al.*, 1985). The most likely explanation of this phenomenon is the effect of the mountain belts and basins in this tectonic block. An analytical method to estimate the effect of 3-D topographic relief on stresses in the crust has

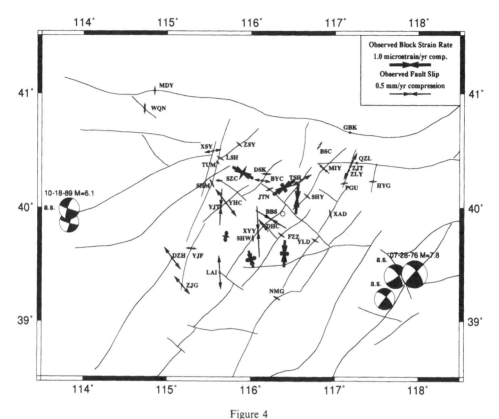

Figure 4
Fault slips, areal geodetic observations in the Beijing area against map-view.

been developed (LIU and ZOBACK, 1992) and can be used to eliminate the topographic effect from the regional stress field. Even though the regional stress field is rather uniform and stable in an historical time frame, numerous seismic focal mechanism and *in situ* stress measurement studies indicated possible stress orientation adjustment before, during, and after large earthquakes within a small area close to the epicenter of an earthquake (CHENG *et al.*, 1982; DING and ZHANG, 1988). However, as the boundary condition input to numerical modeling, the regional stress field is considered as uniform and stable. In our final model of the stress and deformation in the interseismic period, a smaller region with rather uniform state of stress is used (the area in the box of Fig. 3).

3. Tectonic Structures: Fault-block System

There are two sets of fault systems in North China: 1. Fault zones striking NNE; 2. fault zones striking WNW to EW (Figs. 3 and 4). These two sets of faults are almost orthogonal to each other. The tectonic significance has been studied by

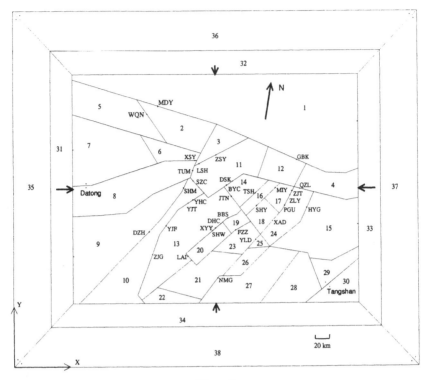

Figure 5
Fault-block model for DDA simulation with boundary conditions (total 38 blocks).

a number of geologists in China (ZHANG *et al.*, 1980; MA *et al.*, 1989, 1990). West of this region the Erdos block (shown as boundary pushing in the DDA model, block No. 31 in Fig. 5) and Yanbei Block (block No. 1 in Fig. 5) are characterized by seismic quiescent and underdeveloped faulting.

There are two pronounced features of the tectonic structure in this region. First, the NNE and NE structures are considerably longer than those of the EW to WNW structures (Figs. 3 and 4). Second, the tectonic activity on faults and structures oriented NE and NNE is characterized by seismic slip. Most of the great earthquakes occurred on the NNE and NE nodal plane (XU *et al.*, 1992). In contrast, faults and structures oriented in the EW and WNW direction are dominated by aseismic slips (GAO and MA, 1993). The structures in the EW direction originated in pre-Mesozoic. NE and NNE structures originated in Mesozoic. Structures in the WNW direction are secondary accompanimental phenomena to EW and NE, NNE structures, so their ages are determined by their parent structure, mostly within pre-Mesozoic to Mesozoic time. Under the current stress regime of ENE compression, the preexisting structures are reactivated in the sense of shear motion.

Structures in the EW to WNW are under left-lateral shear motion and structures striking NE to NNE are under right-lateral shear (SONG et al., 1990) motions.

4. Aseismic Deformation

In this study we will concentrate on discussions of the deformation in the interseismic period simply because we have more robust data and less controversial arguments applicable for this period. Success on modeling in the interseismic period is necessary before attempting to model deformation in the other periods of an earthquake cycle.

From repeated geodetic surveys in this region, the average horizontal deformation rate in a large area can be as large as several tenths of a microstrain per year but it is not statistically significant (HUANG et al., 1989). Vertical deformation has a larger rate than that in the horizontal. Coherent uplift and subsidence have been revealed by modern geodetic surveys in several subregions in North China (XIE, 1993). As well as the conventional triangulation, trilateration, and leveling surveys, cross-fault short baseline and short leveling at permanent sites provide another valuable data set for earthquake studies in this area and these data will set the major constraints in modeling conducted in this paper. Cross-fault surveys frequently reveal higher deformation rates than surveys conducted inside tectonic blocks, an apparent explanation being that the blocks are more rigid inside than at the boundaries. Due to the heterogeneity of the physical properties of faults at different segments and in different orientations, slip rates are also highly diverse, sometimes even in opposite polarities. For example, observations had shown that the Babaoshan fault (the fault on which the station BBS is astride in Fig. 4) is highly active but other faults with the same orientation are rather inactive. According to GAO and MA (1993), NW-oriented faults essentially exhibit steady state creeping while NE-oriented faults are characterized by stick-slip failure. We used a compilation of cross-fault geodetic survey data to constrain our DDA model. A total of 34 cross-fault geodetic survey sites are used in this study (Fig. 4). For most of the sites we have both the baseline length changes and leveling changes, so horizontal shear and the compression/extension of the fault plane can be determined by simple geometric calculations. At some sites only leveling surveys were conducted, therefore only the compression/extension of the fault plane can be determined; no knowledge of the sense of shear can be determined. The cross-fault data are listed in Tables 2 and 3 as well as Figure 4.

Figure 4 shows the strain rates in several blocks derived from triangulation and trilateration surveys over a considerably larger area than the cross-fault surveys. The errors are large for these strain rate results, very few having a magnitude larger than two standard deviations (CHE et al., 1993). Figure 5 shows the area we have chosen for modeling; it encloses the epicentral areas of the Tangshan and Datong earthquakes, even though these events may have some effects on regional aseismic

Table 2

Fault Motion Detected by Short Telluric Network*

Site name	3-lett. code	Longitude (°)	Latitude (°)	Name of fault-crossing	Strike (°)	Dip (°)	Observation period	Exten. rate (mm/yr)	Shear rate (mm/yr)	Confi. code
Tumu	TUM	115.600	40.400	Tumu Fault	15	60	09/85–09/90	−0.05	−0.03	C
Shimen	SHM	115.533	40.217	Shimen Fault	210	65	01/85–12/90	+0.10	+0.06	C
Yanhecheng	YHC	115.700	40.067	Nankou-Yanhecheng Fault	60	70	01/85–12/90	+0.40	−0.06	B
Yijianfang	YJF	115.300	39.683	Zijinguan Fault	30	65	01/85–12/90	−0.12	−0.06	C
Zijingguan	ZJG	115.183	39.433	Zijinguan Fault	30	65	09/85–09/90	+0.27	+0.09	C
Dazhuang	DZH	115.000	39.533	Wulonggou Fault	30	80	01/85–12/90	+0.30	+0.12	C
Baiyangcheng	BYC	116.133	40.200	Nankou-Yanhecheng Fault	45	45	09/85–09/90	+0.12	−0.16	C
Xiaoyouying	XYY	116.050	38.833	Babaoshan Fault	60	50	03/85–09/90	−1.04	−0.53	B
Laishui	LAI	115.617	39.417	Laishui Fault	280	70	03/85–09/90	+0.38	−0.09	B
Jiangtun	JTN	116.267	40.183	Nankou-Sunhe Fault	135	70	01/85–12/90	+1.07	−0.45	D
Beishicheng	BSC	116.817	40.550	Beishicheng Fault	45	25	01/85–12/90	−0.02	−0.10	C
Taoshan	TSH	116.600	40.250	Huangzhuang-Gaoliying	30	70	01/85–12/90	+0.14	−0.23	C
Miyun	MIY	116.883	40.417	Miyun-Erjiayu Fault	15	35	09/85–09/90	−0.17	−0.08	C
Pingu	PGU	117.117	40.217	Chengzhuang-Zhenluoying	15	80	03/85–09/90	+0.09	−0.04	C
Gubeikou	GBK	117.200	40.683	Chongli-Chicheng Fault	285	45	01/85–12/90	+0.00	−0.03	C
Zhenluoying	ZLY	117.133	40.333	Chengzhuang-Zhenluoying	15	72	01/85–12/90	+0.02	−0.02	C
Xiadian	XAD	116.917	39.967	Xiadian Fault	30	75	01/85–12/90	+0.08	+0.08	C
Huangyaguan	HYG	117.433	40.233	Huangyaguan Fault	15	80	03/85–09/90	+1.15	−0.36	D
Jixian	JXN	117.450	40.000	Jixian Fault	105	60	09/85–09/90	+0.21	+0.03	C

* Fault slip rates are from CHE *et al.* (1993). Fault orientations use AKI and RICHARDS (1980) convention. Positive values in fault slip: extension or right-lateral shear; negative values: compression or left-lateral shear.

Table 3

Fault Motion Detected by Short Baseline and Leveling Surveys*

Site name	3-lett. code	Longitude (°)	Latitude (°)	Name of fault-crossing	Strike (°)	Dip (°)	Observation period	Exten. rate (mm/yr)	Shear rate (mm/yr)	Confi. code
Babaoshan	BBS	116.233	39.917	Babaoshan Fault	65	40	01/83–12/90	−0.33	+0.24	C
Dahuichang	DHC	116.117	39.850	Babaoshan fault	18	68	05/71–12/90	−0.15	−0.09	A
Deshengkou	DSK	116.183	40.300	Nankou Fault	50	50	07/82–12/90	−0.08	+0.09	B
Qiangzilu	QZL	117.217	40.433	Miyun-Xinglong Fault	280	87	01/72–12/90	+0.00	−0.03	A
Shangwan	SHW	116.017	39.800	Huangzhuang-Gaoliying	20	48	01/81–12/90	+0.04	+0.08	C
Shizhuangcun	SZC	115.617	40.250	Shizhuang Fault	326	75	08/85–12/90	+0.00	−0.08	A
Wanquan	WQN	114.750	40.900	Shuiguantai Fault	315	80	08/85–12/90	−0.10	+0.09	B
Xiaoshuiyu	XSY	115.533	40.500	Xinglingpu-Yaozitou	51	50	01/74–12/90	+0.10	−0.18	B
Yianjiatai	YJT	115.567	40.000	Yanhecheng Fault	55	40	01/75–12/90	−0.26	−0.33	A
Zhangjiatai	ZJT	117.133	40.367	Zhenluoying Fault	185	60	01/72–12/90	+0.13	−0.32	A
Zhangsanying	ZSY	115.867	40.517	Daxishan Fault	00	60	04/70–12/90	−0.07	−0.06	A

* Fault slip rates are from CHE et al. (1993). Fault orientations use AKI and RICHARDS (1980) convention. Positive values in fault slip: extension or right-lateral shear; negative values: compression or left-lateral shear.

deformation, the central area of the modeling is distant enough to minimize these effects.

DDA Modeling Procedure

Based upon the above discussion of the data and the tool, the fundamentals of the modeling of the DDA method for the interseismic period are described as follows.

1. The Geometric Parameters

This is a two-dimensional plane stress model. We assume that the fault-block system is formed by a layer of linear elastic material and the faults have cut through the entire depth of this layer. This layer of fault-block system coincides with the seismogenic layer in the crust (the uppermost 15 km of the crust in North China (GAO and MA, 1993)). This is a reasonable assumption since the seismogenic layer has a free upper surface and an underlying more ductile layer in which material may flow in a geological time scale. Geophysical deep sounding data indicate that most of the faults we chose in the DDA model cut through the entire seismogenic layer, and some of them even cut through the entire crust (GAO and MA, 1993). Additionally, since our study area is in a strike-slip faulting stress regime (two horizontal principal stresses are the maximum and minimum principal stresses with vertical stress as the intermediate stress), the 2-D approximation would be fairly reasonable.

In choosing the geometry of the model we have required the smallest dimension of a block to be larger than the thickness of the seismogenic layer (15 km). In some areas the faults with lengths less than 15 km are modeled either by abandoning the relatively unimportant one or by a single fault. Additionally we use larger blocks in remote areas, and smaller blocks in areas with more observation data.

We divide the entire area into 38 blocks. Of these 38 blocks 30 (block Nos. 1–30) are model tectonic blocks and 8 are loading and holding blocks. The outermost 4 blocks (block Nos. 35–38) have been fixed at the 4 corners by 8 pins at the corners to eliminate whole body translation. The 4 blocks (block Nos. 31–34) inside the outermost are compressing elements behaving as 4 gigantic flat jacks. The fundamental fault map we used is based upon SONG *et al.* (1990). The model geometry features long NNE faults and short NW faults. Further, as we are more concerned with deformation in the central section, we have divided finer blocks in the central part of the system and larger blocks around the edges to absorb boundary effect.

The axes of the model have been rotated 10 degree counter-clockwise so that the principal direction of the maximum tectonic compression coincides with the x axis and the minimum compression coincides with the y axis in the DDA model (Fig. 5). This simplifies the loading of the boundaries. However, the modeling results (Figs. 6–9) are plotted in traditional map view for the reader's convenience.

2. Physical Parameters

The material parameters have been chosen based upon geophysical and seismological observations in this region. Linear elastic stress-strain relationships are used for blocks. Coulomb's law is applied to the fault planes dividing different blocks as a shear failure criterion. Friction coefficients for faults are chosen according to the age of the faults. Tensile strengths can also be assigned to model tensile failure; for the model here we assigned zero tensile strength for all the faults, which is considered a reasonable approximation for preexisting faults.

We simulate the tectonic stress field in this region (Fig. 5) by applying constant maximum compression in ENE-WSW direction (x axis), and least compression in NNW–SSE direction direction (y axis). Young's modulus of the 4 loading blocks

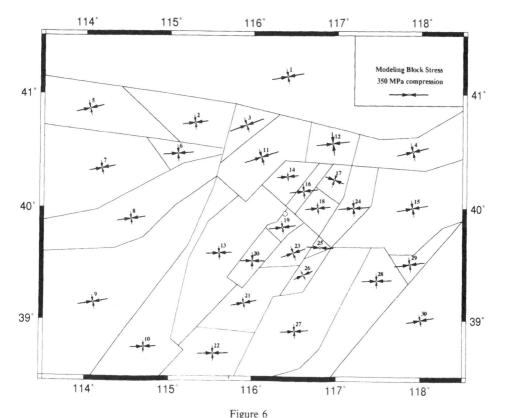

Figure 6
Aseismic principal stress map. Legend shows the scale of 350 MPa.

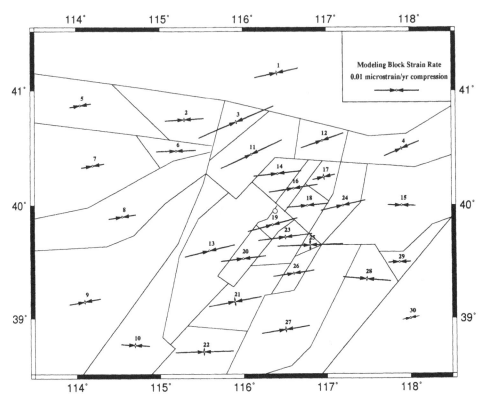

Figure 7
Aseismic principal strain rate map.

is two orders of magnitude lower than that for the tectonic blocks. In the block pair (block Nos. 31 and 33) of x direction (ENE-WSW), the value of compressive loading is 350 MPa, a value of the tectonic stress at a depth of approximately 12 km of the crust in a strike-slip stress regime of a continental area. In this kind of stress regime, the least compressive stress is about one half of the maximum compressive stress (LIU and ZOBACK, 1992), thus the compressive loading (block Nos. 32 and 34) in the y direction (NNW-SSE) is assigned a value of 175 MPa. The mechanical parameters for block interiors include the density of block material, Young's modulus and Poisson's ratio. For the tectonic blocks all these parameters have been assigned uniform values. We used a density of 2.65×10^3 kg/m³, an average value for crustal rocks for all the blocks. Young's modulus is 7.7×10^4 MPa and the Poisson's ratio is 0.25, typical values for the crust. The outermost 4 holding blocks (block Nos. 35–38) have the same values as the tectonic blocks.

For fault parameters, the needed values are the cohesion S_0, and the friction angle ϕ in Culomb's law for shear failure. For NNE-oriented faults, the friction angle used in the modeling is 32° (static coefficient of friction $\mu = \tan \phi = 0.62$,

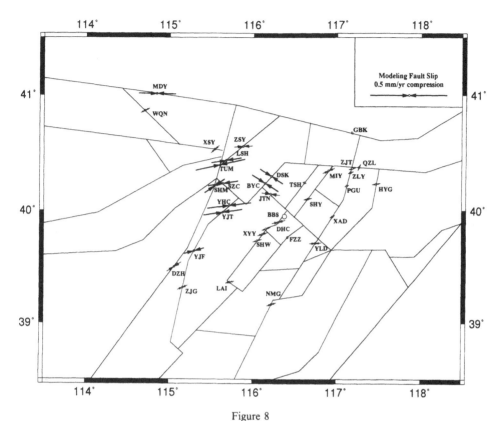

Figure 8
Aseismic fault slip map. The amount of slip is the length change between two ends across the fault.

see JAEGER and COOK, 1979). For NW-oriented faults, the friction angle is 30° and the cohesion is zero, simulating easier aseismic slip on those faults. For the faults dividing block 1 from other blocks, as well as the faults dividing block no. 9 from other blocks, we establish higher cohesion (three orders of magnitude higher than that of the NNE-oriented faults) and higher friction angle ($\phi = 50°$) to simulate the old, inactive and highly healed faults.

Modeling Results and Comparison with Observations

There are four sets of modeling output: the state of stress in each block (Fig. 6); the strain accumulation in each block (Fig. 7); the fault slip at measured sites (Fig. 8); and the block rotation (Fig. 9). Except for the state of stress, these parameters are presented in the form of deformation rates, i.e., the time derivative of the parameters. Since the current DDA model uses a linear elastic constitutive relation,

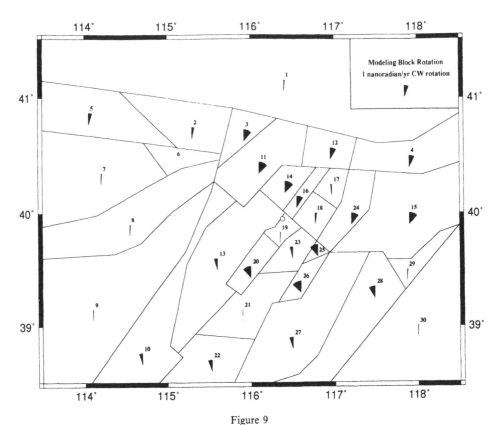

Figure 9
Aseismic block rotation map. Rotation are exaggerated to enhance visualization.

to obtain the strain rate, slip rate and rotation rate, we must use the shortening between two points at two end sides in the maximum compression direction and then convert the spatial shortening to time scale. As we have estimated in previous sections, the representative strain rate in this region is 10^{-8} year^{-1}, the length of the model in the maximum compression is 400 km, consequently the total shortening across the entire region will be 4 mm in one year. Compress the whole region in the model by a certain amount, to derive the state of stress and deformation, dividing the total shortening between those two control points by 4 mm provides the measure of the time scale to secure that much shortening. Consequently, division of the deformation parameters in each block or measuring point by the time scale supplies the deformation rates.

1. State of Stress in Blocks

From Figure 6 we can see that the modeled stress in each block is rather uniform in magnitude, with values close to the loading at the boundaries, however the principal direction varies from block to block, even though the average approximates the direction of regional loading. The variation of principal directions is within a range of about 40 degrees (block No. 3 and block No. 17 as the two extremes), this variation appears to be due to the interaction between blocks through fault-slip inhomogeneity.

2. Strain Rate in Blocks

Figure 7 illustrates the strain rate in each block. if we take 10^{-8} year^{-1} as the regional average strain rate, strain rate varies from block to block about the regional average strain rate. Though the maximum strain is always compressive everywhere, the minimum strain varies from compressive to tensile, but with much smaller magnitude than the maximum strain. The variation of principal direction is about 30 degrees slightly less than the variation of stress orientations.

3. Fault Slips

The modeling result of slips along the two sets of faults in the two nearly orthogonal directions (NNE and NW) is consistent with ENE-WSW tectonic compression. With this current model it is impossible to model those observed slips which are inconsistent with ENE-WSW tectonic compression such as those opposite pairs (e.g., XYY vs. SHW, or ZJG vs. YJF in Figure 4). For those sites there must be some local sources which caused a contradictory sense of slip along the same fault. The detailed comparison of the modeling results with the observed fault slip is shown in Table 4.

4. Block Rotation

Basically we can see that there are two domains of block rotation (Fig. 9). Blocks north of the Beijing area are in a clockwise rotation with the exception of block 17 which is slightly counter-clockwise. Most of the blocks south of Beijing rotate counter-clockwisely. The pure rotation of two adjacent blocks with the same polarity may result in slip on the faults between them with slip opposite to that of tectonic compression slip; for example, block No. 3 and block No. 11 rotate clockwise which may result in a left-lateral shear on the fault between them, however compared with the right-lateral fault slip caused by the compression, this amount of opposite slip is quite insufficient to change the sign of fault slip on that fault. In the other case the rotation may increase the slip caused by tectonic compression, such as the case of the rotation in block No. 13 and block No. 20; such a rotation will enhance the right-lateral shear on the NE trended fault.

Table 4

Comparison of Observed and Modeled Fault Motion

Site name	3-lett. code	Longitude (°)	Latitude (°)	Name of fault-crossing	Obs. extn. rate (mm/yr)	Obs. shear rate (mm/yr)	Mod. extn. rate (mm/yr)	Mod. shear rate (mm/yr)
Tumu	TUM	115.600	40.400	Tumu Fault	−0.05	−0.03	−0.27	+0.14
Shimen	SHM	115.533	40.217	Shimen Fault	+0.10	+0.06	−0.01	+0.15
Yanhecheng	YHC	115.700	40.067	Nankou-Yanhecheng Ft.	+0.40	−0.06	−0.10	+0.25
Yijianfang	YJF	115.300	39.683	Zijinguan Fault	−0.12	−0.06	−0.13	+0.03
Zijingguan	ZJG	115.183	39.433	Zijinguan Fault	+0.27	+0.09	−0.10	+0.01
Dazhuang	DZH	115.000	39.533	Wulonggou Fault	+0.30	+0.12	−0.01	+0.15
Baiyangchen	BYC	116.133	40.200	Nankou-Yanhecheng Ft.	+0.12	−0.16	−0.18	+0.06
Xiaoyouying	XYY	116.050	38.833	Babaoshan Fault	−1.04	−0.53	−0.04	+0.05
Laishui	LAI	115.617	39.417	Laishui Fault	+0.38	−0.09	−0.02	−0.05
Jiangtun	JTN	116.267	40.183	Nankou-Sunhe Fault	+1.07	−0.45	−0.09	−0.08
Taoshan	TSH	116.600	40.250	Huangzhuang-Gaoliying	+0.14	−0.23	−0.02	+0.02
Miyun	MIY	116.883	40.417	Miyun-Erjiayu Fault	−0.17	−0.08	−0.07	+0.05
Pingu	PGU	117.117	40.217	Chenzhuan-Zhenluoying	+0.09	−0.04	−0.03	+0.01
Gubeikou	GBK	117.200	40.683	Chongli-Chicheng Fault	+0.00	−0.03	−0.01	+0.01
Zhenluoying	ZLY	117.133	40.333	Chenzhuan-Zhenluoying	+0.02	−0.02	−0.04	+0.01
Xiadian	XAD	116.917	39.967	Xiadian Fault	+0.08	+0.08	−0.04	+0.01
Huanyaguan	HYG	117.433	40.233	Huangyaguan Fault	+1.15	−0.36	−0.05	+0.02
Babaoshan	BBS	116.233	39.917	Babaoshan Fault	−0.33	+0.24	−0.05	+0.07
Dahuichang	DHC	116.117	39.850	Babaoshan Fault	−0.15	−0.09	−0.03	+0.05
Deshengkou	DSK	116.183	40.300	Nankou Fault	−0.08	+0.09	−0.16	+0.05
Qiangzilu	QZL	117.217	40.433	Miyun-Xinglong Fault	+0.00	−0.03	−0.04	−0.01
Shangwan	SHW	116.017	39.800	Huangzhuang-Gaoliying	+0.04	+0.08	−0.00	+0.05
Shizuangcun	SZC	115.617	40.250	Shizhuang Fault	+0.00	−0.08	−0.174	−0.08
Wanquan	WQN	114.750	40.900	Shuiguantai Fault	−0.10	+0.09	−0.05	−0.04
Xiaoshuiyu	XSY	115.533	40.500	Xinglingpu-Yaozitou	+0.10	−0.18	−0.00	−0.07
Yianjiatai	YJT	115.567	40.000	Yanhecheng Fault	−0.26	−0.33	−0.17	+0.18
Zhangjiatai	ZJT	117.133	40.367	Zhenluoying Fault	+0.13	−0.32	−0.04	+0.01
Zhangsanyin	ZSY	115.867	40.517	Daxishan Fault	−0.07	−0.06	−0.06	+0.10

Conclusion and Assessment of Seismic Potential

From the modeling results we can conclude here by summarizing the following points.

1. Slips on faults with different orientations are generally in agreement with the ENE-WSW tectonic stress field, but the slip pattern of faulting can vary from near orthogonal compression to pure shear along the strike of the faults. This pattern cannot be explained by simple geometric relations between the strike of the fault and the direction of the tectonic shortening. This phenomenon has been observed at many sites of cross-fault geodetic surveys, and our study indicates that this might be caused by the interactions between different blocks and faults.

2. According to the DDA model, if the average aseismic slip rate along major active faults is on the order of several tenths of millimeters per year as observed by the cross-fault geodetic surveys, the typical strain inside a block is on the order of 10^{-8} year^{-1} or less, therefore the rate of 10^{-6} year^{-1}, as reported by observations (even with large error bars) in smaller areas, appears too high to be a representative value of deformation rate in this entire region.

3. It is possible that the sense of slip caused by rigid body rotation in two adjacent blocks is opposite to the slip caused by the tectonic compression. However, the magnitude of slip due to the tectonic compression is considerably larger than that from block rotation. Thus, in general, the slip pattern on faults as a whole agrees with the sense of ENE-WSW tectonic compression in this region. That is to say, the slip caused by regional compression dominates the major part of the entire slip budget.

4. Based on (3), some observed slips in contradiction to ENE tectonic stress field may be caused by more localized sources, and do not have tectonic significance.

In summary, the results of modeling demonstrate that the deformation pattern is consistent with the near east-west tectonic compression in this area. The tectonic shortening in the entire area is accommodated by slip on faults, compressive and shear strain within blocks, as well as translation and rotation of different blocks. Block rotation can occur with a different sense of rotation in different blocks to form varied domains of fault sets (NUR et al., 1989). However, the amount of fault slip caused by block rotation is substantially smaller than that caused by the regional compression itself. It is hard to attribute the left-lateral shear on NE to NNE-oriented faults and the right-lateral shear on NW-oriented faults, (as observed at some cross-fault stations, see Z. CHE et al., 1993) as a fundamental phenomenon with tectonic significance, since it is contradictory to the ENE-WSW compression in this region. Observation of this kind of slip might be caused by some nontectonic, localized sources. Also, localized high strain rates within certain blocks are either related to localized weakened zones in the crust, or simply erroneous blunders. Subsequently if the modeling results agree with the observed slipping rates on faults, then they will not be in agreement with the excessive high strain rates.

Acknowledgments

The authors thank Colleagues in the State Seismological Bureau of China for collecting and sharing the valuable data. One of the authors (L. L.) expresses thanks to Drs. Gen-hua Shi and Max Ma for their generous assistance regarding discussions of the theory and program of the DDA code. The research was supported by the Carnegie Institution of Washington through its Carnegie fellowship.

REFERENCES

AKI, K., and RICHARDS, P. G., *Quantitative Seismology* (Freeman, San Francisco 1980).

ANDERSON, J. G. (1986), *Seismic Strain Rates in the Central and Eastern United States*, Bull. Seismol. Soc. Am. *76*, 273–290.

CHE, Z., *Recent tectonic stress field in Beijing area derived from geodetic data*. In *Seismic Hazard Assessment for the Metropolitan Area of Beijing* (eds. Gao, W. and Ma, J.) (Seismological Press, Beijing 1993) pp. 142–165 (in Chinese with English abstract).

CHE, Z., LIU, T., and YIN, R., *Deformation precursors of the Datong-Yanggao earthquake*. In *Studies on the Datong-Yanggao Earthquake* (eds. Zhang, G. and Ma, L.) (Seismological Press, Beijing 1993) pp. 58–66 (in Chinese with English abstract).

CHE, Y., and YU, J., *Anomalous field feature of groundwater before the Datong-Yanggao earthquake*. In *Studies on the Datong-Tanggao Earthquake* (eds. Zhang, G. and Ma, L.) (Seismological Press, Beijing 1993) pp. 118–125 (in Chinese with English abstract).

CHEN, Y., TSOI, K., CHEN, F., GAO, Z., ZOU, Q., and CHEN, Z., *The Great Tangshan Earthquake of 1976: An Anatomy of Disaster* (Pergamon Press, Oxford 1991).

CHENG, E., LI, G., and CHEN, H. (1982), *On the Direction of the Maximum Compressive Principal Stress Before and After the 1976 Songpan-Pingwu Earthquake (M = 7.2) of the Sichuan Province*, Acta Seismologica Sinica *4*, 136–148 (in Chinese with English abstract).

DING, X., and ZHANG, W. (1988), *State of Modern Tectonic Stress Field in East China Mainland*, Acta Seismologica Sinica *10*, 25–38 (in Chinese with English abstract).

DING, G., *Lithospheric Dynamics of China: Explanatory Notes for the Atlas of Lithospheric Dynamics of China* (Seismological Press, Beijing 1991) (in Chinese).

GAO, X., LI, Z., WANG, H., and LIANG, S. (1992), *Earthquake Prediction Experiment in the Capital Region of China*, Earthquake Research in China *6*, 263–274 (in Chinese with English abstract).

GAO, W., and MA, J., *Seismo-geological Background and Earthquake Hazard in Beijing Area* (Seismological Press, Beijing 1993) (in Chinese).

GENG, L., ZHANG, G., and SHI, Y. (1993), *Preliminary Research on the Relations between Field and Source in Earthquake Preparation*, Earthquake Research in China *9*, 310–319 (in Chinese with English abstract).

HUANG, F., LI, Q., and GAO, Z. (1991), *The Stress Field in the Area from the West of Beijing to the Shanxi-Hebei-Inner Mongolia Border Region*, Acta Seismologica Sinica *13*, 307–318 (in Chinese with English abstract).

HUANG, L., WANG, X., and LIU, T. (1989), *The Horizontal Deformation before and after the Tangshan Earthquake of 1976*, Acta Seismologica Sinica *2*, 520–529 (in Chinese with English abstract).

JAEGER, J. C., and COOK, N. G. W., *Fundamentals of Rock Mechanics* (3rd ed., Chapman and Hall, London 1979).

LINDE, A., AGUSTSSON, K., SACKS, I. S., and STEFANSSON, R. (1993), *Mechanism of the 1991 Eruption of Hekla from Continuous Borehole Strain Monitoring*, Nature *365*, 737–740.

LIU, L., HE, S., NING, C., and LIU, B. (1986), *Installation of Sacks-Evertson Borehole Strainmeter Network in the Beijing-Tianjin Area*, Earthquake Research in China *2*, 102–104 (in Chinese with English abstract).

LIU, L., and ZOBACK, M. D. (1992), *The Effect of Topography on the State of Stress in the Crust: Application to the Site of the Cajon Pass Scientific Drilling Project*, J. Geophys. Res. *97*, 5095–5108.

LIU, L., ZOBACK, M. D., and SEGALL, P. (1992), *Rapid Intraplate Strain Accumulation in the New Madrid Seismic Zone*, Science *257*, 1666–1669.

MA, J., and CHEN, K., *The tectonic deformation model in Beijing Area*. In *Seismic Hazard Assessment for the Metropolitan Area of Beijing* (eds. Gao, W. and Ma, J.) (Seismological Press, Beijing 1993) pp. 202–220 (in Chinese).

MA, X. *et al.*, *Lithospheric Dynamics Atlas of China* (China Cartographic Publishing House, Beijing 1989) (in Chinese with English abstract).

MA, Z., FU, Z., ZHANG, Y., WANG, C., ZHANG, G., and LIU, D., *Earthquake Prediction: Nine Major Earthquakes in China (1966–1976)* (Seismological Press, Beijing 1990) (in Chinese).

MEI, S., FENG, D., ZHANG, G., ZHU, Y., GAO, X., and ZHANG, Z., *Introduction to Earthquake Prediction in China* (Seismological Press, Beijing 1993) (in Chinese).

MEI, S., and ZHU, Y. (1992), *Seismogenic Model of Tangshan Earthquake, Beijing* (Seismological Press, Beijing 1992) (in Chinese).

MORGAN, W. J. (1987), *How Rigid Are the Plates*, Paper presented at the Symposium on Geodynamics, April 22–24, Texas A&M University, Texas.

NUR, A., RON, H., and SCOTTI, O., *Kinematics and mechanics of tectonic block rotation*. In *Slow Deformation and Transmission of Stress in the Earth* (eds. Cohen, S. and Vanicek, P.) Geophysical Monograph *49* (AGU, Washington D.C. 1989) pp. 31–46.

SACKS, I. S., SUYEHIRO, S., LINDE, A. T., and SNOKE, J. A. (1971), *Sacks-Evertson Strainmeter, Its Installation in Japan and Some Preliminary Result Concerning Strain Steps*, Papers Meteorol. Geophys. *22*, 195–208.

SACKS, I. S., SUYEHIRO, S., LINDE, A. T., and SNOKE, J. A. (1982), *Stress Redistribution and Slow Earthquakes*, Tectonophysics *81*, 311–318.

SHI, G., and GOODMAN, R. E., *Discontinuous deformation analysis: A new method for computing stress, strain, and sliding of block systems*. In Proceedings of the 29th U. S. Symposium on Key Questions in Rock Mechanics (1988) pp. 381–393.

SHI, G., and GOODMAN, R. E. (1989), *Generalization of Two-dimensional Discontinuous Deformation Analysis for Forward Modeling*, Int. J. Num. and Analyt. Methods in Geomech. *13*, 359–380.

SHI, G., *Block System Modeling by Discontinuous Deformation Analysis* (Computational Mechanics Publications, Southampton 1993).

SONG, H., HUANG, L., and HUA, X., *Integrated Study of the Crustal Stress Field* (Petroleum Industrial Press, Beijing 1990) (in Chinese).

SU, G., XU, S., MA, S., and WANG, L. (1990), *The Cause of New Tectonic Movement and Migration of Earthquakes in Groups in and around North China*, Acta Geophysica Sinica *33*, 278–290 (in Chinese with English abstract).

WANG, R., HE, G., YIN, Y., and CAI, Y. (1980), *A Mathematical Simulation for the Pattern of Seismic Transference in North China*, Acta Seismologica Sinica *2*, 32–42 (in Chinese with English abstract).

WANG, S., NI, J., MA, Z., ZHANG, Y., SEEBER, L., ARMBRUSTER, J., and ZHANG, L. (1991), *The Characteristics of Fault Plane Solutions and Focal Depths of Strong Earthquakes in North China*, Acta Geophysica Sinica *34*, 42–54 (in Chinese with English abstract).

WEI, J., and SONG, H. (1989), *A Viscoelastic Finite-element Model for Earthquake Migration in the Beijing Region and Its Vicinity*, Acta Seismologica Sinica *2*, 447–455 (in Chinese with English abstract).

XIE, J., *Crustal deformation precursors and its mechanism of the Datong-Yanggao earthquake*. In *Studies on the Datong-Yanggao Earthquake* (eds. Zhang, G. and Ma, L.) (Seismological Press, Beijing 1993) pp. 77–88 (in Chinese with English abstract).

XU, Z., WANG, S., HUANG, Y., and GAO, A. (1992), *Tectonic Stress Field of China Inferred from a Large Number of Small Earthquakes*, J. Geophys. Res. *97*, 11,867–11,877.

ZHANG, G., GENG, L., and SHI, Y. (1993), *A Computer Model for Cyclic Activities of Strong Earthquake in Continental Seismic Zones*, Earthquake Research in China 9, 20–32.

ZHANG, Z., FANG, X., and YAN, H. (1987), *A Mechanical Model of the Formation Mechanism of the Shanxi Graben Zone and the Characteristics of Shanxi Earthquake Zone*, Acta Seismologica Sinica 9, 28–36 (in Chinese with English abstract).

ZHANG, G., and LIANG, B. (1987), *The Application of a Rheologic Model of Rock to the Seismogeny of a Great Earthquake and Research on Seismic Anomalies*, Acta Seismologica Sinica 9, 384–391 (in Chinese with English abstract).

ZHANG, G., and MA, L., Studies on the Datong-Yanggao Earthquake (Seismological Press, Beijing 1993) (in Chinese).

ZHANG, Z. (1985), *Investigation on the Thermal Force Sources of the 1976 Tangshan Earthquake*, Acta Seismologica Sinica 7, 45–56 (in Chinese with English abstract).

ZHANG, Z., FANG, X., and YIAN, H. (1985), *A Mechanical Model of the Formation Mechanism of the Shanxi Graben Zone and the Characteristics of Shanxi Earthquake Zone*, Acta Seismologica Sinica 9, 28–36 (in Chinese with English abstract).

ZHANG, W. *et al.*, Formation and Development of the North China Fault Block Region (Science Press, Beijing 1980) (in Chinese).

ZOBACK, M. L. (1992), *First- and Second-order Patterns of Stress in the Lithosphere: The World Stress Map Project*, J. Geophys. Res. 97, 11,703–11728.

ZHU, C., WANG, L., and LIN, B., *Characteristics of the Datong-Yanggao earthquake sequence*. In Studies on the Datong-Yanggao Earthquake (eds. Zhang, G. and Ma, L.) (Seismological Press, Beijing 1993) pp. 11–18 (in Chinese with English abstract).

(Received April 25, 1995, revised/accepted August 14, 1995)